FARADAY DISCUSSIONS OF THE CHEMICAL SOCIETY
NO. 77   1984

Interfacial Kinetics in Solution

THE FARADAY DIVISION
THE ROYAL SOCIETY OF CHEMISTRY
LONDON

Organising Committee

Professor D. H. Everett (*Chairman*)
Dr. R. Aveyard
Mrs. Y. A. Fish
Dr. W. A. House
Dr. B. H. Robinson
Professor H. Sawistowski
Dr. M. Spiro
Dr. D. A. Young

ISBN: 0-85186-628-X

ISSN: 0301-7249

Printed in Great Britain by J. W. Arrowsmith Ltd, Bristol

A GENERAL DISCUSSION ON
Interfacial Kinetics in Solution

9th, 10th and 11th April, 1984

A GENERAL DISCUSSION on Interfacial Kinetics in Solution was held at the University of Hull on 9th, 10th and 11th April 1984. The President of the Faraday Division, Prof P. Gray, FRS, was in the chair: about 80 fellows of the Faraday Division and visitors from overseas attended the meeting. Among the overseas visitors were:

Prof. G. Astarita, *Italy*
Dr. J. J. M. Binsma, *The Netherlands*
Mr. V. K. Cheng, *Australia*
Prof. J. Christoffersen, *Denmark*
Dr. B. A. W. Coller, *Australia*
Prof. G. Dickel, *West Germany*
Dr. A. R. Flambard, *West Germany*
Dr. V. Friehmelt, *West Germany*
Dr. R. H. Guy, *U.S.A.*
Mr. Z. Kolar, *The Netherlands*
Dr. K. Kontturi, *Finland*
Prof. J. Koryta, *Czechoslovakia*
Prof. M. M. Kreevoy, *U.S.A.*

Prof. I. R. Miller, *Israel*
Dr. E. Nakache, *France*
Prof. W. Nitsch, *West Germany*
Dr. R. D. Noble, *U.S.A.*
Dr. Z. Samec, *Czechoslovakia*
Prof. A. Sanfeld, *Belgium*
Dr. G. Sartori, *U.S.A.*
Dr. A. Steinchen, *Belgium*
Dr. J. Texter, *U.S.A.*
Dr. C. Tondre, *France*
Prof. D. C. Walker, *Canada*
Prof. T. Yasunaga, *Japan*

CONTENTS

- page 7 *Introductory Lecture: Mass Transfer and Reactions at Interfaces*
by P. Meares

- 17 *Amines as Rate Promoters for Carbon Dioxide Hydrolysis*
by D. W. Savage, G. Sartori and G. Astarita

- 33 *Kinetics of Carbon Dioxide Transfer across the Air/Water Interface*
by W. A. House, J. R. Howard and G. Skirrow

- 47 GENERAL DISCUSSION

- 53 *Kinetics and Mechanism of Interfacial Reactions in the Solvent Extraction of Copper*
by W. J. Albery, R. A. Choudhery and P. R. Fisk

- 67 *Separation of Metal Ions by Ligand-accelerated Transfer through Liquid Surfactant Membranes*
by D. T. Wasan, Z. M. Gu and N. N. Li

- 75 *A General Model to Account for the Liquid/Liquid Kinetics of Extraction of Metals by Organic Acids*
by M. A. Hughes and V. Rod

- 85 *The Concept of Interfacial Reactions for Mass Transfer in Liquid/Liquid Systems*
by W. Nitsch

- 97 *Facilitated Transport across Liquid/Liquid Interfaces and its Relevance to Drug Diffusion across Biological Membranes*
by N. Barker, J. Hadgraft and P. K. Wotton

- 105 *Kinetics of Heterogeneous Nitration in Emulsions*
by J. E. Crooks and J. M. Chisholm

- 115 *Use of Microemulsions as Liquid Membranes. Improved Kinetics of Solute Transfer at Interfaces*
by C. Tondre and A. Xenakis

- 127 *Solute Transport and Perturbation at Liquid/Liquid Interfaces*
by R. H. Guy, R. S. Hinz and M. Amantea

- 139 GENERAL DISCUSSION

- 157 *The Variational Principles of Onsager and Prigogine in Membrane Transport*
by G. Dickel

- 169 *Motion Induced by Surface-chemical and Electrochemical Kinetics*
by A. Sanfeld and A. Steinchen

- 181 *Time-dependent Behaviour and Regularity of Dissipative Structures of Interfacial Dynamic Instabilities*
by H. Linde

- 189 *The Contribution of Chemistry to New Marangoni Mass-transfer Instabilities at the Oil/Water Interface*
by E. Nakache, M. Dupeyrat and M. Vignes-Adler

197 *Double Layers at Liquid/Liquid Interfaces*
by Z. Samec, V. Mareček and D. Homolka

209 *Transfer of Alkali-metal and Hydrogen Ions across Liquid/Liquid Interfaces Mediated by Monensin. A Voltammetric Study at the Interface of Two Immiscible Electrolyte Solutions*
by J. Koryta, D. Du, W. Ruth and P. Vanýsek

217 GENERAL DISCUSSION

223 *Ion-exchange Dynamics at the Zeolite/Solution Interface Studied by the Chemical-relaxation Method*
by T. Ikeda, M. Sasaki and T. Yasunaga

235 *Kinetics of Dissolution of Calcium Hydroxyapatite*
by J. Christoffersen and M. R. Christoffersen

243 *Kinetics and Simulation of Dissolution of Barium Sulphate*
by V. K. Cheng, B. A. W. Coller and J. L. Powell

257 *Study of the Dynamic Equilibrium in the CaF_2/Aqueous Solution System using $^{45}Ca^{2+}$ as Radiotracer*
by J. J. M. Binsma and Z. Kolar

265 *High-temperature Dissolution of Nickel Chromium Ferrites by Oxalic Acid and Nitrilotriacetic Acid*
by R. M. Sellers and W. J. Williams

275 *Interfacial Kinetics in Solution. Linear Free-energy Relationships Applicable to Heterogeneously Catalysed Reactions in Solution*
by M. Spiro

287 GENERAL DISCUSSION

309 CLOSING REMARKS
by M. Spiro

313 INDEX OF NAMES

Mass Transfer and Reactions at Interfaces

BY PATRICK MEARES

Chemistry Department, The University, Aberdeen AB9 2UE, Scotland

Received 11th April, 1984

HOMOGENEOUS DIFFUSIVE FLOW

The teaching of the principles of mass transfer in physical chemistry courses commonly starts from either the interdiffusion of a pair of miscible liquids in a binary system or the diffusion of a single solute down its concentration gradient in a solvent. The treatment follows Fick's law with a constant diffusion coefficient and reveals that the choice of reference frame for the mass fluxes is not trivial, and that when the choice is made correctly the binary system is characterized by a single interdiffusion coefficient. Allowance for concentration dependence of the diffusion coefficient can be made with some increase in the mathematical complexity of the treatment.

When more than two components have to be considered the complexity of the mathematical treatment increases substantially and, more seriously perhaps, so does the number of diffusion coefficients needed to describe the system. Fick's law can be generalized to

$$-j_i = \sum_{j=1}^{n-1} D_{ij} \frac{dc_j}{dx} \qquad (1)$$

where j_i is the flux of i relative to the volume-average velocity. Thus a system of n components is described by $(n-1)^2$ diffusion coefficients.

If component n is treated as the solvent, $(n-1)$ coefficients are of the form D_{jj} and resemble in nature and size the diffusion coefficient of j in the solvent n as would be measured in, say, a diaphragm cell. The remainder are cross-coefficients D_{ij} ($i \neq j$) which express the fact that a flux of i can be generated by a concentration gradient of j even in the absence of a gradient of i and that a concentration gradient of i can be generated by a flux of j.

Multicomponent diffusion fluxes are frequently represented by using the notation of irreversible thermodynamics in the form

$$-j_i = \sum_{j=1}^{n-1} L_{ij} \frac{d\mu_j}{dx}. \qquad (2)$$

(For simplicity, the difference between the centre-of-mass velocity and the volume-average velocity has been ignored.)

Since the Onsager reciprocal relations

$$L_{ij} = L_{ji} \qquad (3)$$

apply and reduce the number of independent L_{ij} coefficients to $\frac{1}{2}n(n-1)$, it is evident that, although $D_{ij} \neq D_{ji}$, relations must exist which reduce the number of independent diffusion coefficients to $\frac{1}{2}n(n-1)$. These relations cannot be used without a detailed knowledge of the way in which each chemical potential varies with each of the concentrations. In general this information is not available and would be as tedious to obtain experimentally as it would be to measure all of the $(n-1)^2$ diffusion coefficients.

Multicomponent diffusion will not be further developed here and reference may be made for more information to two excellent texts.[1,2] It must be remembered, however, that the cross-coefficients D_{ij}, although usually (except in macromolecular systems) smaller than the straight D_{jj} coefficients, are frequently significant and can lead to unknown or unexpected phenomena.[3]

MASS FLOW WITH REACTION

Faced with so many experimental, theoretical and interpretive problems in the study of multicomponent diffusion in systems without phase boundaries, the physical chemist might be tempted to look for a simpler background against which to study reaction mechanisms. Such problems, however, cannot be evaded in many of the most important operations of the chemical industry. Furthermore, the objective of many industrial processes is to bring about a separation and consequently at least two phases are involved and molecular transport takes place between them. Thus it has come about that most of the work on mass transfer up to and across interfaces, with and without chemical reaction, has been carried out by chemical engineers and is frequently expressed in a terminology not wholly familiar to physical chemists.

Despite the complexity of the molecular processes involved, it is essential to characterize the mass-transfer behaviour of any experimental system in an unambiguous, although empirical, way if the kinetics and mechanism of concurrent chemical reactions are to be inferred from their effects on the rates of the processes taking place in the system.

When a substance i is being transported across an interface and then consumed by a first-order reaction with rate constant k_i, the rate of change of concentration of i in a volume element distant x from the surface may be written[4]

$$\frac{\partial c_i}{\partial t} = D\frac{\partial^2 c_i}{\partial x^2} - u\frac{\partial c_i}{\partial y} - k_i c_i. \quad (4)$$

Here $u\partial c_i/\partial y$ allows for convection with velocity u parallel to the interface and the third term on the right-hand side deals with the consumption of i by reaction.

By solving eqn (4) with and without the reaction term one may obtain an expression for the mass-transfer coefficient of i with, k_M, and without, k_M°, reaction. Provided the surface concentration and hydrodynamic conditions are kept constant, a reaction factor ϕ is defined by

$$\phi = k_M/k_M^\circ. \quad (5)$$

Since ϕ is a function of the velocity constant k_i, it is the experimental quantity upon which information about the reaction depends. The solution of the continuity eqn (4) can be carried out only when the mass transfer in the non-reacting system has been correctly modelled. Most of this introductory paper is devoted to an elementary discussion of this problem.

The difficulty of characterizing the mass-flow conditions is emphasized by the wide variety of experimental arrangements that have been used by the authors whose work is included in this Discussion. Stirred and unstirred systems have been used to study gas/liquid, liquid/liquid and solid/liquid interfaces, and the interfaces themselves are sometimes planar and sometimes spherical. Many more complex

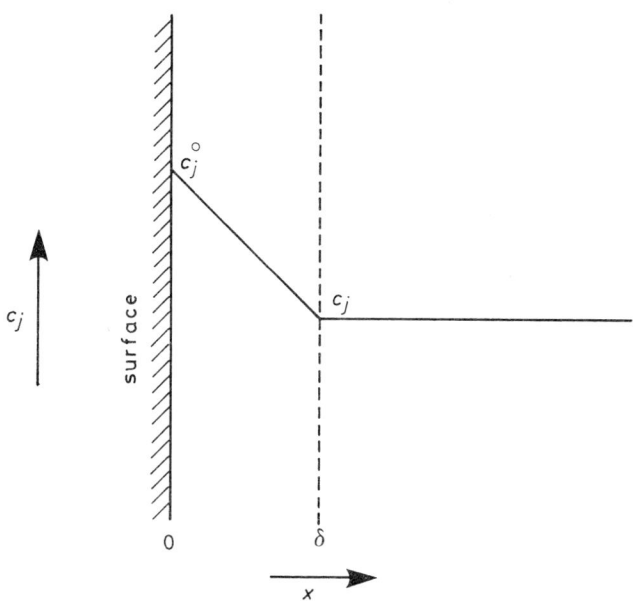

Fig. 1. Concentration profile at a solid/liquid interface in the presence of a stagnant liquid film of thickness δ.

arrangements and surface geometries are to be found in industrial mass-transfer plants.

THE STAGNANT-FILM MODEL

Three models are in common use to describe mass transfer at interfaces. They are the stagnant-film model, the penetration model and the turbulent-boundary-layer model.[4]

The stagnant-film model is probably the best known and was introduced by Nernst at the beginning of this century. It is assumed that the resistance to mass transfer up to a phase boundary in a liquid or a gas lies wholly within a thin layer adjacent to the surface and that all regions of the phase further from the surface than the thickness of this layer can be regarded as well stirred and of uniform composition. The fluid in immediate contact with a stationary surface is at rest and a pair of fluids in contact at an interface are at rest there relative to one another. Thus it is appropriate to model the resistant boundary layer as a stagnant film and to assign to it a thickness δ (fig. 1).

Transport across this film takes place by diffusion only and can usually be regarded as confined to the direction normal to the surface. Fluxes can be expressed by using Fick's law and ignoring effects due to cross-coefficients D_{ij}. The flux j_i relative to the volume-average velocity given by Fick's equation must be corrected to find the flux J_i relative to the fixed surface. Thus

$$J_i = j_i + c_i \sum_{i=1}^{n} J_i \bar{V}_i \qquad (6)$$

where $\bar{V}_i$ is the partial molar volume of i.

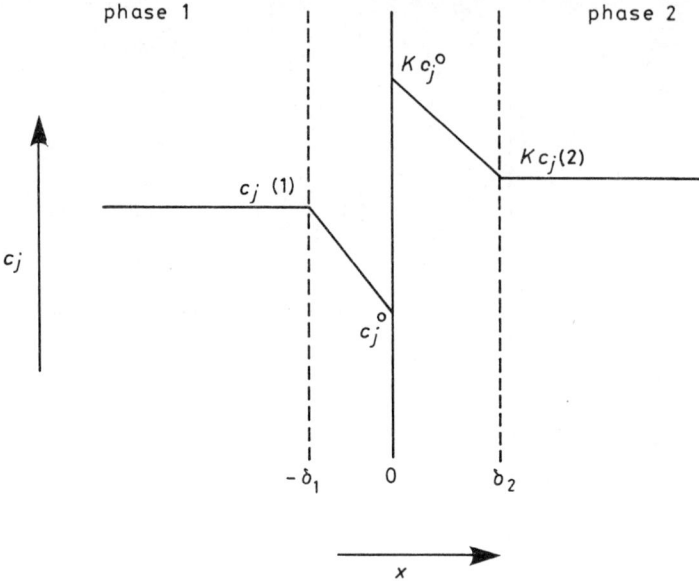

Fig. 2. Concentration profile at a liquid/liquid interface with stagnant films of thicknesses δ_1 and δ_2 and distribution coefficient K.

The stagnant-film model is easy to visualize and set down but it can only be used quantitatively in the absolute sense in a few special cases. There are two difficulties; the stagnant-layer thickness cannot be measured independently and, except at some fluid/solid interfaces, the concentrations at the surface will not be known.

Where two fluid phases are in contact, only the bulk concentrations are known and it is necessary to elaborate the model to include two stagnant films, one on each side of the interface (fig. 2). Then one may assume also that there is no resistance to transfer at the interface itself, i.e. that the fluids in contact there are at partition equilibrium. This assumption is discussed more fully later. The total resistance to mass transfer is then given by the sum of the two stagnant-film resistances, which act in series.

The values of δ for the two films have to be estimated from measurements of mass-transfer coefficients of suitable substances under the same hydrodynamic conditions as will be used in the kinetic studies. Alternatively, the hydrodynamic conditions may be varied in a systematic way that permits extrapolation to $\delta \to 0$, i.e. to conditions of perfect mixing in each phase. Finally, in order to estimate the resistance to mass transfer from the two sets of bulk concentrations and stagnant-film thicknesses, the equilibrium distribution or partition coefficients K of the materials in contact across the phase boundary must be known. This equilibrium might follow the linear law of Henry at a gas/liquid interface or for the partition of a solute between two solvents, but in a case where liquid ion-exchangers or ionophores are involved a quadratic or higher-order ion-exchange equilibrium isotherm is needed.

The stagnant-film model has been particularly useful in work on membrane transport and in the study of electrode kinetics. In membrane science the membrane phase can be regarded as a stagnant phase and transport within it treated as due to molecular-diffusion processes, perhaps coupled with chemical reactions. The transport equations in the membrane are derived in terms of the surface concentrations. It does not affect the use of the model that these equations are frequently far more complex than Fickian diffusion equations. Usually it can be arranged that either the hydrodynamic conditions, and hence the values of δ, are the same at both membrane/solution interfaces or the solutions can be chosen so that the resistance to mass transfer is negligible at one side relative to the other. Then only one value of δ is important. The stagnant film can be assigned an effective thickness by studying the transport of a suitable passively transported solute, and the assumption of equilibrium at the membrane faces is usually justified.

The use of Fick's law and a characteristic value of δ leads to a linear increase of flux with diffusion coefficient. However, there is no distinct plane at which the bulk changes from being well stirred to stagnant. The effectiveness of the stirring declines gradually as the surface is approached. The greater the value of D the greater will be the distance from the wall at which the diffusion flux becomes comparable with the convective eddy flux. A monotonous increase of δ with $D^{1/3}$ is expected but in view of the relatively small range spanned by the diffusion coefficients of a series of solutes in a single solvent, this variation of δ is a less serious problem than many others met in multicomponent transport systems.

PRECISELY KNOWN HYDRODYNAMICS

Where the phase at one side of the interface is effectively solid it may be possible to extrapolate mass-transfer data obtained under various conditions so as to estimate the flux that would be obtained under perfect stirring conditions, *i.e.* zero film thickness.[5]

Such extrapolation procedures are essentially empirical. However, when one phase takes the form of a disc, such as a metal electrode, which is rotated within a liquid then it is found[6] that the stagnant-layer thickness has the same value over the whole surface of the disc. Further, if the rate of rotation is restricted to the regime of laminar motion in the fluid, the mass-transfer coefficient can be related to the rate of rotation and other measurable parameters of the system. An unambiguous way of extrapolating out the liquid-phase mass-transfer resistance is then indicated by the theory.

An important advance has been made by extending the rotating-disc technique from metal electrodes and catalysts and from rate-of-dissolution studies to the rotating diffusion cell,[7] which has been used by several authors contributing to this Discussion.

Exact treatments of mass transfer across a phase boundary have been carried out also for several practical cases involving laminar flow over plates, spheres and cylinders given stationary concentrations at the interface.[4] They represent a refinement of the stagnant-film model in these special cases. They are of particular value for studies on rates of absorption of gases by fluids.[8]

THE PENETRATION MODEL

The penetration model differs fundamentally from the stagnant-film theory. Whereas the latter is directed towards steady-state mass transfer, the former deals with transient conditions. If the fluid at the interface is constantly replaced by fresh fluid from the bulk in a time comparable to that required to establish a steady concentration gradient across the stagnant film, it is found that the rate of mass transfer of a substance into the fluid should vary with the square root of its diffusion coefficient in the fluid.[8] The model can be used only in combination with various empirical quantities, especially the characteristic length of time during which an element of fluid remains in contact with the interface. This model is not developed more fully here because it does not appear to be appropriate in the papers which are to follow except perhaps in the studies of liquid membranes.[9] The liquid in contact with the moving emulsion droplets is subject to continual renewal, and permeation into each volume element of the fluid occurs only during the short period required for a droplet to pass through it.

THE TURBULENT-BOUNDARY-LAYER MODEL

When the velocity and its gradient in shearing flow near to an interface are large enough for the momentum forces to exceed the viscous forces the flow ceases to be laminar and becomes irregular and turbulent. Macroscopic volume elements of the liquid then 'diffuse' through the mass rather in the way that single molecules diffuse in steady conditions. The transporting effect of this irregular motion is referred to as eddy diffusion; it is normally far more effective than molecular diffusion in equalizing concentrations throughout the fluid.

The theory of turbulent flow close to interfaces[10] is highly complex and yet relatively primitive in terms of the level of understanding achieved. One may imagine that close to a stationary interface turbulence is damped and that mass transfer across the interface will generate steep concentration gradients in a thin layer, the turbulent boundary layer, between the surface and the fully turbulent regions in the bulk. This thin layer consists of the stationary film at the interface across which transfer is by molecular diffusion and an intermediate layer in which eddy diffusion and molecular diffusion both play a role. An empirical quantity, the eddy diffusivity E, is then assigned to take care of the turbulent contribution to the flux, giving

$$J_i = -(D_i + E) \, dc_i/dx. \tag{7}$$

In the nature of eddy flow it seems likely that E, which is a function of the hydrodynamic conditions, has about the same value for all components.

It is unlikely that one would choose to disentangle the kinetic characteristics of a chemical reaction under the deliberate complication of turbulent conditions. None of the experimental contributions to this Discussion have done this but, as will be described, turbulence may arise at the interface for reasons other than the use of agitation at high Reynolds numbers.

SURFACE RESISTANCE AND ADSORBED MONOLAYERS

The foregoing discussion has been based on the assumption that the compositions and hence properties in each of the two phases would, at equilibrium, be uniform up to the interface and that during the non-equilibrium state of mass transfer between the phases there is no relative motion of the phases actually at the interfacial layer of molecules apart from the steady mass-transfer flux normal to the interface. The layers of molecules on either side of the interface are assumed to be in thermodynamic equilibrium with only a small energy of activation required for molecular exchanges between them, *i.e.* there is no barrier to mass flow in the interface itself.

Clearly these assumptions are a great oversimplification. The properties of the system, such as density, concentration, dielectric constant *etc.* change abruptly at the interface. The molecules there are subjected to strong force fields which affect their properties, cause orientation and give rise to well understood phenomena such as interfacial electric-potential differences, adsorption and interfacial tension. When, as in the cases of interest here, one at least of the phases contains several components, their mole fractions will change as the surface is approached even at equilibrium. This surface activity normally leads to an accumulation near the surface of solutes which lower the surface tension. The thermodynamic interconnection of surface concentrations and surface pressure or surface tension γ is made through the Gibbs isotherm in terms of the surface excess Γ_i:

$$\Gamma_i = -\partial \gamma / \partial \mu_i. \tag{8}$$

When the surface excess is large, *i.e.* for highly surface-active solutes, it is believed that the interfacial layer may consist almost wholly of a monomolecular film of the adsorbed solute.

To the extent that the diffusion coefficients of the solutes are concentration-dependent, their values in the interfacial layer may differ, as a result of the surface excess concentrations, from those in the bulk. Far more importantly, a close-packed monolayer actually forming the interface may represent a substantial barrier to mass transfer. Thus in the case of the stagnant-film model, the total resistance R will be given by

$$R = R_1 + R_2 + R_s \tag{9}$$

where R_1 and R_2 are the two stagnant-film resistances and R_s the surface resistance.

In the absence of powerfully surface-active substances R_s is usually negligible compared with R_1 and R_2 even under conditions of good stirring, but the effect of surfactants can be very marked for gas and vapour transfer at gas/liquid interfaces.[11] Perhaps the most widely known is the effect of a monolayer of cetyl alcohol in retarding evaporation from reservoirs.

The surface resistance to solute transfer at a liquid/liquid interface is usually negligible compared with that due to the stagnant films adjacent to the interface but surfactants may have a marked effect on another interfacial influence on mass transfer yet to be discussed.

SPONTANEOUS INTERFACIAL TURBULENCE

It has been recognized for more than a century that, when a solute diffuses across an oil/water interface, the interface may become unstable, turbulent motion may develop spontaneously and an emulsion may form of one phase dispersed in the other. Such phenomena are commonly grouped together as Marangoni effects.[4,10-13] There is a fascinating and substantial literature on this subject and only a few simple principles can be mentioned here.

The true Marangoni effects are driven by fluctuations in the interfacial tension. They should be distinguished from convection driven by an unstable density distribution which can arise, for example, as a result of a substance transferring upwards across a phase boundary if the liquid left behind at the interface is then more dense than the bulk liquid lying below. Spontaneous density convection can also arise in multicomponent systems where a flow of one component generates a concentration gradient of another and, in the process, inverts the gravitationally stable density profile.[3]

The Marangoni effect can arise at a gas/liquid or liquid/liquid interface; turbulent effects develop at plane and at spherical interfaces. Fig. 3 and 4 show how such effects can arise at a plane interface where the transported substance lowers the interfacial tension. A local downward fluctuation in the interfacial tension due

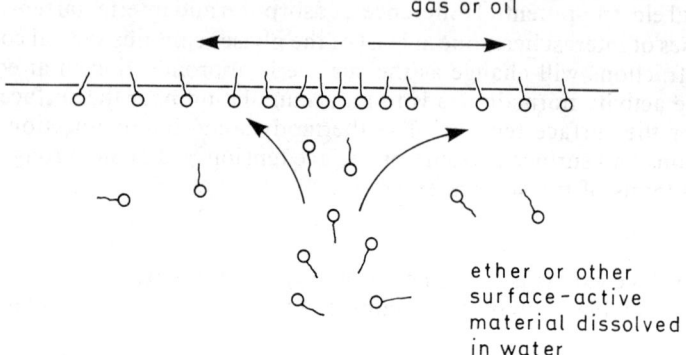

Fig. 3. Diagrammatic representation of surface-active material spreading from a a locally high concentration in the surface and dragging some of the underlying water with it. This brings up more surfactant to the surface and amplifies the disturbance (from Davies and Rideal).[11]

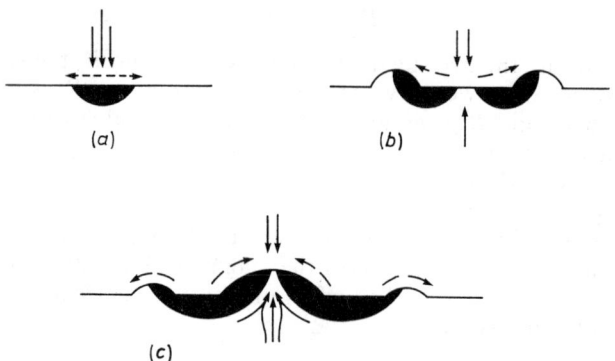

Fig. 4. Diagrammatic representation of interfacial turbulence creating a surface ripple (from Ellis and Biddulph).[24] (a) A spot of very low surface tension is formed. (b) The spot spreads violently, forming an annulus and exposing the bulk liquid. (c) A large ripple is formed as the central motion reverses.

to a local concentration fluctuation leads to an outward movement of the surface from the site of the fluctuation which drags up from the bulk still more of the surface active substance. Thus instead of dying away the disturbance is amplified. The balance of forces is such that, at the surface, a line across which the interfacial tension changes produces a ripple as it spreads outwards. The mutual interferences of many such ripples may produce a regular cellular pattern on a plane surface (plate 1).[12] In the case of a drop of one phase dispersed in another, violent kicking of the drop may appear.[14-16] The precise theory of such effects is complex and incomplete although major advances have been made in recent years through the interest now manifest in dissipative structures.[17]

The importance of spontaneous turbulence in the study of interfacial reactions is that such turbulence greatly increases the rate of mass transfer across the interface. In practical terms, such an effect would be extremely beneficial in a process such as liquid extraction but on the other hand the unsuspected existence of interfacial

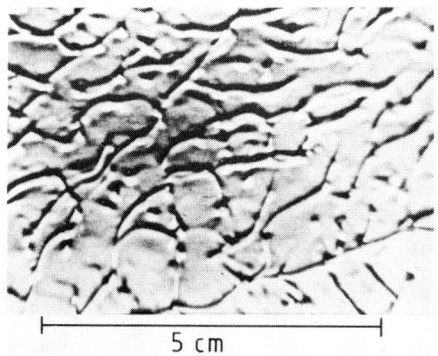

Plate 1. Water desorbing from a 10 mm deep pool composed of equal volumes of water and 1,4-dioxane (from Berg et al.[25]).

turbulence in the kinetic study of a reaction could lead to erroneous conclusions. It is essential therefore to have a reliable set of criteria to enable the likelihood of interfacial instability to be predicted. The author was highly conscious of this problem while reading the preprinted papers for this Discussion. It is helpful that several of the main recent contributors to the theory of Marangoni effects are contributing also to the Discussion and may throw light on their relevance to some of the experimental situations described in the other papers.

The first major theoretical analysis of the Marangoni effect[18] showed that instability for the transfer of a component that lowers the surface tension is likely to occur when transfer is out of the phase of higher viscosity or lower diffusion coefficient, i.e. for a gas/liquid interface desorption is liable to be unstable but adsorption is not. The absorption of carbon dioxide, however, which raises the surface tension of an aqueous solution, can give rise to instability especially when it is absorbed into a solution of monoethanolamine,[19] and the same is true for the absorption of ammonia by acetic acid solutions.[20]

It is frequently found that the presence of the interface of highly surface-active solutes reduces the spreading tendency of the surface fluctuations of the concentration of the transferable solute. The adsorbed film of surfactants also introduces a surface elasticity and viscosity which damp the fluctuations and so preserves stability.

Important new studies of the stability criteria[21-23] have served to extend and largely confirm the results of Sternling and Scriven.[18] Especially pertinent here is the extension from purely diffusional and hydrodynamic effects of earlier work to include chemical reactions at interfaces.[22]

CONCLUSION

In order to study the kinetics and mechanism of interfacial reactions it is essential to be able to disentangle the influences of reaction steps and mass transfer on the rate of the overall processes that can be observed in an experiment. The more precise is the knowledge of the mass-transfer restraints in the system the more fundamental are the conclusions that can be drawn regarding the chemical reactions. For this reason it is preferable to construct experimental systems with very well defined hydrodynamic situations as in rotating-disc or well developed laminar-flow systems. Alternatively, the use of carefully controlled conditions which can be characterized by studying the fluxes of inert substances in terms of the stagnant-film model or the penetration model is acceptable. It is not too serious that these model theories do not apply exactly to the experimental circumstances because much of the error cancels when one divides k_M by k_M^o in order to find the reaction factor ϕ. It is important, however, that substances occurring in the reacting system that were absent from the characterizing system do not induce interfacial instability and turbulence, which would greatly increase the rate of mass transfer across the interface and frustrate the attempt to derive the contribution of the reaction to the observed phenomena.

[1] D. D. Fitts, *Non-equilibrium Thermodynamics* (McGraw-Hill, New York, 1962).
[2] E. L. Cussler, *Multicomponent Diffusion* (Elsevier, Amsterdam, 1976).
[3] R. P. Wendt, *J. Phys. Chem.*, 1962, **66**, 1940.
[4] T. K. Sherwood, R. L. Pigford and C. R. Wilke, *Mass Transfer* (McGraw-Hill, New York 1975), chap. 5 and 8.
[5] H. D. Spriggs and N. N. Li, in *Membrane Separation Processes*, ed. P. Meares (Elsevier, Amsterdam, 1976), chap. 2.
[6] B. G. Levich, *Physico-chemical Hydrodynamics* (Prentice-Hall, Englewood Cliffs, N.J., 1962).

[7] W. J. Albery, A. M. Couper, J. Hadgraft and C. Ryan, *J. Chem. Soc., Faraday Trans. 1*, 1974, **70**, 1124.
[8] P. V. Danckwerts, *Gas–Liquid Reactions* (McGraw-Hill, New York, 1970), chap. 5.
[9] D. J. Wasan, Z. M. Gu and N. N. Li, *Faraday Discuss. Chem. Soc.*, 1984, **77**, 67.
[10] J. T. Davies, *Turbulence Phenomena* (Academic Press, New York, 1972), chap. 9.
[11] J. T. Davies and E. K. Rideal, *Interfacial Phenomena* (Academic Press, New York, 1961), chap. 7.
[12] J. C. Berg, in *Recent Developments in Separation Science*, ed. N. N. Li (C.R.C. Press, Cleveland, Ohio, 1972), vol. II, pp. 1–31.
[13] B. G. Levich and V. S. Krylov, *Annu. Rev. Fluid Mech.*, 1969, **1**, 293.
[14] J. T. Davies and D. A. Haydon, *Proc. 2nd Int. Congr. Surface Activity* (Butterworths, London, 1957), vol. I, pp. 417–425.
[15] M. V. Ostrovsky and R. M. Ostrovsky, *J. Colloid Interface Sci.*, 1983, **93**, 392.
[16] T. S. Sørensen, *J. Chem. Soc., Faraday Trans. 2*, 1980, **76**, 1170.
[17] P. Glansdorff and I. Prigogine, *Thermodynamic Theory of Structure, Stability and Fluctuations* (Wiley-Interscience, London, 1971).
[18] C. V. Sternling and L. E. Scriven, *AIChE J.*, 1959, **5**, 514.
[19] P. V. Danckwerts and A. T. da Silva, *Chem. Eng. Sci.*, 1967, **22**, 1613.
[20] A. J. M. A. Oyekan and H. Sawistowski, *Chem. Eng. Sci.*, 1971, **26**, 1772.
[21] M. Hennenberg, P. M. Bisch, M. Vignes-Adler and A. Sanfeld, *J. Colloid Interface Sci.*, 1979, **69**, 128; 1980, **74**, 495.
[22] W. Dalle Vedove and A. Sanfeld, *J. Colloid Interface Sci.*, 1981, **84**, 318; 328.
[23] J. Reichenbach and H. Linde, *J. Colloid Interface Sci.*, 1981, **84**, 433.
[24] S. R. M. Ellis and M. Biddulph, *Chem. Eng. Sci.*, 1966, **21**, 1107.
[25] J. C. Berg, M. Boudart and A. Acrivos, *J. Fluid Mech.*, 1966, **24**, 721.

Amines as Rate Promoters for Carbon Dioxide Hydrolysis

By D. W. Savage* and G. Sartori

Corporate Research, Exxon Research and Engineering Co., Annandale, New Jersey 08801, U.S.A.

and G. Astarita

Chemical Engineering Department, University of Delaware, Newark, Delaware 19711, U.S.A.

Received 17th January, 1984

Amines act as homogeneous catalysts for the carbon dioxide hydrolysis reaction, so that they are very effective rate promoters for carbon dioxide absorption in carbonate solutions. Experimental data show that the rate-promotion effect is a very conspicuous one, to the point where the catalysed reaction can be regarded as essentially instantaneous in comparison with diffusion phenomena. Possible mechanisms of this effect are discussed.

The rate-enhancement effect is in addition to the effect that amines have on the capacity of carbonate solutions. The relationship between the rate and capacity effects is discussed.

In the hot carbonate process for CO_2 removal from gases, the chemical sink for carbon dioxide is the bicarbonate ion; CO_2 is chemically consumed by the following overall reaction:

$$CO_2 + CO_3^{2-} + H_2O \rightarrow 2HCO_3^-. \tag{1}$$

The chemistry of the process in the case where potassium (or sodium) carbonate is the only reactive species originally present in the liquid phase is well understood. The slow step of the overall sequence resulting in reaction (1) is CO_2 hydrolysis:

$$CO_2 + OH^- \rightarrow HCO_3^- \tag{2}$$

and the kinetic constant of reaction (2) has been determined[1] for ionic strengths up to the highest ones used in industrial operation and for temperatures up to 110 °C. The vapour–liquid equilibrium (VLE) behaviour has also been satisfactorily modeled; a correlation is available for the physical solubility of CO_2 in concentrated carbonate solutions, and the mass-transfer rates can be predicted accurately from the available physicochemical information.[2]

The hot carbonate process has many attractive features, but the *rate* of mass transfer is comparatively low, owing to the small value of the hydroxide ion concentration in reaction (2). Therefore, rate promoters have been in common industrial use for a long time. The industrially important rate promoters fall into two categories: inorganic (usually weak acids) and organic (amines, usually amino-alcohols). Rate promoters, in addition to their effect on the mass transfer rates, may also influence the VLE behaviour of the system.

This paper is concerned mainly with the analysis of the rate-promotion effect of amines, and in particular of a new class of sterically hindered amines. The chemistry of hindered amines has been discussed recently by Sartori and Savage[3].

Rate promotion has been discussed in the literature, and two different mechanisms have been considered. In the first mechanism the promoter acts as a homogeneous catalyst for reaction (2); this is the commonly accepted mechanism in the case of arsenious acid.[4-6] In the other mechanism the reaction steps are separated by diffusion steps; this has been called the 'shuttle mechanism'[7] and has been proposed to explain the low-temperature behaviour of amines as rate promoters.[8,9] A paper has recently been published[10] where rate promotion is discussed in general terms; the two mechanisms have been shown to be only quantitatively but not qualitatively different. In this paper, the rate-promotion effect of hindered amines is discussed following the lines of the general argument developed in.[10]

We have performed mass-transfer rate and VLE experiments with a variety of amine promoters. We report results for two amines; diethanolamine (DEA), an amine in common commercial use as a rate promoter, and a sterically hindered diamine (HDA). A hindered amine is defined structurally as a primary amine in which the amino group is attached to a tertiary carbon atom, or a secondary amine in which the amino group is attached to a secondary or tertiary carbon atom. Some examples of sterically hindered amines have been given.[3] Hindered amines are characterized by a low tendency to form carbamates owing to the bulkiness of the substituent attached to the amino group. HDA contains a sterically hindered secondary amino group and an unhindered primary amino group. The latter serves mainly to increase the solubility of the diamine in the hot potassium carbonate solution.

Before discussing the mass-transfer rate experiments a discussion will be given of the thermodynamic (VLE) behaviour of amine-promoted carbonate solutions in the presence of CO_2. The thermodynamic model is needed to calculate the driving forces for mass transfer and in the interpretation of the chemical kinetics.

THERMODYNAMICS

The thermodynamic analysis given here is based on a general methodology for developing the VLE behaviour of gas-treating systems described in ref. (4). Consider an aqueous solution originally made up of m mol dm^{-3} of K_2CO_3, and Rm mol dm^{-3} of an amine rate promoter, RNH (for reasons that will become apparent later, we exclude from consideration the case where the amine is tertiary). As CO_2 is absorbed, the composition of the solution will change; let there be ym mol dm^{-3} of chemically combined CO_2 in the solution (the original solution corresponding to $y=0$).

From a thermodynamic viewpoint we are interested in the prediction of the composition of the liquid phase, and of the corresponding equilibrium vapour pressure of CO_2, as functions of the degree of chemical saturation y. It should be borne in mind that, while in the case of an unpromoted solution ($R=0$) the value of y is restricted to the range 0–1 (except at very high CO_2 partial pressures where physical solubility of CO_2 is significant), for a solution promoted by an alkaline species such as an amine the upper bound to the value of y becomes $1+R$, since the amine itself provides an additional chemical sink for CO_2 through its ability to transform into the protonated form RNH_2^+.

In addition to its free (RNH) and protonated (RNH_2^+) forms, the amine may also be present in the carbamate ion form, $RNCOO^-$. We neglect the possibility that a significant amount of alkylate may form through reaction of alcoholic functional groups of the amine, since the range of pH values of interest is too low. Also, we will not consider in detail the additional complications arising in the case of polyamines.

Should the amine be entirely in the form of free amine, the concentrations of the relevant non-volatile components in the liquid phase would be given by

$$[CO_3^{2-}] = m(1-y) \tag{3}$$

$$[HCO_3^-] = 2my \tag{4}$$

$$[RNH] = Rm \tag{5}$$

$$[RNH_2^+] = [RNCOO^-] = 0. \tag{6}$$

However, the amine can become converted to its protonated and carbamate forms by the occurrence of the following reactions:

$$HCO_3^- + RNH \to CO_3^{2-} + RNH_2^+ \tag{7}$$

$$HCO_3^- + RNH \to RNCOO^- + H_2O. \tag{8}$$

Let f be the fraction of the amine converted (so that $1-f$ is the fraction present as free amine). Thus

$$[RNH] = Rm(1-f) \tag{9}$$

$$[HCO_3^-] = m(2y - Rf) \tag{10}$$

$$0 \leq f \leq \min(1, 2y/R). \tag{11}$$

Let g be the fraction of amine converted to the carbamate form. Thus

$$[RNCOO^-] = mRg \tag{12}$$

$$[CO_3^{2-}] = m[1 - y + R(f-g)] \tag{13}$$

$$[RNH_2^+] = Rm(f-g) \tag{14}$$

$$0 \leq g \leq f. \tag{15}$$

Eqn (9)–(15) give all the relevant concentrations in terms of the two parameters f and g. The value of the latter is in turn determined by the equilibrium condition for reactions (7) and (8). In particular, let A be the equilibrium constant for reaction (7), i.e. the ratio of the second dissociaton constant of carbonic acid to the protonation constant of the amine. As the alkalinity of the amine increases, so does the value of A; a value of unity corresponds to an amine with $pK_a = 10.3$ at room temperature.

The equilibrium constant A is concentration-based and therefore in principle its value will depend on the composition of the liquid phase. There is precedent in the literature for using a lumped parameter such as ionic strength as a measure of the composition. However, the ionic strength of a carbonate solution does not change very significantly as y changes, and therefore it is reasonable to regard A as independent of y, although it will depend on m. The equilibrium condition for reaction (7) thus becomes

$$A = \frac{[1 - y + R(f-g)](f-g)}{(1-f)(2y - Rf)}. \tag{16}$$

Let B be the product of the equilibrium constant of reaction (8) times the molarity m. The value of B is the appropriate dimensionless measure of the stability of the carbamate:

$$B = \frac{g}{(2y - Rf)(1-f)}. \tag{17}$$

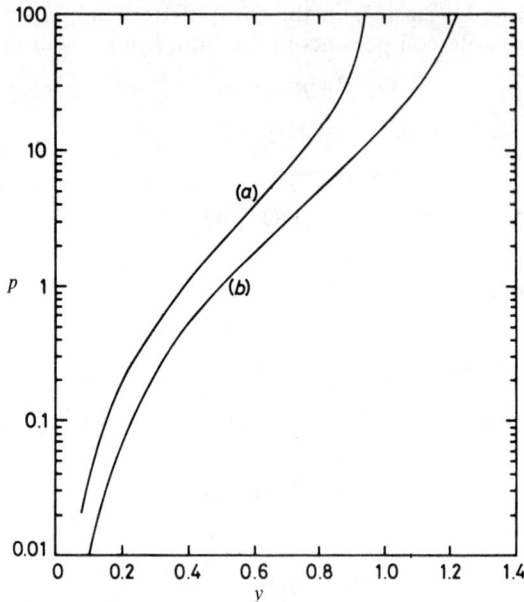

Fig. 1. Dimensionless CO_2 equilibrium partial pressure, p, plotted against CO_2 loading in solution, y (in mol CO_2 per mol initial K_2CO_3). (a) Unpromoted; (b) promoted ($A = 1$, $B = 1$, $R = 0.3$).

The system of eqn (16) and (17) for the two unknowns f and g is equivalent to a fourth-order polynomial equation and thus possesses four roots. However, only one of the roots will be such that f and g are real and satisfy the constraints (11) and (15). Once f and g have been calculated, the concentrations of the non-volatile components are known. If K is the equilibrium constant for reaction (1) and H is the Henry's-law constant for the physical solubility of CO_2, the equilibrium partial pressure p^* is given by

$$p^* = \frac{H[HCO_3^-]^2}{K[CO_3^{2-}]}. \qquad (18)$$

Substitution of eqn (10) and (13) into (18) gives the dimensionless equilibrium partial pressure p as

$$p = \frac{p^* K}{Hm} = \frac{(2y - Rf)^2}{1 - y + R(f - g)}. \qquad (19)$$

Limiting degenerate cases of course arise when either A or B is much larger or much smaller than unity. Results of the calculation are therefore presented with respect to a base case where $A = 1$, $B = 1$ and $R = 0.3$. The base case corresponds to an amine with a room temperature pK_a of 10.3, and a moderately stable carbamate.

Fig. 1 shows the beneficial effect of promoter addition on the capacity of the solution (capacity may be considered to be the swing in CO_2 loading corresponding to a change in CO_2 partial pressure e.g. $p = 0.1$–100). The effect is of particular importance if the partial pressure of CO_2 in the raw gas corresponds to a value of $p >$ ca. 1.0.

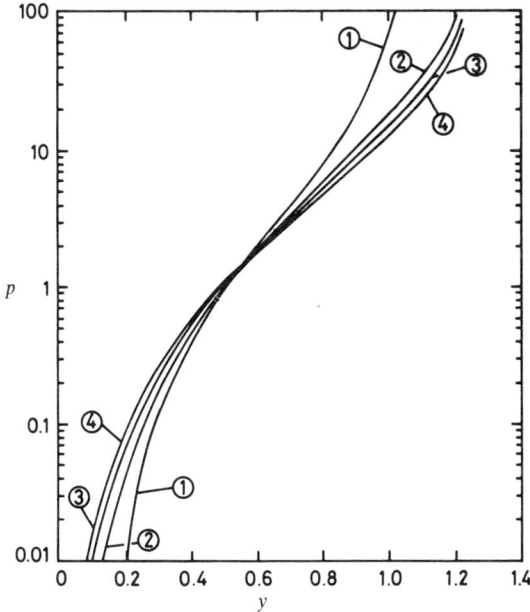

Fig. 2. As fig. 1. $A = 1$, $R = 0.3$. (1) $B = 100$, (2) $B = 3$, (3) $B = 1$, (4) $B = 0.3$.

Fig. 2 shows the effect of the carbamate stability constant, B, on the VLE behaviour. As the carbamate stability decreases, the capacity of the solution increases. However, from this viewpoint, at values of $B <$ ca. 0.3 the capacity increase becomes negligible; the curve for $B = 0$ being almost indistinguishable from that corresponding to $B = 0.3$. Conversely, there is a very significant loss of capacity as B is increased above the value of unity; the curve for $B = 100$ (an amine with a very stable carbamate) actually results in a capacity less than that of an unpromoted solution.

Fig. 3 shows the effect of carbamate stability on the fraction of free amine. As will be discussed later, the effective rate promoter is considered to be the free amine, and therefore it is important to know what fraction of the total promoter is present in the rate-effective form. Again, as the carbamate stability constant decreases, the fraction of unconverted amine increases. With B values of 0.3 or less, a substantial fraction of amine is unconverted even at y values approaching the upper bound of $1 + R$.

Fig. 4 shows the effect of amine alkalinity on capacity. The effect is not very marked. However, if the value of A becomes less than ca. 0.3, the capacity decreases significantly. As the amine alkalinity decreases, more and more amine is present in the form of free amine. Therefore, from a purely thermodynamic viewpoint it would appear that an amine slightly less alkaline than the carbonate, and with a B value of 0.3 or less, is possibly the best organic rate promoter in terms of capacity enhancement and amount of free amine.

The thermodynamic model discussed above does not strictly apply to a diamine, HNRR′NH. For a diamine the following converted terms are possible: singly protonated ($^+H_2NRR'NH$, $NHRR'NH_2^+$), doubly protonated ($^+H_2NRR'NH_2^+$), singly carbamated ($^-OOCNRR'NH$, $HNRR'NCOO^-$), doubly carbamated ($^-OOCNRR'NCOO^-$) and self-neutralizing, ($^-OOCNRR'NH_2^+$, $^+H_2NRR'NCOO^-$).

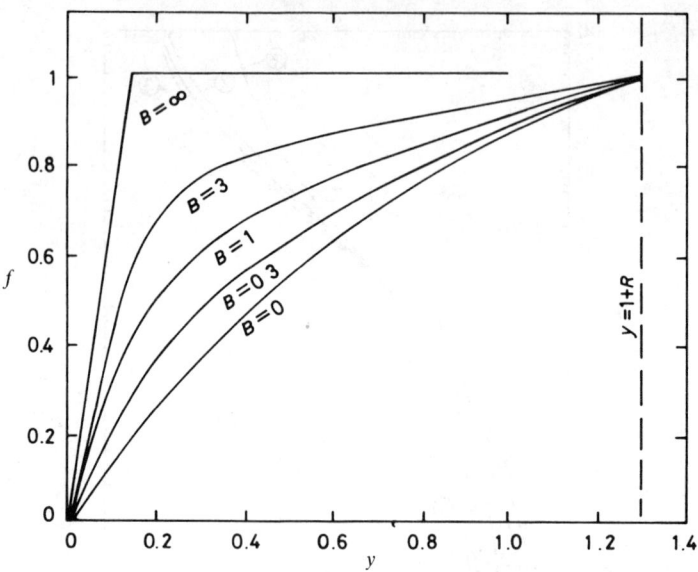

Fig. 3. Fraction of amine converted, f, plotted against CO_2 loading in solution, y. $A = 1$, $R = 0.3$.

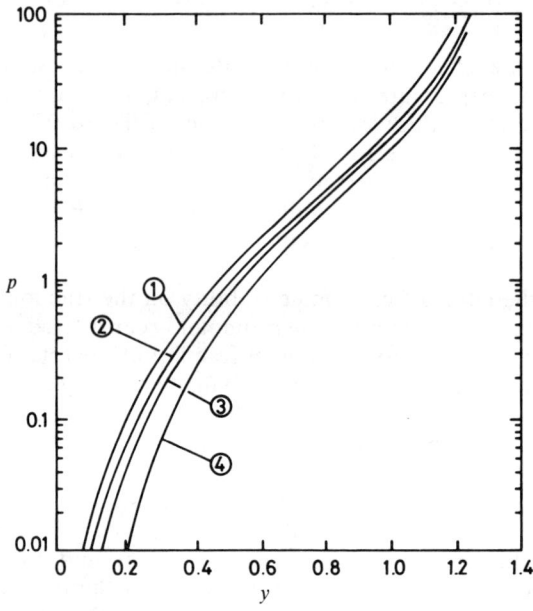

Fig. 4. As fig. 1. $B = 1$, $R = 0.3$. (1) $A = 0.3$, (2) $A = 1$, (3) $A = 3$, (4) $A \to \infty$.

However, consider a diamine where the two nitrogen atoms are separated by a comparatively short and hence rather stiff organic backbone, such as the HDA used in this work. The two amphoteric forms would require severe bending of this backbone, and therefore their formation can be neglected. The doubly protonated form and the doubly carbamated form are likely to be present at concentrations

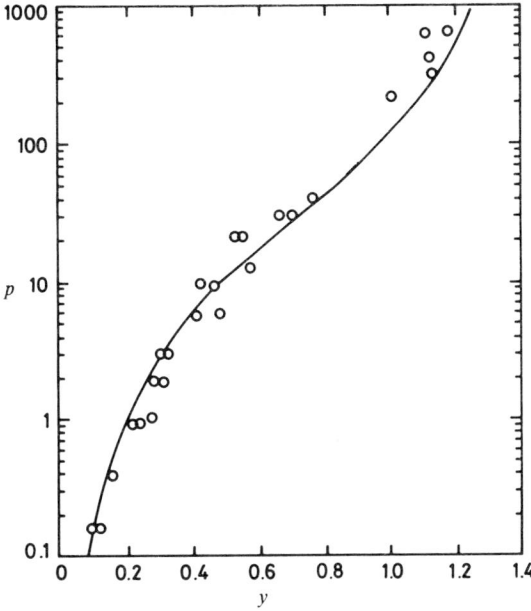

Fig. 5. As fig. 1. The solid curve is theoretical for $A = 1$, $R = 0.3$ and $B = 0.3$. The data points are for hindered amine (HDA), $R = 0.3$.

well below those of the singly protonated and singly carbamated forms, owing to the proximity of the two nitrogen atoms. The equilibrium equations between the two singly protonated forms, and between the two singly carbamated forms, are linear, so that either form is simply proportional to the sum of the two:

$$[^+H_2NRR'NH] \propto [^+H_2NRR'NH] + [HNRR'NH_2^+]$$

and therefore the equations given above hold for the sum of the concentrations of the two protonated and the two carbamated forms. Of course, the possibility of double protonation results in the fact that the upper bound to y is $1 + 2R$ rather than $1 + R$; the plot of log p^* against y will have a high-pressure branch going from $y = 1 + R$ to $y = 1 + 2R$. This, however, occurs at unrealistically high values of p^* for any realistically assumed value of the equilibrium constant for the formation of the doubly protonated form.

The considerations above imply that the model is a reasonable approximation, and this is borne out by comparison with experimental VLE data. Fig. 5 shows data for a K_2CO_3 solution containing 0.3 mol of HDA per mol of K_2CO_3; the curve is calculated from the model. Fig. 6 is a similar plot for a DEA-promoted solution. The cyclic capacity of the HDA-promoted solution is significantly larger than that of the DEA-promoted solution provided the partial pressure of CO_2 in the feed gas is high enough for the high-pressure branch of the plot of p against loading to be utilized.

RATE EXPERIMENTS

The experimental determination of rates of absorption (and desorption) of CO_2 into (or from) aqueous solutions of K_2CO_3 containing the amine promoters has

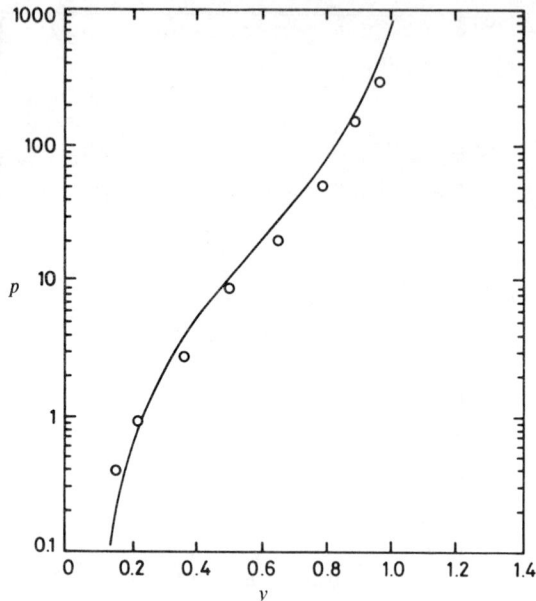

Fig. 6. As fig. 1. The solid curve is theoretical for $B = 1.5$. The data points are for DEA, $R = 0.6$.

been carried out on a one-sphere absorber. The experimental technique has been described elsewhere.[1] The data were reduced to an apparent mass-transfer coefficient k_L defined as

$$k_L = \frac{NH}{p_i - p^*} \qquad (20)$$

where N is the mass transfer flux, H is the Henry's-law constant for CO_2 in the solution, p_i is the gas–liquid interface partial pressure of CO_2 and p^* is the equilibrium vapour pressure of CO_2 corresponding to the bulk liquid composition.

The value of N was measured in the rate experiment. The value of H was assumed to be the same as in an unpromoted solution; the latter had been determined independently.[2] The value of p_i was the difference between the total pressure of the gas and the vapour pressure of water over the solution; the latter had been measured independently in VLE experiments. Finally, the value of p^* was obtained from smooth curves drawn through the VLE data (see fig. 5 and 6). Fig. 7 is a typical rate plot obtained from one of the runs with DEA as the rate promoter.

From the smooth curves drawn through the data in plots such as in fig. 7, the values of a promotion factor E_p have been derived. E_p is defined as follows:

$$E_p = (k_L)_{\text{promoted}} / (k_L)_{\text{unpromoted}}. \qquad (21)$$

The value of $(k_L)_{\text{unpromoted}}$ was known from previous experiments carried out with K_2CO_3 solutions.[1] With 'conventional' amines such as DEA, the rate-promotion factor is approximately independent of fractional saturation (see fig. 8). The values of E_p are rather large; in fact they are at least as large as values reported in the literature,[8,9] in spite of the fact that the latter were obtained under conditions where $(k_L)_{\text{unpromoted}}$ is essentially equal to the value k_L° one would observe in the absence of any chemical reaction. In other words, although at high temperature even

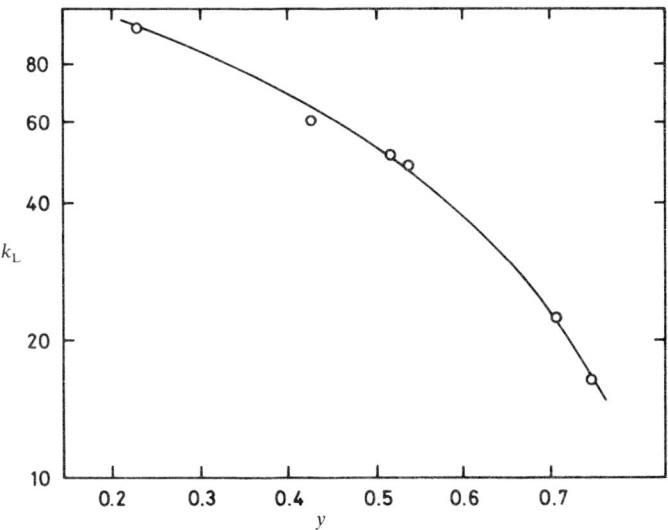

Fig. 7. Mass-transfer coefficient, k_L (cm min^{-1}) plotted against CO_2 loading in solution, y, for a one-sphere absorber at 90 °C.

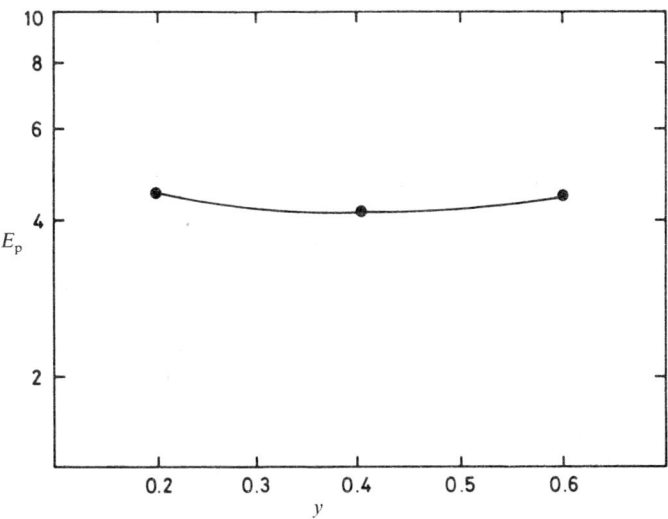

Fig. 8. Absorption promotion factor, E_p, plotted against CO_2 loading in solution, y, for a one-sphere absorber at 90 °C.

unpromoted carbonate solutions show very significant rate enhancement over physical absorption, the addition of amines still has a large promotion effect.

Our desorption rate data are much less extensive than the absorption data. The desorption runs indicate that gas-phase resistance is controlling; indeed the desorption rates for both DEA- and HDA-promoted solutions superimpose in a plot of rate against equilibrium vapour pressure of CO_2. Therefore the only conclusion which can be drawn from the desorption data is that the rate-promotion effect of

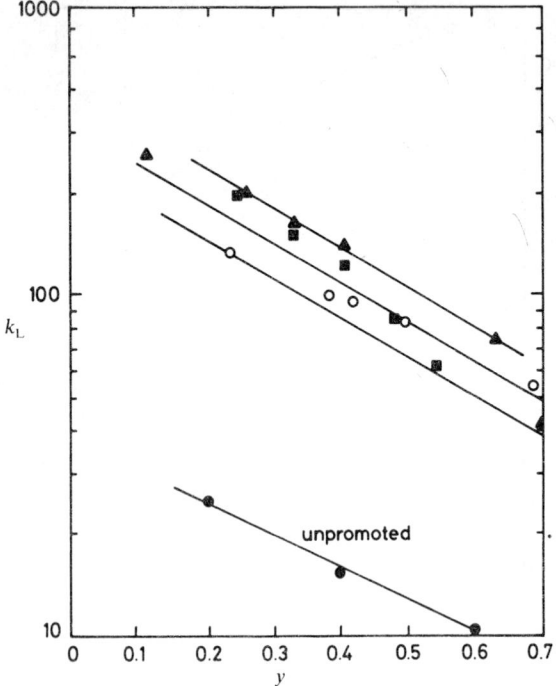

Fig. 9. As fig. 7. Flow rate (cm^3 min^{-1}) as follows: ■, 60; ▲, 60/130; ○, 130; ●, unpromoted.

amines is strong enough to make the gas-phase resistance mass-transfer controlling. This conclusion is supported by the experimental result that desorption rates observed in parallel experiments on a continuously stirred tank reactor (CSTR) are higher than those on the sphere unit. The CSTR, which contains a propeller in the vapour space, has of course better gas-phase mass-transfer properties. An estimate of the gas-phase mass-transfer coefficient for desorption runs allows estimation of an upper bound for a reliable estimate of E_p at ca. 3. Therefore one can only conclude that rate promotion in desorption is large enough to make $E_p > 3$.

Absorption data from three runs for HDA-promoted solutions are shown in fig. 9. Taking the data as a set, all data points fall within ±30% of the mean line. The differences from run to run are believed to be real, reflecting minor variations in solution composition, flow rate and amine manufacturer.

It is important to notice that the values of E_p corresponding to the HDA data in fig. 9 are of the order of 8, as compared to 4–5 for the DEA data in fig. 7. The actual measured rates show an even larger difference between HDA and DEA promoters. Owing to the difference in VLE behaviour, at any given value of y the value of $p - p^*$ is larger for HDA-promoted than for DEA-promoted solutions.

In the interpretation of absorption-rate data, such as that reported in fig. 7 and 9, it is useful to first consider whether the observed behaviour is consistent with the instantaneous-reaction (I-model) or fast-reaction (F-model) regimes of mass transfer. For the exact definitions and analysis of I-models and F-models, see ref. (4).

Available absorption-rate data strongly support an I-type model, at least for the more efficient hindered amine promoter. The evidence is as follows. (i) Absorption rates are much less than proportional to the physical driving force, $p_i - p^*$. Fig. 10

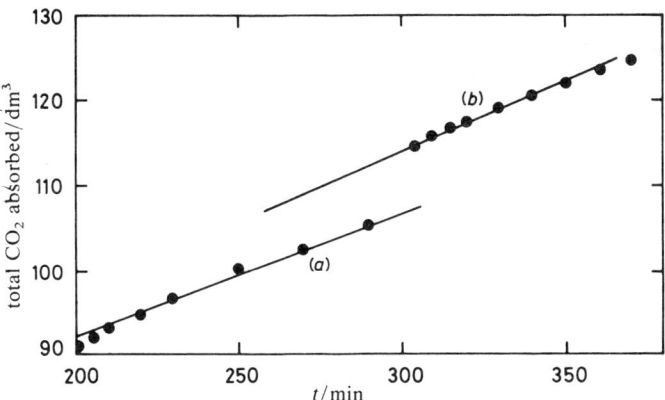

Fig. 10. Total CO_2 absorbed (dm^3) plotted as a function of time, t, for a one-sphere absorber. (a) Slope = 0.17 dm^3 min^{-1}, (b) slope = 0.14 dm^3 min^{-1}. At the point shown by the arrow the pressure was increased from 50 to 105 kpag.

is a plot for a run where the driving force was increased abruptly by a factor of ca. 2; the corresponding increase of the rate of absorption is only ca. 20%. (ii) Rates of absorption on the one-sphere absorber seem to be strongly influenced by the value of the liquid flow rate, as an I-model would predict. (iii) Values of k_L are much larger for promoted than for unpromoted solutions. Since the latter are well correlated by an F-model,[2] it is likely that an I-model prevails in systems containing highly efficient rate promoters.

INTERPRETATION OF DATA

The discussion below follows an approach to the interpretation of rate-promotion behaviour presented in ref. (10). We recall that, given any promoter, Prom, a sequence of chemical reactions is presumed to take place which can be written as

$$CO_2 + Prom \rightarrow Int \qquad (22)$$

$$Int + R + Prom \rightarrow S \qquad (23)$$

where Int is some intermediate, R is the non-volatile reactant and S is the chemical sink for CO_2. The thermodynamic analysis presented in the first part of this work shows that, over a rather wide range of y values, the main reaction taking place during absorption is essentially the transformation of the carbonate ion to the bicarbonate ion. Therefore, R can be identified with CO_3^{2-} and S with HCO_3^-. It is quite likely that Prom is the free (unconverted) amine, although for the case of a diamine such as HDA forms in which one of the nitrogen atoms is converted could conceivably act as promoters. What the intermediate, Int, may be is open to speculation.

There are two mechanisms of rate promotion which have been considered in the literature: the shuttle mechanism (SM) and homogeneous catalysis (HC). In the SM-model, reactions (22) and (23) are separated by diffusion steps, with reaction (22) taking place within the concentration boundary layer and reaction (23) in the bulk of the liquid. In the HC-model, both reactions (22) and (23) take place in the concentration boundary layer.

An absolute upper limit for a SM rate promotion can be calculated on the basis of the following two hypotheses: (i) Reaction (22) proceeds in the I-regime and is essentially irreversible. (ii) All the amine in the solution is present in the form of 'promoter' (presumably free amine). Both hypotheses are very approximate, and both lead to a gross overestimate of E_p. Reaction (22) is unlikely to be almost irreversible, since its reverse is required to take place in the bulk of the liquid. Furthermore, as little as 20% of the total amine may be present in the 'promoter' (free amine) form, as shown by the thermodynamic analysis given earlier. In view of these considerations, actual values of E_p are expected to be significantly less than the value $E_{p,max}$ obtained from the above two assumptions.

The analysis of the SM in the literature[8] refers to the case where there is no rate enhancement in the unpromoted solution, and therefore cannot be applied directly to the case at hand. The relevant differential equations for the film-theory model, taking into account the unpromoted rate enhancement, are, however, easy to integrate. The result is

$$E_{p,max} = \left(1 + \frac{k_L^{\circ 2} C_p^2 H^2}{DkK'(p_i - p^*)^2 [CO_3^{2-}]/[HCO_3^-]}\right)^{1/2} \quad (24)$$

where C_p is the promoter concentration, D is the diffusivity, k is the kinetic constant for direct hydrolysis of CO_2 via reaction with the hydroxide ion and K' is the equilibrium constant for the following reaction:

$$H_2O + CO_3^{2-} \rightleftharpoons HCO_3^- + OH^- \quad (25)$$

and $[CO_3^{2-}]$ and $[HCO_3^{2-}]$ are bulk-liquid values. The value of the product kK is well known.[1]

Fig. 11 is a plot of $E_p/E_{p,max}$ for DEA-promoted solution based on the hypothesis that C_p equals the total amine concentration. Observed values are much too large; it appears that in the solutions considered there is not enough free amine to allow for the observed rate promotion via an SM.

An even stronger argument for rejecting the SM hypothesis is the fact that quite significant rate promotion is observed also in desorption. The discussion in ref. (10) clearly shows that a SM cannot be active in both absorption and desorption.

We conclude therefore that, at high temperatures, the rate promotion of amines takes place through a homogeneous catalysis (HC) mechanism. The analysis based on the HC hypothesis is straightforward. Since the experimental data suggest an I-model, and the HC hypothesis implies that the overall reaction takes place in the interface region, the I-model equations apply directly. This is, of course, an absolute upper bound: all resistance due to chemical kinetics has been eliminated, and absorption proceeds at a rate governed simply by the diffusion of reaction products from the interface to the bulk of the liquid. The upper bound for the rate of absorption is based on the HC mechanism:

$$(N_{max})_{HC} = k_L^{\circ}(y_i^* - y_0)m \quad (26)$$

where y^* is the value of y corresponding to equilibrium with p_i and y_0 is the bulk-liquid value of y.

Fig. 12 is a plot of $N/(N_{max})_{HC}$ for absorption into DEA- and HDA-promoted solutions. The ratio $N/(N_{max})_{HC}$ is less than unity for both promoters, and as a first-order approximation appears to be independent of fractional saturation. It should be borne in mind that $(N_{max})_{HC}$ was calculated by assuming that all diffusivities are equal; a correction for unequal diffusivities would result in a value

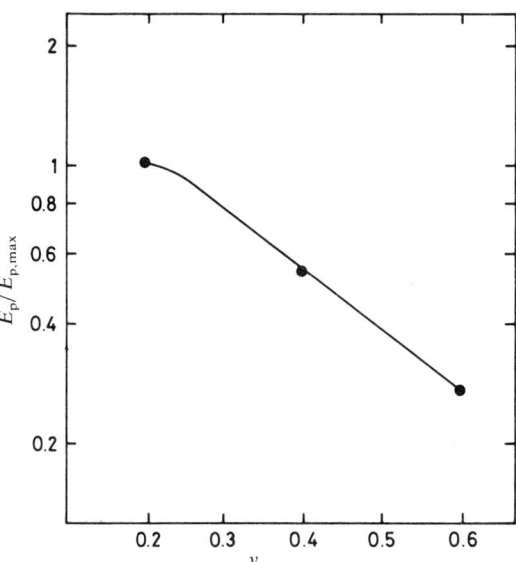

Fig. 11. Normalized promotion factor, $E_p/E_{p,max}$, plotted against CO_2 loading in solution, y.

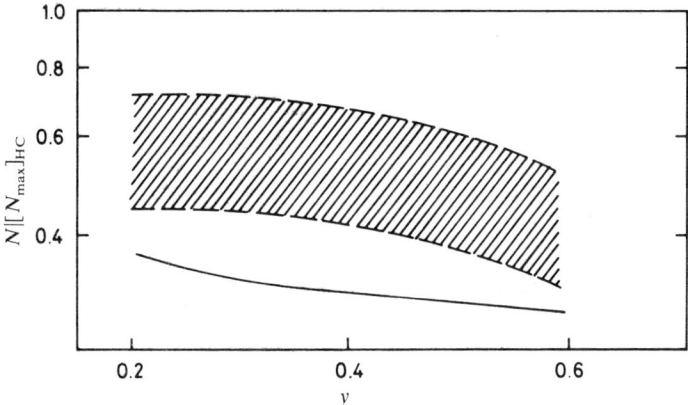

Fig. 12. Normalized absorption rate, $N/[N_{max}]_{HC}$, plotted against CO_2 loading in solution, y. (---) DEA data, 90 °C. Shaded area: HDA data, 90 °C. (Note: 70 and 90 °C data almost coincide.)

of $(N_{max})_{HC}$ ca. 20% lower than the one shown. This seems to indicate that the HDA-promoted data are, within experimental accuracy, in agreement with the I-model predictions: the rate-promotion effect is so large that the CO_2 hydrolysis is practically instantaneous at the conditions of the experiments.

Supporting evidence for an I-model based on HC is obtained by considering the fact that temperature seems to have no effect on the value of $N/(N_{max})_{HC}$. Once the chemical-kinetic resistance has been eliminated, any further increase in the magnitude of the kinetic constants does not result in additional rate increases.

The fact that our data are well correlated by an I-model implies, unfortunately, that no actual values of kinetic constants can be extracted from them; however, a lower bound can be estimated. An I-model is justified if the following condition is satisfied:[4]

$$(Dk_{app})^{1/2} > k_L. \tag{27}$$

For the conditions of our experiments, this results in a lower bound estimate for the apparent pseudo-first-order kinetic constant of $10^6 \, s^{-1}$, an extremely (incredibly) large value.

It is important to point out that, when a rate promoter is active enough to make the rate of CO_2 hydrolysis essentially instantaneous as compared with diffusional phenomena (*i.e.* when an I-model applies), the mass-transfer rates actually observed will be determined uniquely by the equilibrium behaviour of the system. In fact, given a value of the partial-pressure driving force for absorption, $p_i - p^*$, the rate of mass transfer is proportional to the corresponding liquid-phase driving force $y_i^* - y_0$ [see eqn (25)], where y_i^* is related to p_i by the equilibrium condition, and so is y_0 to p^*. Consideration of fig. 2 shows that, as the capacity of a solution is increased, so is the value of $y_i^* - y_0$ corresponding to any given $p_i - p^*$. Therefore, for any two promoters which both yield an essentially instantaneous hydrolysis reaction, the one exhibiting the larger capacity will also exhibit larger rates.

POSSIBLE CHEMICAL MECHANISM

Since actual values of apparent kinetic constants for the catalysed rate of hydrolysis could not be extracted from our data, any attempt to interpret the underlying chemical mechanism is very difficult. The literature on reactions between CO_2 and amines is abundant, and a recent paper gives a good review.[11] However, most of the information available is concerned with systems where the amine acts as a reagent, rather than as a catalyst for CO_2 hydrolysis. The only analogue of the system considered in this work is the case of the reaction of CO_2 with tertiary amines. In fact, even if the tertiary amine acts as a reagent, it does so simply by providing the basicity required to neutralize the HCO_3^- ion, which is the actual chemical sink for CO_2. (Since a tertiary amine cannot form a carbamate, it could in a sense be regarded as being infinitely hindered.) However, there are several indications in the literature[11-13] that the rate of CO_2 hydrolysis in tertiary amine solutions is significantly larger than one would calculate from the known value of the kinetic constant for direct hydrolysis:

$$CO_2 + OH^- \rightarrow HCO_3^-. \tag{28}$$

Furthermore, the rate appears to be proportional to the concentration of free amine and not to the concentration of hydroxide ions. This seems to indicate that tertiary amines act as homogeneous catalysts for reaction (28), *via* a mechanism of the type given by reactions (22) and (23), with P being the unconverted amine. Blauwhoff *et al.*[11] and Donaldson and Van Nguyen[12] hypothesize that the intermediate I may be a hydrated form of the amine; Yu[13] hypothesizes that it may be a zwitterion-type of weakly bonded amine–CO_2 intermediate.

Drawing from the analogy with tertiary amines, the fact that HDA, with $B \approx 0.3$ and hence a high concentration of free amine (see fig. 3), is a good rate promoter makes sense. However, it is not clear why HDA should be a more effective catalyst by several orders of magnitude than the tertiary amines.

We acknowledge the contributions of A. L. Bisio, J. N. Begasse, R. E. Noone and W. C. Yu to this work.

NOMENCLATURE

DEA = diethanolamine
HDA = hindered diamine
m = original K_2CO_3 concentration in mol dm^{-3}
R = original molar ratio of amine to K_2CO_3
y = concentration of chemically combined CO_2 in mol dm^{-3}
f = fraction of amine converted
g = fraction of amine converted to carbamate
A = equilibrium constant for reaction (7)
B = equilibrium constant of reaction (8) × m
K = equilibrium constant for reaction (1)
H = Henry's-law constant for CO_2
p^* = equilibrium partial pressure of CO_2
$p = p^* K / Hm$
$k_L = NH/(p_i - p^*)$
N = mass-transfer flux
p_i = partial pressure of CO_2 at interface
$E_p = (k_L)_{promoted}/(k_L)_{unpromoted}$
k_L° = value of k_L in the absence of chemical reaction
K' = equilibrium constant for reaction (24)
y^* = value of y corresponding to equilibrium with p_i
y_0 = bulk liquid value of y

[1] D. W. Savage, G. Astarita and S. V. Joshi, *Chem. Eng. Sci.*, 1980, **35**, 1513.
[2] S. V. Joshi, G. Astarita and D. W. Savage, *Chem. Eng. Progr., Symp. Ser.*, 1981, **77**, 63.
[3] G. Sartori and D. W. Savage, *Ind. Eng. Chem., Fundam.*, 1983, **22**, 239.
[4] G. Astarita, D. W. Savage and A. L. Bisio, *Gas Treating with Chemical Solvents* (J. Wiley, New York, 1983).
[5] D. Roberts and P. V. Danckwerts, *Chem. Eng. Sci.*, 1962, **17**, 961.
[6] G. Astarita, G. Marrucci and F. Gioia, *III Europ. Symp. Chem. React. Eng.*, 1964, supplement to *Chem. Eng. Sci.*, 1965.
[7] G. Astarita, D. W. Savage, *Adv. Transport Proc.*, 1983, **3**, 340.
[8] A. L. Shrier, P. V. Danckwerts, *Ind. Eng. Chem., Fundam.*, 1969, **8**, 415.
[9] F. Leder, *Chem. Eng. Sci.*, **26**, 1381.
[10] G. Astarita, D. W. Savage, J. M. Longo, *Chem. Eng. Sci.*, 1981, **36**, 581.
[11] P. M. M. Blauwhoff, G. F. Versteeg and W. P. M. Van Swaaij, *Chem. Eng. Sci.*, 1983, **18**, 1411.
[12] T. L. Donaldson and V. N. Van Nguyen, *Ind. Eng. Chem., Fundam.*, 1980, **19**, 260.
[13] W. C. Yu, *Ph.D. Thesis* (University of Delaware, 1984).

Kinetics of Carbon Dioxide Transfer across the Air/Water Interface

By William A. House* and John R. Howard

Chemistry Laboratory, Freshwater Biological Association, East Stoke, Wareham, Dorset BH20 6BB

and Geoffrey Skirrow

Department of Inorganic, Physical and Industrial Chemistry, University of Liverpool, Donnan Laboratories, Grove Street, Liverpool L69 3BX

Received 29th November, 1983

The transport rate of CO_2 into aqueous solution may be augmented by a chemically enhanced flux caused by the hydration of CO_2. The importance of hydration reactions in controlling CO_2 transport depends upon the solution composition and temperature as well as the hydrodynamic conditions at the air/water interface. The transfer has been examined under conditions approaching laminar flow at low Reynolds number. Factors considered to be important in determining the transport rate under these conditions include temperature, pH, alkalinity, solution flow rate and the gas-phase composition.

The transfer rate is expressed in the form

$$\frac{V \, d \sum [CO_2]}{dt} = K_L A ([CO_2]_s - [CO_2]_b)$$

where A is the surface area, V is the solution volume and $\sum [CO_2]$ is the total inorganic carbon concentration. The subscripts s and b refer to the surface and bulk solution concentrations, respectively, and K_L is the gas transfer velocity. The CO_2 concentration at the solution surface is determined by assuming equilibrium between the gas-phase CO_2 and the solution. In the absence of chemical enhancement, the transfer velocity is independent of the pH and solution alkalinity and depends only on the temperature and hydrodynamic conditions at the interface. However, at high pH values the hydration reactions between OH^- and CO_2 cause deviations from the physical transfer model and K_L may be predicted from the solution of a second-order non-linear differential equation subject to appropriate boundary conditions and electroneutrality at all points within the diffusion layer. A numerical model is presented which permits K_L to be calculated for various degrees of chemical enhancement.

Experimental results for a range of bicarbonate concentrations [$(2-40) \times 10^{-3}$ mol dm^{-3}] and CO_2 gas-phase compositions are given. The effects of temperature (20–35 °C) and flow rate (100–500 cm^3 min^{-1}) have also been investigated. The carbonate alkalinity ([HCO_3^-] + 2[CO_3^{2-}]) has a pronounced effect on K_L at high pH. The results are interpreted using the numerical model and compared with calculations based on the assumption of instantaneous hydration equilibrium. The effect of carbonic anhydrase on catalysing the hydration reactions is also briefly discussed.

When disequilibrium exists between a gas phase and a solution, then net gas transfer between the two phases will occur. For CO_2 transport, physical dissolution is accompanied by hydration and ionisation reactions,[1] *i.e.*

$$CO_2 + H_2O \underset{k_{-1}}{\overset{k_1}{\rightleftharpoons}} H_2CO_3 \qquad (I)$$

$$CO_2 + OH^- \underset{k_{-2}}{\overset{k_2}{\rightleftharpoons}} HCO_3^- \qquad (II)$$

$$H_2CO_3^* \overset{K_1}{\rightleftharpoons} HCO_3^- + H^+ \quad [\text{ref. (2)}] \qquad (III)$$

$$HCO_3^- \overset{K_2}{\rightleftharpoons} CO_3^{2-} + H^+ \quad [\text{ref. (3)}] \qquad (IV)$$

where K_1 and K_2 are the dissociation constants of carbonic acid and $H_2CO_3^* \equiv CO_2 + H_2CO_3$. Reactions (III) and (IV) are instantaneous but the hydration reactions (I) and (II) are relatively slow and, in certain conditions, may limit the mass-transport rate of CO_2 into solution. For example, at 25 °C $k_1 = 0.026$ s^{-1}, i.e. $t_{1/2} = 26.7$ s. The purposes of this work are to ascertain the factors controlling interfacial CO_2 transport and to develop a model to describe the transfer between gas mixtures and $KHCO_3 + K_2CO_3$ solutions having carbonate alkalinities, a_C ($=[HCO_3^-]+2[CO_3^{2-}]$), similar to those of natural freshwaters. Although the major ion components in hard waters are calcium and bicarbonate, the susceptibility of pure solutions of $Ca(HCO_3)_2$ to heterogeneous nucleation and precipitation of calcite at high pH prevents their use over a wide pH range. The addition of inorganic phosphate (ca. 20 μmol dm^{-3}) may confer some metastability on these solutions, but even so it is difficult to attain and maintain the pH values characteristic of biologically productive freshwater canals[4] and therefore of interest in this work.

Air/water CO_2 exchange is important in a number of geochemical and biological problems. The potential climatological and other consequences of an increasing concentration of atmospheric CO_2 has stimulated discussion on transfer routes and rates.[5,6] Much of this has centred around transport to the sea, since this is likely to be a major CO_2 sink. Because of their much smaller areal extent, freshwater systems will make only a trivial contribution to atmospheric CO_2 buffering, and interest in transfer to them arises instead from its biological implications for particular local systems. Additionally, many freshwater systems show not only compositions and alkalinities which differ considerably from those of the open sea, but also wide diurnal and seasonal pH ranges, matched only in the marine environment by localised waters such as rock pools. For these reasons, knowledge of the factors which control interphase CO_2 transport for a range of chemical and physical environments is needed.

This investigation concentrates on CO_2 transfer from the gas phase to bicarbonate solutions showing approximately laminar flow. This model system should give information about the importance of reactions (I) and (II) in controlling the transfer and enable predictions to be made of gas transfer in our field system, the Leeds to Liverpool canal.[4] The complicating effects of turbulence in the solution phase were minimised by using a channel with dimensions such that the gas/liquid surface area was sufficient to permit measurable transfer over several hours but yet ensure laminar flow conditions for moderate (ca. 1 cm s^{-1}) flow rates.

EXPERIMENTAL

APPARATUS

The trough (fig. 1) was constructed from 15 mm thick Perspex and was provided with a gas-tight removable lid. The solution was recirculated (via inlet and exit ports B and D) by means of an impeller pump (Gorman–Rupp) and the volume flow rate measured with a

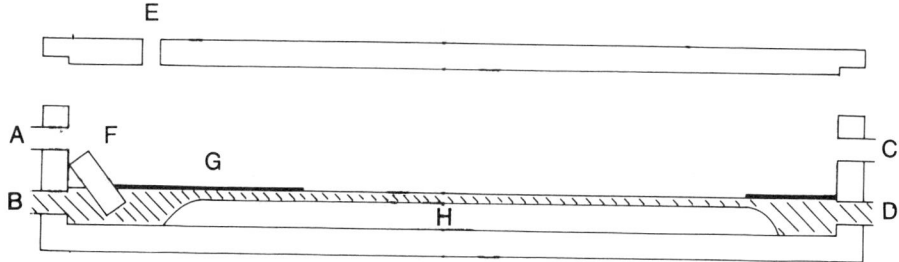

Fig. 1. Recirculating trough used in the CO_2 transfer experiments. Key: (A) gas inlet, (B) solution inlet, (C) gas outlet, (D) solution outlet, (E) aperture for electrode and platinum resistance thermometer, (F) baffle, (G) floats and (H) platform.

Platon flowmeter (0–800 cm^3 min^{-1}). This rate was adjusted to within ±10 cm^3 min^{-1} using a Variac (MC401) to control the pump driver voltage. Baffle F and the entrance well ensured even distribution of the entering solution over the trough width. The gas mixture (see below) was vented *via* ports A and C at *ca.* 1 dm^3 min^{-1}. An aluminium-sheet air thermostat provided temperature control to ±0.2 °C. The trough was mounted on a platform *ca.* 6 cm above the base of the thermostat cabinet and air was circulated underneath the platform *via* two 75 W heaters and a cooling coil. This arrangement produced a uniform temperature within the cabinet. Air and solution temperatures were measured using two four-wire platinum resistance thermometers [conforming to BS 1904 grade II (DIN 43760)] in conjunction with a Kelvin double bridge.

$CO_2 + N_2$ gas mixtures were prepared by mixing the individual pure-gas components using a volumetric mixing apparatus designed to allow flexibility in the selection and control of the composition. Fine-control pressure regulators provided a constant pressure source for each gas and a needle valve controlled the nitrogen flow rate from its source at 1 dm^3 min^{-1}. A Platon Flostat type MN controlled the CO_2 flow in the range 1–60 cm^3 min^{-1} and automatically compensated for upstream pressure variations. Gas flow rates were measured using conventional 'plumb-bob' float flowmeters which were calibrated using bubble meters. Low CO_2 flows (<6 cm^3 min^{-1}) were determined from the pressure difference developed across a 8 cm length of 0.2 mm i.d. stainless-steel tube using a dibutyl phthalate manometer. The CO_2 and N_2 streams were mixed in a bead-packed vessel and the resulting mixture brought to the experimental temperature and saturated with water before entering the trough. Gas-chromatographic checks of the gas composition agreed with those expected from the measured flow rates to within 4% over the CO_2 partial-pressure range 0.003–0.02 atm.*

MATERIALS AND PROCEDURE

Solutions were prepared using A.R. grade potassium carbonate and potassium chloride with singly distilled water. Carbonic anhydrase was obtained from bovine erythrocyte (Sigma Chemical Co.). Carbonate alkalinities of the solutions used (0.002–0.04 mol dm^{-3}) were periodically checked to within 0.1% by Gran titration[7] with standard HCl.

The solution (300 cm^3) was introduced into the trough, the pH electrode and platinum resistance thermometer probe inserted into the solution *via* aperture E and through an opening in float G (see fig. 1) and the liquid circulation commenced. The pH was continuously monitored using a Radiometer PHM64 instrument and a combination electrode (GK2321C). This approach had obvious advantages over sampling methods, although care was needed in electrode calibration. Performance checks before and after each experiment were essential. The solution was monitored for at least 30 min before initiating the gas flow. The digital

---

* 1 atm ≡ 101 325 Pa.

Table 1. Characteristic parameters for laminar flow

flow rate at 25° C/cm³ min⁻¹	$\bar{v}$/cm s⁻¹	Re	h_v/cm	h_f/cm
113	0.41	83	2.1	38
329	1.19	240	6.0	110
545	1.98	400	10.0	184

output from the meter was input at 1 or 2 min intervals to a microcomputer and stored until the termination of the run. Subsequently they were used to calculate solution compositions and gas transfer velocities for particular pH values.

HYDRODYNAMIC CONSIDERATIONS

Preliminary experiments enabled the hydrodynamics of the liquid flow to be characterised, and on the basis of these adjustments were made to the baffle F, floats G and platform H (see fig. 1).

For laminar flow in a rectangular tube (i.e. beneath the first float), the Reynold's number, Re, is[8]

$$Re = \frac{4R\bar{v}}{\nu} \quad (1)$$

where R is the hydrodynamic radius (i.e. the cross-sectional area divided by the wetted perimeter), $\bar{v}$ is the mean velocity of the flow and ν is the kinematic viscosity. Laminar flow is expected when $Re < 2000$. Calculated values for the final design (fig. 1) using 300 cm³ of solution and flow rates in the range 113–545 cm³ min⁻¹ (table 1) fall below this limit. A flow rate of 329 cm³ min⁻¹ was adopted for most experiments.

The 'entrance' length, h, over which the parabolic profile is developed (Levich[9]) is given by

$$h \approx 0.1 \, r \, Re \quad (2)$$

where r is the half-width of the profile. For a rectangular cross-sectioned tube, parabolic profiles develop in both the vertical and horizontal planes leading to 'entrance' lengths h_v and h_f, respectively (table 1). The length of the float G on the inlet side (10 cm) ensured proper development of the vertical profile before gas exchange occurred. The floats also prevented exchange in the 'mixing wells' immediately after the baffle F and near the exit port D. Table 1 shows that $h_f \gg h_v$, thus indicating that plug flow across the trough may be assumed at moderate flow rates. The profiles were also examined by photographing the pattern which developed following injection of a dye at port D. Well developed parabolic profiles were observed in the vertical plane.

THEORETICAL

For given solution conditions, the flux (J) across the interface will be proportional to the surface area (A) and the difference in the CO_2 partial pressures of the gas and liquid phases (p_{CO_2} and P_{CO_2}, respectively). The CO_2 concentrations in the uppermost surface layer (in equilibrium with p_{CO_2}) and in the bulk solution are given, respectively, by $[CO_2]_s = \alpha p_{CO_2}$ and $[CO_2] = \alpha P_{CO_2}$, where α is the solubility coefficient, and for a solution of volume V

$$\frac{d\sum[CO_2]}{dt} = \frac{JA}{V} = \frac{K_L A \alpha}{V}(p_{CO_2} - P_{CO_2})$$

$$= \frac{K_L A}{V}([CO_2]_s - [CO_2]_b) \quad (3)$$

where $\sum [CO_2]$ is the sum of the inorganic carbon species and K_L is the mass transfer velocity. A similar, equivalent, equation can be deduced from Fick's first law on the assumption that as CO_2 passes through the boundary layer it behaves as an unreactive gas (K_L is then independent of the solution pH). In fact, because of the hydration processes K_L will not be constant but is expected to increase with increasing pH. Even so, eqn (3) can be used to obtain experimental values for K_L, i.e. K_L^{exptl}. The results can be interpreted according to the severity of the hydration augmentation of the exchange rate. Three exchange possibilities can be envisaged: (i) no hydration reaction during boundary-layer passage, (ii) hydration equilibrium attained at all points within the layer (chemical-equilibrium model) and (iii) partial hydration during boundary-layer passage (finite-reaction model). These are examined below, where (i) is shown to be a special case of (ii).

CHEMICAL-EQUILIBRIUM MODEL (CEM)

If hydration equilibrium is reached at all points within the boundary layer, then the theoretical maximum flux of CO_2 across the interface is attained and may be calculated [see e.g. ref. (10)]. In the following, the boundary layer is assumed to extend from the surface ($x = 0$) to $x = \delta_{eff}$ where $[CO_2] = [CO_2]_b$. δ_{eff} is the effective thickness of the diffusion boundary layer. From Fick's law at steady state we have

$$\sum_i D_i \frac{d^2 c_i}{dx^2} = -\frac{d \sum [CO_2]}{dt} \tag{4}$$

where i refers to each of the CO_2, HCO_3^- and CO_3^{2-} species. The total flux J_T (defined as positive when the solution phase is gaining inorganic carbon) is given by

$$\sum_i D_i \frac{dc_i}{dx} = -\sum_i J_i = -J_T. \tag{5}$$

From the charge-balance equation for a bicarbonate/carbonate solution when the pH is not too extreme, i.e. $[H^+]$ and $[OH^-]$ are small in comparison with $[HCO_3^-]$ and $[CO_3^{2-}]$, it follows that

$$z \frac{d[M^{z+}]}{dx} \approx \frac{d[HCO_3^-]}{dx} + \frac{2 d[CO_3^{2-}]}{dx} \tag{6}$$

where M^{z+} is the balancing metal ion of charge z. If it is assumed that

$$\frac{d[M^{z+}]}{dx} = 0$$

then

$$\frac{d[HCO_3^-]}{dx} = -\frac{2 d[CO_3^{2-}]}{dx}. \tag{7}$$

The further assumption that $D_{HCO_3^-} = D_{CO_3^{2-}}$ leads to

$$J_{HCO_3^-} = -2 J_{CO_3^{2-}}. \tag{8}$$

Thus from eqn (5) and (8) the total flux is

$$J_T = J_{CO_2} + \tfrac{1}{2} J_{HCO_3^-} \tag{9}$$

or

$$J_T = -D_{CO_2}\frac{d[CO_2]}{dx} - \tfrac{1}{2}D_{HCO_3^-}\frac{d[HCO_3^-]}{dx}. \tag{10}$$

Integration of eqn (10) between the surface and the limit when $[CO_2]=[CO_2]_b$ at $x = \delta_{eff}$ produces

$$J_T = \frac{D_{CO_2}}{\delta_{eff}}([CO_2]_s - [CO_2]_b) + \frac{D_{HCO_3^-}}{2\delta_{eff}}([HCO_3^-]_s - [HCO_3^-]_b). \tag{11}$$

Under conditions such that chemical reactions contribute insignificantly to the total flux, eqn (11) reduces to

$$J_T = J_{CO_2} = \frac{D_{CO_2}}{\delta_{eff}}([CO_2]_s - [CO_2]_b). \tag{12}$$

Comparison of eqn (3) and (12) shows the transfer velocity to be defined by

$$K_L^{exptl} = D_{CO_2}/\delta_{eff}. \tag{13}$$

The concentration gradients of CO_2, H^+, HCO_3^- and CO_3^{2-} may be computed from the chemical-equilibrium model by determining J_T using eqn (11) and by integration of eqn (10) from the surface to some point within the film.

The chemical augmentation of the flux may be quantified by defining a relative flux augmentation parameter, J_R:[10]

$$J_R = (J_T - J_{CO_2})/J_{CO_2}. \tag{14}$$

FINITE-REACTION MODEL (FRM)

The possibility that the hydration reaction (II) may be rate-limiting is now considered. Combination of Fick's second law with the rates of reactions (I) and (II) when steady-state flux conditions are attained at all points within the layer yields a non-linear differential equation:

$$D_{CO_2}\frac{d^2[CO_2]}{dx^2} = [CO_2](k_1 + k_2[OH^-]) - [HCO_3^-](k_{-1}[H^+] - k_{-2}). \tag{15}$$

Use in eqn (15) of the reduced form: $T^\dagger = x/\delta_{eff}$ gives

$$\frac{D_{CO_2}}{\delta_{eff}^2}\frac{d^2[CO_2]}{dT^{\dagger 2}} = [CO_2](k_1 + k_2[OH^-]) - [HCO_3^-](k_{-1}[H^+] + k_{-2}). \tag{16}$$

It is necessary to solve eqn (16) with the following initial conditions:

$$[CO_2]_{T^\dagger=0} = [CO_2]_s$$
$$\frac{d[CO_2]_{T^\dagger=0}}{dT^\dagger} = -\frac{J_T \delta_{eff}}{D_{CO_2}} \tag{17}$$

so that the boundary condition

$$[CO_2]_{T^\dagger=1} = [CO_2]_b \tag{18}$$

is fulfilled.

Eqn (16) incorporates $[HCO_3^-]_{T^\dagger}$. This may be evaluated by integration of eqn (10) between the surface and position $T^\dagger$ and making $[HCO_3^-]_{T^\dagger}$ the subject:

$$[HCO_3^-]_{T^\dagger} = [HCO_3^-]_s - \frac{2D_{CO_2}}{D_{HCO_3^-}}\left(\frac{J_T \delta_{eff} T^\dagger}{D_{CO_2}} - [CO_2]_s + [CO_2]_{T^\dagger}\right). \quad (19)$$

The values of $[OH^-]$ and $[H^+]$ in eqn (16) are determined numerically from the charge-balance condition within the boundary layer given the values of $[CO_2]_{T^\dagger}$ and $[HCO_3^-]_{T^\dagger}$. Thus the method of solution requires a choice of δ_{eff} and an initial value of $[HCO_3^-]_s$. A numerical solution of eqn (16) is then sought for the chosen value of $[HCO_3^-]_s$. If the solution does not meet the boundary condition [eqn (18)], then $[HCO_3^-]_s$ is adjusted in an iterative manner until this condition is met. In practice, it is convenient to define a reduced parameter $\beta = [HCO_3^-]_s/[K^+]$ in applying the iterative adjustment. The relative flux augmentation parameter [eqn (14)] and transfer velocity may then be calculated as for the chemical-equilibrium model.

Two distinct numerical methods were used to solve the differential equation. Both are available in the Numerical Algorithms Group (NAG) library. Most of the computations were performed using the variable-step Adams method.[11] Some results were also checked using the variable-order, variable-step Gear method.[11] The K_L value was found to be independent of the tolerance parameter used in the NAG routine. This parameter controls the error in the integration and in the determination of the position where $[CO_2]_{T^\dagger} = [CO_2]_b$. The relationship between the tolerance parameter and accuracy in the solution, i.e. in $T^\dagger$, cannot be guaranteed. For this reason, various tolerance parameters were used to estimate the accuracy of the final K_L. In all the calculations, β was adjusted iteratively to a relative accuracy of 10^{-9}.

RESULTS AND DISCUSSION

In this analysis the thermodynamic constants K_1 and K_2, the Henry's law solubility coefficient of CO_2 and K_W were taken from ref. (2), (3), (12) and (13), respectively. The rate constants and diffusion coefficients are given in table 2.

STANDARD-CONDITION EXPERIMENTS

The reproducibility and calculation procedures were tested under predefined experimental conditions. Solution alkalinity and flow velocity were chosen to be similar to those of the freshwater system being studied (carbonate alkalinity, 0.004 mol dm^{-3}; linear flow rate, ca. 1 cm s^{-1}). For convenience, the temperature was chosen as 25 °C and the gas-phase partial pressure of CO_2 controlled at 0.008 ± 0.0005 atm. These conditions caused the solution to absorb CO_2. The microcomputer stored the pH values at 1 min intervals. These were subsequently used in a 7-point quadratic smoothing routine to generate dpH/dt at 0.1 pH intervals in the pH range 10–7.6. Values of d$\sum[CO_2]$/dt were then calculated from the relationship

$$\frac{d\sum[CO_2]}{dt} = \frac{d\sum[CO_2]}{dpH} \times \frac{dpH}{dt}$$

in which d$\sum[CO_2]$/dpH is the CO_2 buffer capacity of the bicarbonate solution.[16] The apparent equilibrium constants were calculated using the Davies equation[17] for the activity coefficients. For the $KHCO_3 + K_2CO_3$ solutions used, no correction for ion-pairing was needed. The experimental transfer velocity, K_L^{exptl}, for each pH was computed using eqn (3).

Table 2. Constants used in the data analysis

$T/°C$	k_1/s^{-1} [ref. (14)]	k_2 /dm^3 mol^{-1} s^{-1} [ref. (14)]	D_{CO_2} /10^{-9} m^2 s^{-1} [ref. (15)]
20	0.0173	5900	1.785
25	0.0257	8500[a]	2.043
30	0.0373	12 400	2.320
35	0.0513	17 387	2.614

[a] Ref. (1).

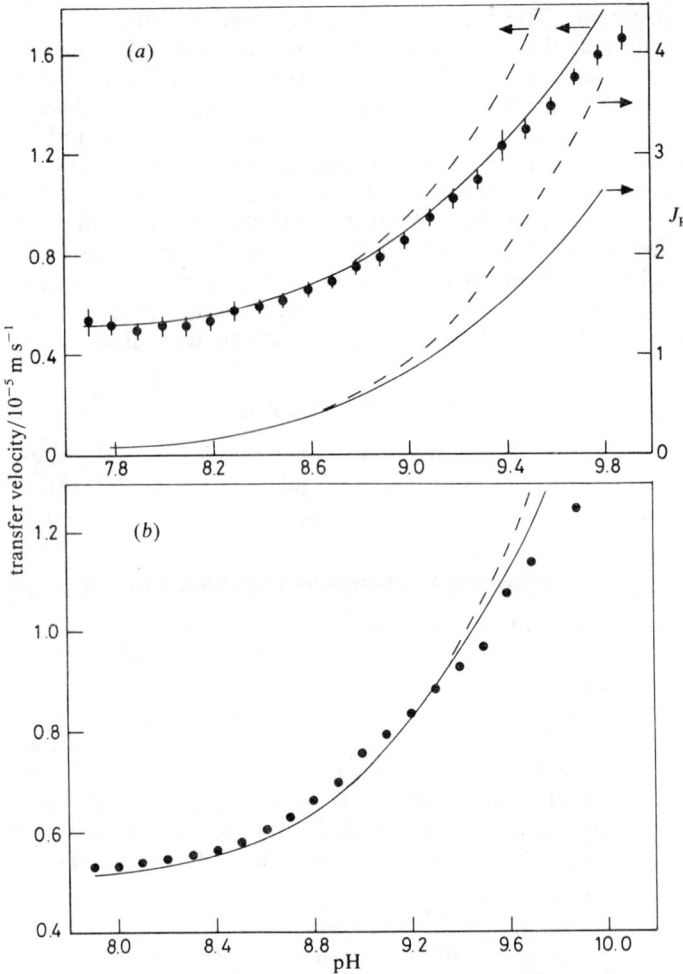

Fig. 2. (a) Results from experiments in standard conditions (alkalinity 0.004 mol dm^{-3}, 25 °C, flow rate 329 cm^3 min^{-1}, $p_{CO_2} = 0.008 \pm 0.0005$ atm). ●, Mean of 4 experiments with standard deviations shown; (---) predicted by CEM; (———) predicted by FRM. (b) Alkalinity 0.002 mol dm^{-3} but otherwise standard conditions. ●, Mean of 2 experiments; (---) predicted by CEM; (———) predicted by FRM.

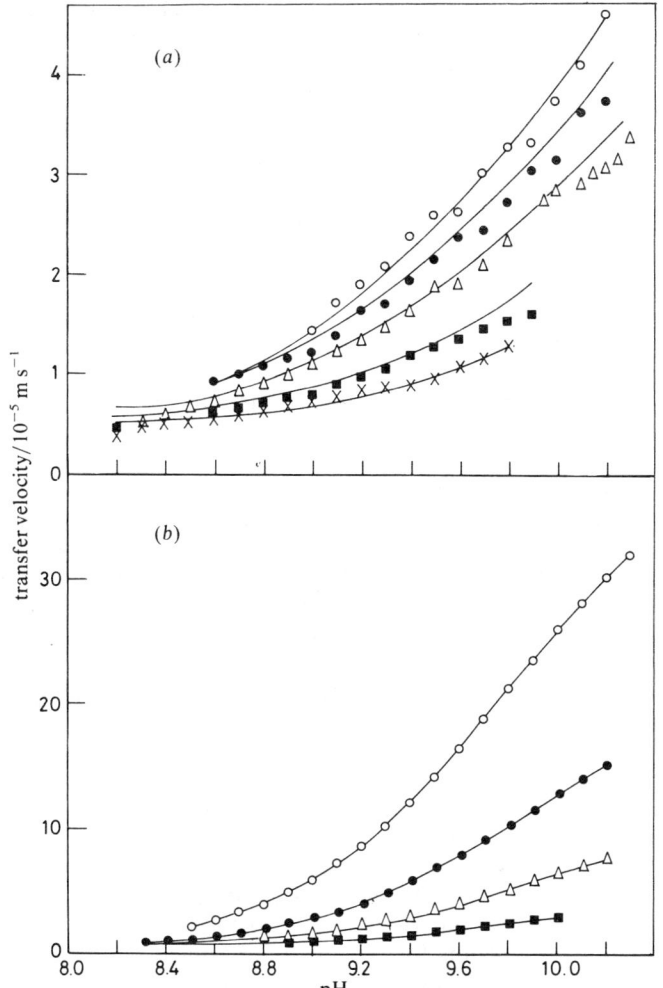

Fig. 3. (a) Alkalinity dependence of K_L^{exptl} at 25 °C, flow rate 329 cm³ min⁻¹, p_{CO_2} = 0.008 ± 0.0005 atm. ×, 0.002; ■, 0.004; △, 0.01; ●, 0.02; ○, 0.04 mol dm⁻³. Solid lines are predicted from FRM. (b) Alkalinity dependence of K_L predicted by CEM at 25 °C. ■, 0.004; △, 0.01; ●, 0.02; ○, 0.04 mol dm⁻³.

The mean results from four experiments are shown in fig. 2(a) together with the standard deviations. Transfer velocities decreased smoothly with descending pH and reached a plateau at pH < 8.2. The standard experiment was repeated at regular intervals to check the equipment performance.

EFFECTS OF ALKALINITY AND IONIC STRENGTH

The dependence of K_L^{exptl} on the solution alkalinity at 25 °C was investigated for the alkalinity range 0.002–0.04 mol dm⁻³ using K_2CO_3 solutions and a gas-phase partial pressure of CO_2 of 0.008 ± 0.0005 atm. All the experiments were performed with a flow rate of 329 cm³ min⁻¹. The results are shown in fig. 3(a). It was expected that the hydration reactions and the effects of chemical augmentation by HCO_3^- and CO_3^{2-} ions would be less important at the lower alkalinities. For alkalinities below 0.001 mol dm^{-3} the slowness of transfer made measurements impractical.

The results summarised in fig. 2(b) are the mean of two experiments for an alkalinity of 0.002 mol dm^{-3} and were analysed using both the CEM and FRM. A program for use with the CEM minimised the quantity

$$\gamma = [\sum (K_L^{exptl} - K_L)^2]^{1/2}/n \qquad (20)$$

(where n is the number of data points) by systematically adjusting δ_{eff}. The agreement between the experimental and the CEM K_L values obtained with the optimum δ_{eff} ($\delta_{eff} = 413$ μm; $\gamma = 0.012$) is shown in fig. 2(b). This value of δ_{eff} was used in the FRM calculations.

Fig. 2(b) shows that the transfer velocities are in good agreement with both models for pH < 9.4, but in the higher pH region (where the two model predictions diverge) the FRM gives better agreement. At pH 9.4 the predicted relative flux augmentation is 0.94 (FRM) and 0.97 (CEM). In view of the agreement between the two models at an alkalinity of 0.002 mol dm^{-3}, the effective diffusion-layer thickness was fixed at 413 μm. This was considered to be a satisfactory compromise between using the more complex FRM to perform the optimisation and using a set of less reliable data for the transfer into 0.001 mol dm^{-3} alkalinity solution, where better agreement with the CEM at higher pH is expected.

Interpretations based on both models have been compared with observed K_L^{exptl} data for the higher alkalinity, 0.004 mol dm^{-3}, and the comparisons are shown in fig. 2(a). Flux-augmentation predictions based on the two models are also given. For this alkalinity J_R(CEM) has risen to 1.99 at pH 9.4, and in general the difference between the two transfer-velocity predictions increases with increasing alkalinity. The particularly large increases in the transfer velocity with increasing alkalinity predicted by the CEM should be noted [see fig. 3(b)]. For example, at the alkalinity 0.04 mol dm^{-3} and pH 9.5, the CEM predicts a transfer velocity of 14.3×10^{-5} m s^{-1} compared with K_L^{exptl} of 2.69×10^{-5} m s^{-1}. The corresponding FRM prediction is 2.50×10^{-5} m s^{-1}. Fig. 3(a) summarises the overall agreement between the calculated (FRM) and experimental results for the alkalinity range examined.

The effects of ionic strength are less dramatic. Transfer velocities were measured using 0.004 mol dm^{-3} alkalinity solutions with concentrations of KCl in the range 0.01–0.1 mol dm^{-3}. An increase in K_L of 0.15×10^{-5} m s^{-1} for the highest ionic strength solution at high pH values decreased to an undetectable amount below pH 8.1. This result was found to be in good agreement with the theoretical predictions from the CEM.

EFFECT OF THE GAS-PHASE COMPOSITION

The applied CO_2 partial pressure was expected to have a large effect on K_L^{exptl} particularly in the high pH region, where chemical augmentation fluxes are significant. The lowest partial pressure that could be used was determined by the slowness of the transfer reaction. All the experiments were performed at 25 °C and 0.004 mol dm^{-3} alkalinity with partial pressures in the range of 0.03–0.0004 atm. K_L^{exptl} values obtained with p_{CO_2} close to 0.008 atm are shown in fig. 4(a), together with the K_L (FRM) predictions. As p_{CO_2} decreases, the rate-limiting effect of the hydration reaction increases, as shown by the changes in J_R in table 3.

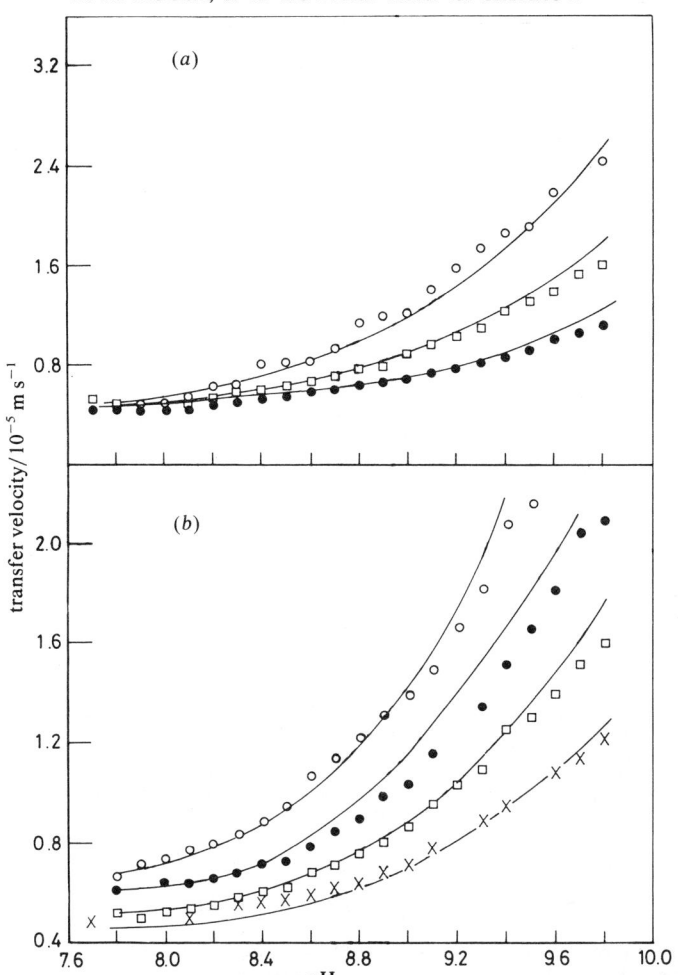

Fig. 4. (a) p_{CO_2} dependence of K_L^{exptl} at 25 °C, flow rate 329 cm³ min⁻¹, alkalinity 0.004 mol dm⁻³. ●, 0.0179; □, 0.008; ○, 0.0027 atm. Solid lines are predicted by FRM. (b) Effect of temperature on K_L^{exptl} for alkalinity 0.004 mol dm⁻³, flow rate 329 cm³ min⁻¹, $p_{CO_2} = 0.008 \pm 0.0005$ atm. ×, 20; □, 25; ●, 30; ○, 35° C. Solid lines are predicted from FRM.

Table 3. Comparison of results from the theoretical models with the experimental data to illustrate the effect of varying p_{CO_2} (carbonate alkalinity = 0.004 mol dm⁻³, 25 °C)

pH	p_{CO_2}	chemical-equilibrium model		finite-reaction-rate model		K_L^{exptl} /10⁻⁵ m s⁻¹
		K_L/10⁻⁵ m s⁻¹	J_R	K_L/10⁻⁵ m s⁻¹	J_R	
9.8	0.0179	1.4	1.7	1.25	1.5	1.12
9.8	0.008	2.3	3.8	1.80	2.6	1.60
9.8	0.0027	6.2	11.6	2.53	4.1	2.43
8.2	0.0179	0.53	0.07	0.53	0.07	0.51
8.2	0.008	0.58	0.16	0.57	0.15	0.55
8.2	0.0027	0.70	0.42	0.64	0.29	0.63

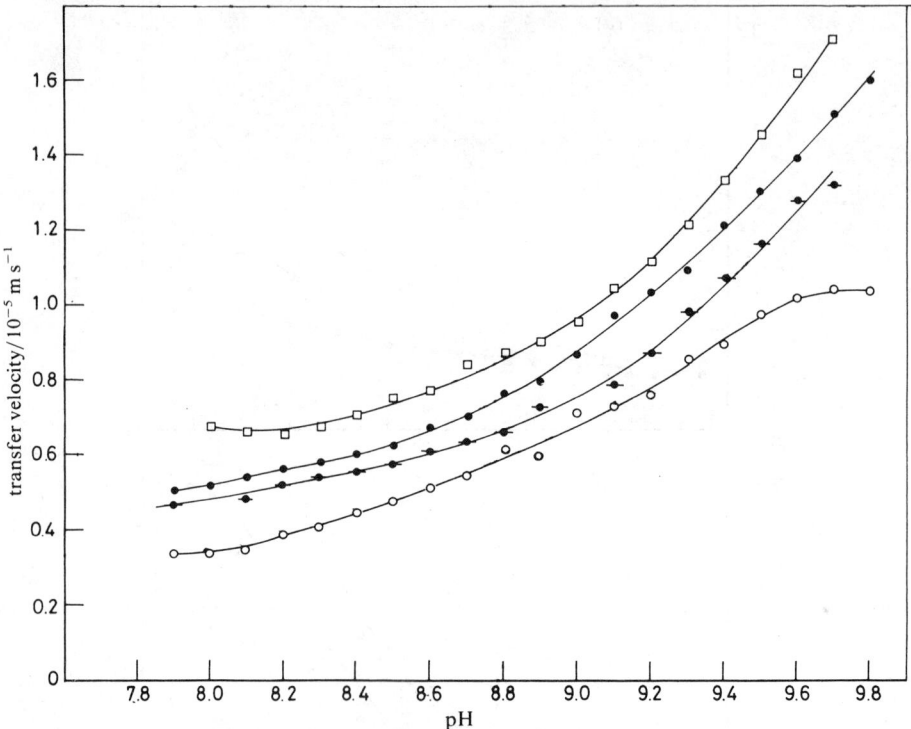

Fig. 5. Flow-rate dependence of K_L^{exptl} under otherwise standard experimental conditions. ○, 113; -●-, 221; ●, 329; □, 545 cm^3 min^{-1}.

EFFECT OF TEMPERATURE

Duplicate measurements of the transfer velocity at temperatures in the range 20–35 °C for 0.004 mol dm^{-3} alkalinity solutions and $p_{CO_2} = 0.008 \pm 0.0005$ atm are shown in fig. 4(b). The constants used at the selected temperatures are given in table 2. The value of $\delta_{\text{eff}} = 413$ μm was retained throughout the temperature range. Observed K_L^{exptl} data and K_L (FRM) predictions agree reasonably at 20, 25 and 35 °C. Agreement at 30 °C is less satisfactory and may reflect inadequacy in the selection of constants. These results support the assumption that δ_{eff} is not sensitive to small changes in temperature.

EFFECT OF FLOW RATE

The use of volume flow rates in the range 113–545 cm^3 min^{-1} (but otherwise standard conditions) showed a five-fold increase in flow to have a small but distinct effect on K_L^{exptl} (see fig. 5). Below pH 9 this amounted to a difference of ca. 0.28×10^{-5} m s^{-1} in K_L^{exptl}. Interpretation in the pH range 7.80–8.60 using the CEM showed J_R to be <0.4 for all the flow rates and good agreement between the CEM and FRM was obtained. Using the CEM, the optimum values of δ_{eff} (for the pH range 7.8–8.6) were found to be 612, 467, 427 and 352 μm for flow rates of 113, 221, 329 and 545 cm^3 min^{-1}, respectively; the minimum values of γ [eqn (20)] were 0.024, 0.007, 0.006 and 0.009, respectively. The value of δ_{eff} at 329 cm^3 min^{-1} is in agreement

GENERAL DISCUSSION

Dr. C. Tondre (*University of Nancy, France*) said: I have two questions that I would like to address to Prof. Astarita, one concerning the thermodynamic aspects of the work presented and a second related to the kinetic aspects. Concerning the first point it seems that the stability constant of the carbamate [eqn (8) and (17)] is an adjustable parameter in his treatment. One could probably obtain a good estimate of this parameter (at least for DEA) by the use of the product of the equilibrium constants of the individual reactions into which eqn (8) can be split:

$$H_2CO_3 \rightleftharpoons CO_2 + H_2O \tag{1}$$

$$HCO_3^- + H^+ \rightleftharpoons H_2CO_3 \tag{2}$$

$$CO_2 + RNH \rightleftharpoons RNCOO^- + H^+. \tag{3}$$

The equilibrium constants for reactions (1) and (2) are well known, and in the case of DEA the equilibrium constant for reaction (3) has been obtained from stopped-flow experiments[1] and confirmed by ^{13}C n.m.r. measurements.[2] Has Prof. Astarita tried to measure the stability constant of HDA by such techniques?

I come now to the kinetic part of my question: the authors seem to admit that the main contribution to CO_2 absorption is not carbamate formation but rather the base-catalysed hydration reaction, and they mention a lower-bound estimate for the kinetic constant of $10^6 \, s^{-1}$. Whereas the contribution of base-catalysed hydration reactions is probably significant for tertiary amines,[3-5] the values obtained for the second-order rate constants are quite low (of the order of 3 to 5 $dm^3 \, mol^{-1} \, s^{-1}$) and for primary or secondary amines such a contribution is expected to be negligible compared with carbamate formation and pure hydration reactions.[1,6] Carbamate formation has been suggested[7] possibly to take place in two steps: (i) formation of a zwitterion by the attack of CO_2 on the nitrogen atom and (ii) a proton-removal step from the zwitterion, which can be base-catalysed. Could the influence of the amine on this last step be an explanation for the base-catalysis observed? Could the carbamate itself not be a sink for CO_2? Is it a question of pH?

[1] D. Barth, C. Tondre and J. J. Delpuech, *Int. J. Chem. Kinet.*, 1983, **15**, 1147.
[2] D. Barth, P. Rubini and J. J. Delpuech, *Nouv. J. Chim.*, 1983, **7**, 563.
[3] T. L. Donaldson and Y. N. Nguyen, *Ind. Eng. Chem., Fundam.*, 1980, **19**, 260.
[4] D. Barth, C. Tondre, G. Lappai and J. J. Delpuech, *J. Phys. Chem.*, 1981, **85**, 3660.
[5] P. M. Blauwhoff, G. F. Versteeg and W. P. Van Swaaij, *Chem. Eng. Sci.*, 1983, **38**, 1411.
[6] D. E. Penny and T. J. Ritter, *J. Chem. Soc. Faraday Trans. 1*, 1983, **79**, 2103.
[7] P. V. Danckwerts, *Chem. Eng. Sci.*, 1979, **34**, 443.

Dr. W. A. House (*Freshwater Biological Association, Dorset*) said: The chemical promoters have two effects: they catalyse the CO_2 hydration reaction and increase the capacity of the carbonate solutions for CO_2. Does Prof. Astarita know if the catalytic effect is maintained at low concentrations (1 mmol dm^{-3} or even 1 μmol dm^{-3}) of promoter?

Prof. H. Linde (*Academy of Sciences of the G.D.R., Berlin*) (*communicated*): In connection with the papers of Prof. Astarita and Dr. House I would like to consider the following problems. The promotion effect is strong enough in comparison with the hydration reaction to make the rate of adsorption and desorption controlled by mass transfer, *i.e.* strongly influenced by the value of the liquid flow rate. Is it possible that differences in the flow rate of the interface itself caused by interfacial dynamic effects have an additional and different influence on the transfer rate?

Usually interfacial dynamic surface renewal depends strongly on its amplification by Marangoni instability due to driving forces and the resulting direction of mass- and/or heat-transfer and/or chemical reaction, as well as on damping of surface renewal by adsorption layers by the mechanism of Gibbs–Marangoni elasticity.

For very clean water surfaces the amplification of surface renewal by Marangoni instability functioning as an intensification effect is possible.

For water surfaces contaminated with even small amounts of surface-active agents or with absorption layers, additional damping effects of the forced surface renewal have to be taken into consideration under conditions of slow forced convection (see Dr. House's paper). If the flow velocity and surface dimensions are small enough, as in the experiment reported in fig. 1 of the paper, at least a part of the water surface is expected to be stagnant because of unfavourable contamination by adsorption layers. For extended water surfaces with very slow forced convection only, this stagnant layer, stemming from natural and man-made contamination is to be expected and causes the damping effect.

Accordingly, stronger forced convection and extended surfaces are at the conditions required to break the adsorption layer and to reproduce the adsorption rate appropriate to the level of surface renewal.

Prof. G. Astarita (*University of Naples, Italy*) said: There is one general point which needs to be clarified with regard to several of the questions which were asked. In our experiments the concentration of amine was never >30% of that of carbonate, and was often significantly less than that; therefore, the capacity of the solution is essentially due to the carbonate, and the main effect of the amine is to act as a rate promoter and not as a reactant. We have done some experiments with lower amine concentrations, although none at concentrations so low as to have no effect on capacity, and certainly nothing as low as 1 mmol dm^{-3}. We have some indication that the catalytic effect would not be observable at extremely low amine concentrations.

There are several reasons for our interpretation of the data in terms of homogeneous catalysis, rather than a shuttle mechanism, and these are discussed in the paper. The most relevant is that the shuttle mechanism, even with simplifying assumptions leading to a gross overestimate of the rate promotion, cannot account for the magnitude of the observed rates.

As for the question of bifunctionality, we have also performed experiments with hindered monoamines, which exhibit very similar results. The amine used in the industrial process, for which results have been presented, has a second amino group added mainly to increase its solubility in concentrated carbonate solutions.

It is true that, in the thermodynamic modelling, the value of B is treated as an adjustable parameter. However, the best fit with the data is obtained with $B = 0.3$, which is in good agreement with an independent estimate based on n.m.r. data. It is also true that carbamate stability data for 'conventional' amines do exist in the literature. However, the essentially novel feature of hindered amines is the very fact that they form carbamates which are remarkably less stable than those of conventional amines.

As stated in the paper, the data are such that actual values of kinetic constants cannot be extracted from them, and therefore any discussion of the underlying chemical mechanism is highly speculative. It is also possible that with hindered amines zwitterion formation is the first step; however, the following steps should lead to the formation of the bicarbonate, which is the chemical sink for carbon dioxide.

The value of H represents the ratio (at equilibrium) of the partial pressure to the concentration of physically dissolved carbon dioxide; therefore in no way is it related to the capacity of the solution. We have assumed it to be the same as in unpromoted solution, and we believe this to be a reasonable assumption. Furthermore, the results of the analysis are very insensitive to the value assumed for H, since instantaneous reaction conditions are approached.

In reply to Prof. Linde's communicated remark I note that our experiments were conducted on a one-sphere liquid–gas device. The gas–liquid interface appeared by visual inspection as very smooth and regular, and therefore gross interfacial instabilities can be ruled out. Of course we have no way to ascertain that no microscopic instabilities did occur. However, the data are in reasonable agreement with those one would calculate for instantaneous reaction conditions on the assumption that no such instabilities are present, and therefore if they are present their effect must be no larger than the experimental inaccuracy.

To answer Dr. House, I would merely say that we have no direct experimental evidence on the catalytic effect at very low amine concentrations.

Dr. W. A. House (*Freshwater Biological Association, Dorset*) (*communicated*): In response to Prof. Linde's communicated comment, I emphasise that, as shown in fig. 5 of our paper, the effect of flow rates in the range 0.41–1.98 cm s^{-1} was measured. As expected, the flow rate does affect the film model in a reduction in δ_{eff}, the diffusion boundary layer thickness. We have not analysed our results using a surface renewal-model because of the defined hydrodynamics in our experimental arrangement. Well developed parabolic profiles were observed in the vertical plane of the trough and plug flow in the horizontal plane. A slide of the parabolic profile was shown to participants at the Discussion.

There is no evidence for Marangoni instabilities caused by adsorption of surface-active agents in our experiments. Presumably such effects would have been detected

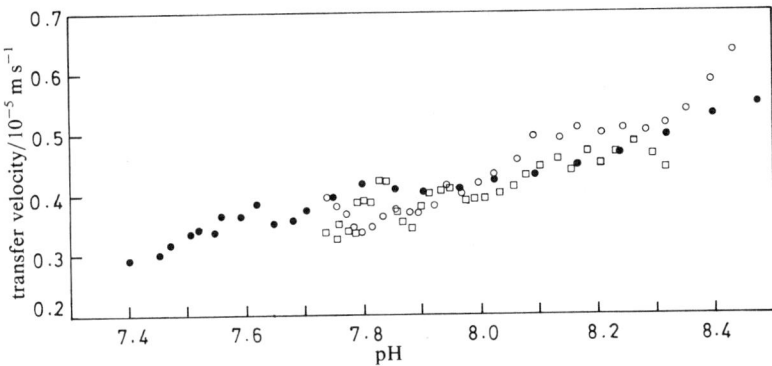

Fig. 1. Comparison of the transfer velocity measured using the recirculating trough apparatus with various hard waters. Flow rate = 329 cm^3 min^{-1}, P_{CO_2} = 0.008 ± 0.0002 atm. □, Synthetic Ca(HCO$_3$)$_2$ solution, alkalinity 0.002 51 mol dm^{-3}; ●, natural hard-water sample [bridge no. 20 on the Leeds to Liverpool canal, see ref. (4) of our paper], total alkalinity 0.002 16 mol dm^{-3}; ○, natural hard-water sample (R. Frome, Dorset, SY 868 868), total alkalinity 0.003 92 mol dm^{-3}.

in the photographs of the dye flowing in the trough. Experiments were also performed using clean $Ca(HCO_3)_2$ solutions and were compared with results of transfer experiments using natural water from a canal and a chalk stream. The results have not been published, but are shown in fig. 1 of this Discussion. The results obtained from the two systems are in close agreement. The freshwater samples would be expected to contain surface-active agents and any additional resistance they caused would have reflected in changes in K_L. However, we have made no systematic study of the effect of surface-active agents or insoluble monolayers under our experimental conditions.

Dr. M. Spiro (*Imperial College, London*) (*communicated*): In their theoretical treatment of the chemical-equilibrium model (CEM), House et al. included the assumption that the trace diffusion coefficients D of HCO_3^- and CO_3^{2-} are equal. This can be tested by making use of the Nernst relation[1] for any ion i

$$D_i^\circ = (RT/F^2)(\lambda_i^\circ/z_i^2)$$

where R is the gas constant, T the absolute temperature, F the Faraday constant and z_i the charge number of the ion. The values of the limiting molar conductance, λ_i°, are known for both ions in aqueous solution at 25 °C, being 45.4^2 and 138.5^1 S cm^2 mol^{-1} for HCO_3^- and CO_3^{2-}, respectively. Hence

$$D^\circ(HCO_3^-) = 1.20_9 \times 10^{-9} \text{ m}^2 \text{ s}^{-1}$$

and

$$D^\circ(CO_3^{2-}) = 0.922 \times 10^{-9} \text{ m}^2 \text{ s}^{-1}$$

a difference of 31%. Both values will be smaller at finite ionic strengths.[1]

Fortunately the assumption of equal diffusion coefficients is not essential to the theory of the CEM. Let us therefore write

$$D(HCO_3^-) = \xi D(CO_3^{2-}).$$

According to Stokes' law, the ξ value of 1.31 will not vary much with temperature. Then Fick's first law, together with eqn (7) of the paper, leads to the following revised equations:

$$J(HCO_3^-) = -2\xi D(CO_3^{2-}) \tag{8'}$$

$$J_T = J(CO_2) + \left(1 - \frac{1}{2\xi}\right) J(HCO_3^-) = J(CO_2) + 0.62 J(HCO_3^-). \tag{9'}$$

The change from 0.5 to 0.62 in eqn (9) will affect the bicarbonate terms in subsequent equations such as (11) and (19). The optimum δ_{eff} value derived from the data in fig. 2(b) of the paper should therefore be increased accordingly.

[1] R. A. Robinson and R. H. Stokes, *Electrolyte Solutions* (Butterworths, London, 2nd edn, 1959), pp. 317 and 463.
[2] G. F. Cassford and A. D. Pethybridge, unpublished work.

Dr. W. A. House (*Freshwater Biological Association, Dorset*) (*communicated*): Our assumption given after eqn (7), $D(HCO_3^-) = D(CO_3^{2-})$, was based on the rigid-sphere model in which the tracer diffusion coefficients are inversely proportional to the square root of the molecular weights. I agree with Dr. Spiro's comment that this restriction in the theory could have been avoided by using the Nernst relationship for $D^\circ(HCO_3^-)$ and $D^\circ(CO_3^{2-})$ although there is uncertainty about the effects of ionic strength on the diffusion coefficients. As Dr. Spiro shows, the effect of the assumption is to alter the magnitude of the chemical augmentation term in eqn (9). The non-augmented flux, $J(CO_2)$, remains unaltered. Repeat calculations with the revised diffusion coefficients for $D(HCO_3^-)$ and $D(CO_3^{2-})$ at an alkalinity of 0.002 mol dm^{-3} and otherwise standard conditions, show that the effect on K_L is $<0.1 \times 10^{-5}$ m s^{-1} below pH 8. The additional augmentation flux will thus have a small effect on the evaluation of δ_{eff} determined for the various flow rates (see fig. 5 of our paper) and over the pH range 7.80–8.60.

The general question of the choice of appropriate diffusion coefficients is difficult because of the range of experimental values reported in the literature at any temperature. A major source of error in estimating δ_{eff} concerns $D(CO_2)$. Differences of up to 20% are obtained between the best theoretical predictions of $D(CO_2)$ and experimental values in the temperature range 6.5–65 °C.[1] It is also worth noting that reliable experimental values of the limiting molar conductance of CO_3^{2-} are not available for temperatures other than 25 °C and until recently[2] were not available for HCO_3^-. They can, however, be estimated using the Walden rule and the limiting molar conductances at 25 °C.

[1] A. Akgerman and J. L. Gainer, *J. Chem. Eng. Data*, 1972, **17**, 372.
[2] G. E. Cassford, *Ph.D. Thesis* (University of Reading, 1983), p. 130.

Mr. V. K. Cheng (*Monash University, Australia*) (*communicated*): Perhaps I was confused by the drawing in fig. 1 of Dr. House's paper. Could he clarify why the platform H is considerably smaller in size than that of the trough. Surely the intrusion made by the platform in the water trough will induce turbulent flow at the edge of the platform. If the water level above the platform is not sufficiently high, the movement of the gas/water interface may be turbulent as well. Was there any competition between the maintenance of constant temperature and laminar fluid flow when the size of the platform was varied?

The F.R.M. model assumes the coupling of the slow rate-determining steps (I and II) at the interface and the mass transfer of products through the interfacial diffusion layer. Reaction II would suggest that the rate of hydration increases with pH. As a result, at lower pH the rate-determining characteristic of steps I and II and the deviation between the fitting of data with the F.R.M. and C.E.M. would be expected to be more marked. How do the accumulation of products from the various elementary steps and the mass-transfer characteristics, such as the concentration gradient of the reactants and products, relate to each other by the numerical models and by intuitive expectations?

Dr. W. A. House (*Freshwater Biological Association, Dorset*) (*communicated*): The main reason for introducing platform H was to ensure mixing and thus even distribution of the solution in the 'well region' upon entering the trough from tube B. The platform H is smaller than the trough to produce the depression at the entrance. The position of the baffle F was found to be critical in ensuring that plug flow developed across the trough. No experiments were done with different sizes of platform. The 'well region' also provided a convenient section in the mixed solution for the temperature probe and pH electrode. With the flow rates used in these experiments, namely $2.0 > \bar{v}/\text{cm s}^{-1} > 0.4$, the dye profile was photographed and found to be parabolic in the vertical plane. At higher flow rates some turbulence caused by the baffle and platform did develop, and so the hydrodynamic conditions were unsuitable for gas-transfer experiments.

The hydration reactions occur at the air/water interface and as CO_2 passes through the diffusion boundary layer. Thus concentration gradients in CO_2, HCO_3^-, CO_3^{2-} and H^+ occur across the boundary layer. At the lower pH values (typically pH <8), the chemical augmentation flux is a small contribution to J_T. The rate of the hydration reaction (II) depends on both $[OH^-]$ and $[CO_2]$ and the dehydration reaction upon $[HCO_3^-]$ [see eqn (16) of our paper]. Consequently the reaction moves to chemical equilibrium within the film with decreasing pH. As the pH decreases from 10 to 8 $[CO_2]$ and $[HCO_3^-]$ in the solution increase. For a fixed gas-phase composition the concentration gradients of HCO_3^- and CO_3^{2-} across the boundary layer decrease, leading to a large reduction in the second term in eqn (11).

Kinetics and Mechanism of Interfacial Reactions in the Solvent Extraction of Copper

By W. John Albery,* Riaz A. Choudhery and Peter R. Fisk†

Department of Chemistry, Imperial College, London SW7 2AY

Received 13th December, 1983

The rotating-diffusion-cell technique for studying reactions at liquid/liquid interfaces has been applied to the study of the kinetics and mechanism of the extraction of Cu^{2+}, using an oxime ligand. The technique has been extended by using a ring electrode in the ring–disc configuration to measure the flux of Cu^{2+} when it is stripped from the organic phase into strong acid. Both extraction and stripping reactions have been studied at various acid concentrations. In the extraction direction the rate of reaction is first order in Cu^{2+} and first order in oxime at low concentrations of both reactants; at higher concentrations a limiting rate is observed. In the stripping direction a zero-order reaction is observed. This behaviour is interpreted in terms of an interfacial reaction in which the rate-limiting step is the attachment of the first oxime to Cu^{2+}.

The study of the kinetics of systems reacting at the interface between two liquids requires first an interface of defined area, and secondly control of the mass transfer bringing the reactants to the interface and taking the products away. Without control of the area and the mass transfer the measurement of rate constants and the deduction of mechanism is fraught with difficulty. We have therefore developed the rotating-diffusion-cell method[1-4] for the study of such systems. In this method the liquid/liquid interface is established by surface tension on the pores of a Millipore filter. This defines both the location and the area of the interface. The filter disc is then rotated. The hydrodynamics of a rotating disc are known, and the transport of material to and from the disc surface has been calculated by Levich.[5] The apparatus is illustrated in fig. 1. The use of a baffle ensures that one obtains the correct hydrodynamics and transport on both sides of the disc.[1] The rate of reaction can be followed by analysing the composition of the solution in the outer or in the inner compartment. We have shown that a particularly sensitive method is the use of a pH-stat to titrate acidic or basic products of the reaction, thereby providing a continuous record of the rate of reaction and its variation with rotation speed.[1-4] In this paper we describe a further development of the technique, in which the filter-paper disc is surrounded by a ring electrode to provide a second method, by which one can obtain a continuous measurement of the flux of the chemical species leaving the disc surface. The ring–disc electrode is a well established technique in electrochemistry.[6]

Both the pH-stat and the ring–disc methods have been applied to a study of the kinetics and mechanism of the solvent extraction of copper using the oxime ligand 'Acorga P50'. Over half a million tonnes of copper are produced by this method *per annum*. The reaction scheme is as follows:

$$Cu^{2+}_{aq} + 2HL_{org} \rightleftharpoons CuL_{2,org} + 2H^{+}_{aq}$$

† Present address: Ciba-Geigy Industrial Chemicals, Manchester.

where HL is

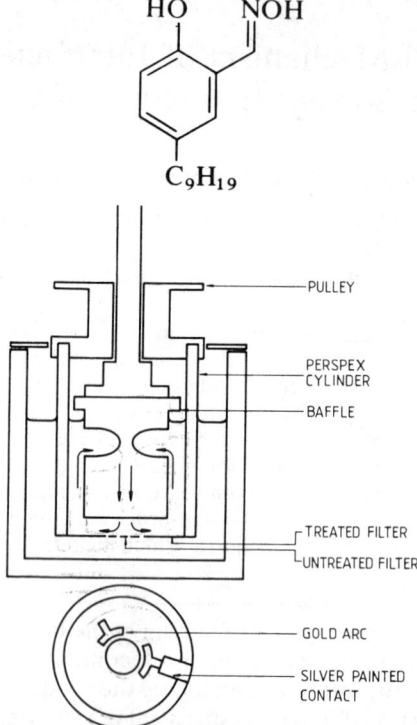

Fig. 1. The rotating diffusion cell. The lower diagram shows the ring–disc arrangement on the lower side of the filter paper.

In our studies the organic phase is n-heptane. In the extraction direction, where the copper is being transferred into the organic phase, the reaction is followed by the pH-stat titrating the H^+ as it is released. We have studied this reaction over a range of pH from 2 to 4. The stripping reaction in the opposite direction takes place when the pH is less than 1. It is therefore impossible to use the pH stat method. Instead the ring–disc technique is used to monitor Cu^{2+} entering the aqueous phase by measuring the current from the reduction of Cu^{2+} on the ring electrode. These kinetic studies allow us to conclude that the reaction takes place on the interface between the two liquids, and to determine the mechanism and rate-limiting step of the reaction.

EXPERIMENTAL

GENERAL

The rotating diffusion cell and apparatus have been described previously.[1] The Millipore filter disc is treated with clearing solution (33% n-hexane, 33% 1,4-dioxane and 34% 1,2-dichloroethane) to collapse the pores, leaving a small untreated disc in the centre; it is on this small central disc that the reaction takes place. We have improved the procedure for making the disc by mounting the filter on a lathe and then painting on the clearing solution using a paint brush and a cylindrical template. In this way good well centred discs can be made. The 'ring' electrodes were made by evaporating gold on to the Millipore filter surface through a mask. Instead of making a complete ring, we found it easier to make an arc that subtended ca. 70°. Two such arcs were evaporated around each disc, so that there was a replacement if the first arc failed. Electrical contact was made by a strip made of evaporated gold strip and then of painted silver running to the outer circumference of the disc and then up the side of the rotating cylinder as shown in fig. 1. This strip was insulated from the solution by painting it with Perspex cement. The counter-electrode was made of platinum gauze and the reference electrode was a saturated calomel electrode (SCE). To reduce Cu^{2+} the ring electrode was potentiostatted at -0.4 V; all potentials are reported with respect to SCE.

The collection efficiency[7] of the ring–disc depends on the geometry. The radius of the disc, r_1, and the internal and external radii of the arc, r_2 and r_3, respectively, were measured by using a travelling microscope on the lathe. Measurements along at least 10 diameters were made. The angle subtended by the arc was measured by making an enlarged photograph of the assembly. The collection efficiency was calculated in two ways. In the first method average values of the radii were calculated and the usual formula[7] was then applied. In the second method the formula was applied to each segment in turn and the overall collection efficiency found by summing the contributions from each segment. No significant difference was found between the results from the two methods.

When the pH is less than 4 a given flux produces a smaller change and therefore special precautions need to be taken in using the pH-stat to follow the reaction. A Burr-Brown 3627 amplifier was fitted to the top of the glass electrode to act as an impedance converter, thereby reducing pick-up in the leads to the electrode.

All chemicals were of AnalaR grade, except for the Acorga P50 and 4-nonylphenol, which were kindly supplied by Dr R. F. Dalton of ICI.

RESULTS AND DISCUSSION

RING–DISC TECHNIQUE

The new ring–disc technique was tested in two ways. First the cell was filled with 0.2 mol dm^{-3} K_2SO_4; an aliquot of $Fe(CN)_6^{3-}$ was added to the inner compartment and the flux measured by the arc electrode and by measuring $Fe(CN)_6^{3-}$ spectrophotometrically in samples taken from the outer compartment. In the second method 1,4-diaminobenzene was added to the inner compartment; the flux of this compound could be simultaneously measured by oxidising it on the arc electrode and by titrating it with the pH-stat. In both experiments good agreement was found between the flux measured by the arc electrode and that measured either spectrophotometrically or by the pH-stat.

RESULTS FOR EXTRACTION

Fig. 2 shows typical results for the flux, j, for the extraction reaction as the oxime concentration, the metal concentration and the pH are varied. At low concentrations of oxime or metal the reaction is first order in both oxime and metal. However, as the concentration of oxime or metal is raised the reaction rate reaches a limit. This type of behaviour is typical of a surface reaction where the surface becomes saturated. The results at pH 3 show that the rate is slower by a factor of 3 compared with the rate at pH 4. Experiments at pH 2 showed that the extraction rate was decreased by a further factor of 2 compared with the rate at pH 3.

The Levich equation for the rotating-disc electrode states that the thickness of the hydrodynamic boundary layer, Z_D, is given by[5]

$$Z_D = 0.643 \nu^{1/6} D^{1/3} W^{-1/2} \tag{1}$$

where ν is the kinematic viscosity, D is the diffusion coefficient and W is the rotation speed in Hz.

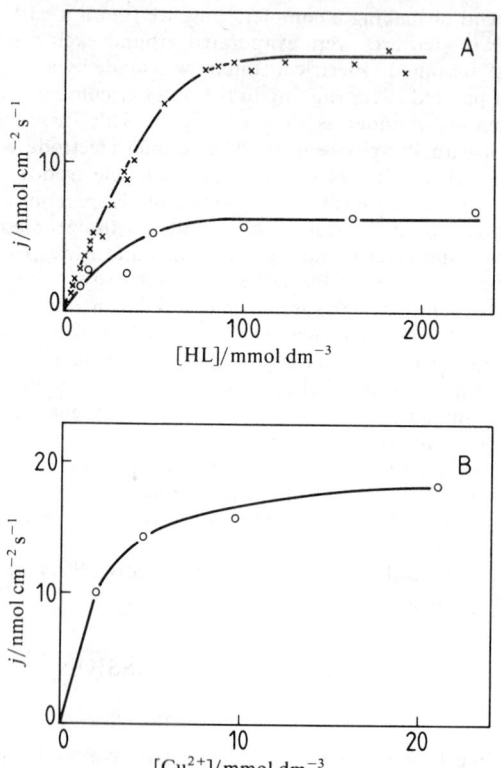

Fig. 2. Typical results for the extraction reaction at a rotation speed of 2.9 Hz. In A $[Cu^{2+}]$ was 10 mmol dm^{-3} and the pH was as follows: ×, 4.0; ○, 3.0. In B [HL] was 67 mmol dm^{-3} and the pH was 4.0.

We have shown[1-4] that the flux, j, is usually given by an equation of the form

$$j^{-1} = j_0^{-1} + \alpha Z_D / Dc \qquad (2)$$

where j_0 is the flux at infinite rotation speed, α is the area of the pores divided by the geometric area and c is the concentration of a reactant.

The second term in eqn (2) describes the effect on the flux of transport through the diffusion layers. By plotting j^{-1} against $W^{-1/2}$ one can extrapolate to find j_0, which describes the flux at infinite rotation speed where mass transport in the diffusion layers is insignificant.

Fig. 3 shows typical results plotted in this way for the extraction reaction at pH 4 for a series of different oxime concentrations. Reasonable straight lines are found.

We have added the product, CuL_2, to the inner compartment to see if the extraction rate was inhibited by the product. Addition of CuL_2 up to a concentration of 60 mmol dm^{-3} made no significant difference to the rate.

RESULTS FOR STRIPPING

Fig. 4 shows typical results for the stripping reaction in 3.0 mol dm^{-3} H_2SO_4. These results have been plotted using a modified form of eqn (2):

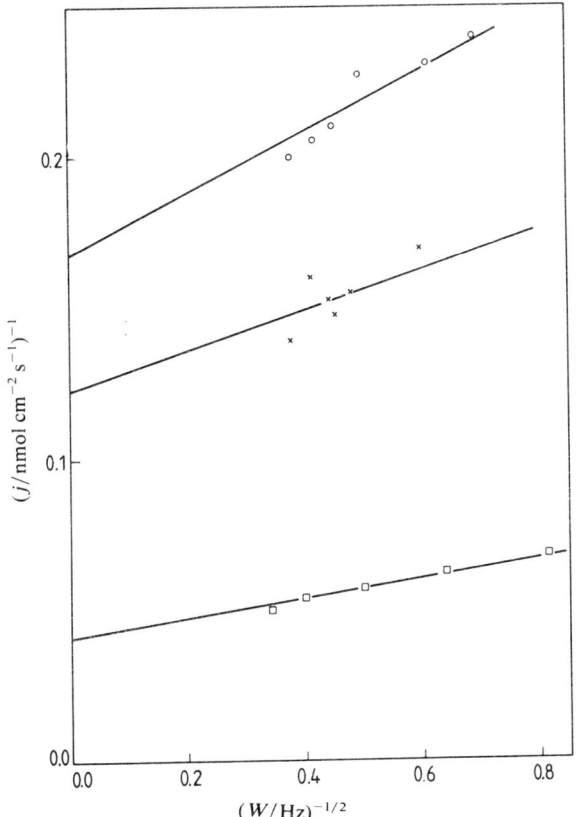

Fig. 3. Typical results for the extraction reaction plotted according to eqn (2). The pH was 4.0 and $[Cu^{2+}]$ was 10 mmol dm^{-3}. [HL] in mmol dm^{-3} was as follows: ○, 13; ×, 20; □, 100.

$$c/j = c/j_0 + \alpha Z_D/D \qquad (3)$$

where c is the bulk concentration of CuL_2. It can be seen that as c is varied the gradients of the different lines are the same, and this agrees with eqn (3). If the reaction at the interface was first order in CuL_2, then all the lines would have a common intercept. This is not found. We may conclude first that the kinetics of the reaction at the interface are significant, and secondly that the order of the reaction with respect to CuL_2 is less than unity.

Similar plots are found for the results in 1.5 mol dm^{-3} H_2SO_4. Fig. 5 shows intercepts from these experiments plotted against c. The linear variation with c corresponds to a rate-limiting step at the interface that is zero order in CuL_2. It is satisfactory that the intercepts of the plots for the two different acid concentrations are in good agreement and correspond to the diffusion of CuL_2 through the filter. It is clear that the zero-order rate depends on the acid concentration; it is faster in 3 mol dm^{-3} H_2SO_4, where the fluxes are close to transport control. The variation of the stripping rate with the concentration of H^+ is shown in fig. 6. The kinetic step at the interface is roughly first order in acid.

We have also found that the stripping reaction in 1.5 mol dm^{-3} H_2SO_4 is inhibited to some extent by the addition of free oxime. Fig. 7 shows results for two different

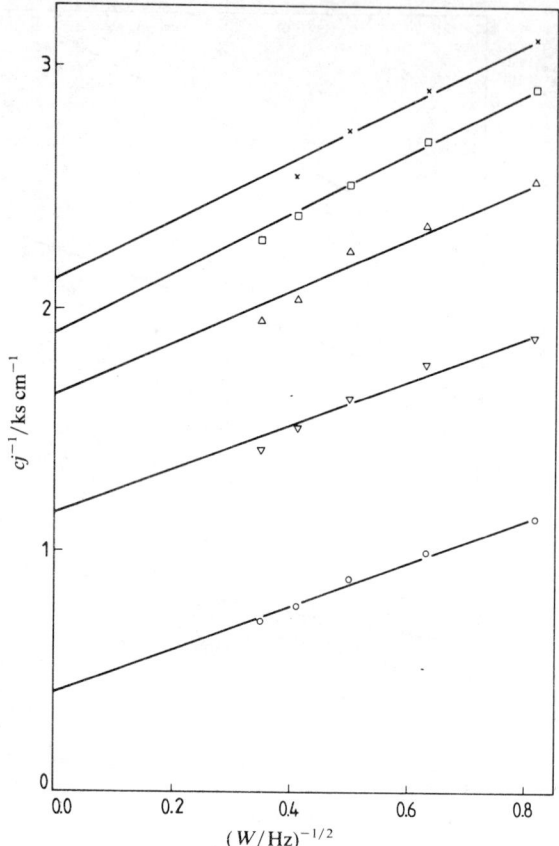

Fig. 4. Typical results for the stripping reaction in 3 mol dm^{-3} H$_2$SO$_4$, plotted according to eqn (3). The [CuL$_2$] in mmol dm^{-3} were as follows: ○, 10; ▽, 36; △, 65; □, 100; ×, 161.

values of c; only modest effects are observed even for concentrations of L as large as 300 mmol dm^{-3}.

LOCATION OF THE REACTION

It has been suggested that the reaction takes place entirely in the aqueous phase[8] by a mechanism in which a limited amount of ligand would dissolve in the aqueous phase and there complex the copper before returning to the organic phase. In the stripping direction CuL$_2$ would dissolve in the aqueous phase and then dissociate. Using stopped-flow spectrophotometry we have measured k_R, the second-order rate constant for complex formation, to be *ca.* 10^5 dm^3 mol^{-1} s^{-1}. This value implies that the reaction length would be smaller than the diffusion length of the rotating disc. Under these conditions the flux for extraction would obey the following equation:

$$j = (k_R D[\text{Cu}^{2+}])^{1/2}[L]. \quad (4)$$

The flux for stripping would be

$$j = (k_D D)^{1/2}[\text{CuL}_2]. \quad (5)$$

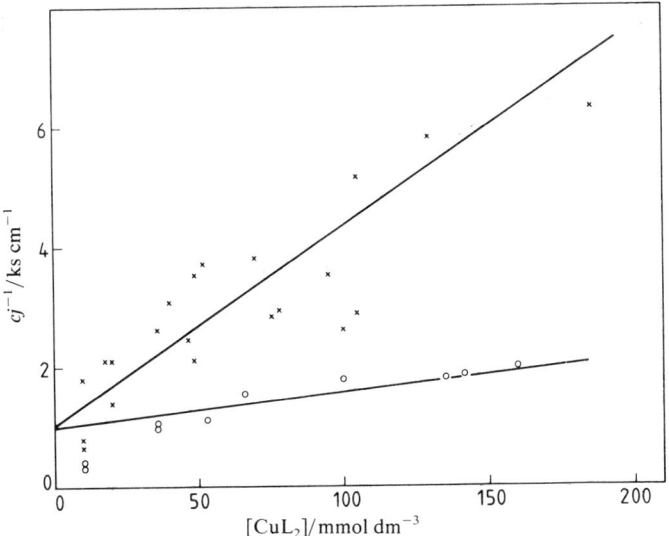

Fig. 5. Intercepts of plots, similar to those shown in fig. 4, plotted against [CuL$_2$]. The concentration of H$_2$SO$_4$ in mol dm^{-3} was as follows: ×, 1.5; ○, 3.

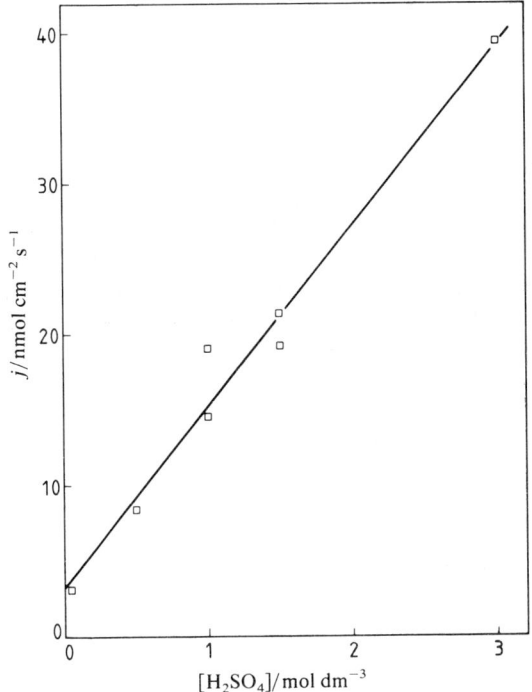

Fig. 6. Rate of the stripping reaction at a rotation speed of 4 Hz plotted against [H$_2$SO$_4$]; the [CuL$_2$] was 100 mmol dm^{-3}.

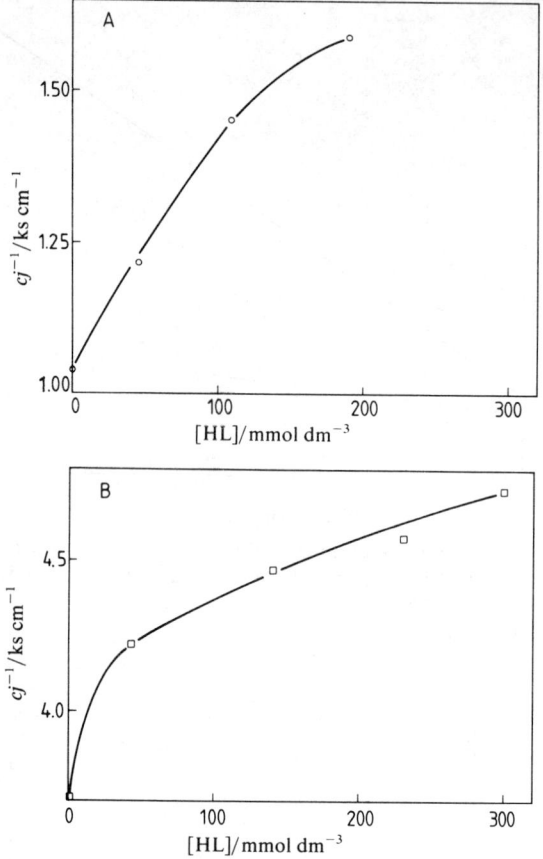

Fig. 7. Effect on the rate of the stripping reaction of the addition of free ligand to the organic phase. The rotation speed was 2.5 Hz and [H$_2$SO$_4$] was 1.5 mol dm^{-3}. In A [CuL$_2$] was 10 mmol dm^{-3} and in B it was 100.

The orders with respect to the reactants in these equations do not agree with those found above. This model does not explain the limiting rates observed in the extraction and the zero-order rate with some inhibition by HL in stripping. We therefore conclude that the reaction does not take place in the aqueous phase, but does take place on the interface.

THE KINETIC MODEL

We propose the following model for the interfacial reaction:

$$\begin{array}{lcccccc}
\text{organic} & 2\text{HL} & \text{HL} & \text{HL} & & & \text{CuL}_2 \\
 & \;\;\;\;K_L\;\Big\updownarrow & \;\;\;\;k_1\;\Big\updownarrow\;\;k_{-1} & \;\;\;\;K_L\;\Big\updownarrow & & & \\
\text{interface} & \rightleftharpoons \text{HL} & \rightleftharpoons \text{CuL}^+ & \rightleftharpoons \text{HL, CuL}^+ & \underset{k_{-2}}{\overset{k_2}{\rightleftharpoons}} \text{CuL}_2 & \underset{k_{-3}}{\overset{k_3}{\rightleftharpoons}} & \\
\text{aqueous} & \text{Cu}^{2+} & \text{Cu}^{2+} & \text{H}^+ & \text{H}^+ & 2\text{H}^+ & 2\text{H}^+ \\
\end{array}$$

We assume that at equilibrium each of the species HL, CuL$^+$ and CuL$_2$ absorb on the interface according to a Langmuir isotherm and that the constant for HL is K_L.

Steady-state analysis of the model gives the following expression for the flux j, where j is positive for extraction and negative for stripping:

$$j\left(\frac{k_2 x_{HL}}{k_{-1}} + 1 + \frac{k_{-2} x_{HL}}{k_L l_0 k_3}\right) = n x_{HL}^2 \left(K_1 k_2 m_0 - \frac{k_{-2} c_0}{K_3 (K_L l_0)^2}\right) \quad (6)$$

$\ddagger 1 \quad \ddagger 2 \quad \ddagger 3$

where $K_i = k_i/k_{-i}$, n is the number of sites on the interface in mol cm^{-2}, x_{HL} is the fraction of sites occupied by HL, m_0 is the surface concentration of Cu^{2+} in the aqueous phase, c_0 is the surface concentration of CuL$_2$ in the organic phase and l_0 is the surface concentration of HL in the organic phase. The fraction, x_{HL}, is given by:

$$x_{HL}\left(\frac{1}{K_L l_0} + 1 + K_1 m_0 + \frac{c_0}{K_3 K_2 l_0}\right) = 1 + \frac{j}{k_{-1} n} - \frac{j}{k_3 n}. \quad (7)$$

empty   HL   CuL$^+$ CuL$_2$

As indicated, the left-hand side of eqn (6) contains three terms corresponding to the three transition states. If one of the transition states is clearly rate limiting, then its term will dominate the left-hand side of eqn (6). The right-hand side of this equation describes the thermodynamics of the system. It will be zero at equilibrium. The first term will dominate for irreversible extraction giving a positive value of j; the second term will dominate for irreversible stripping giving a negative value of j. Turning to eqn (7) for x_{HL}, as indicated, the left-hand side describes the competition for the sites, when it is governed by thermodynamics. The terms in j on the right-hand side describe how the thermodynamic distribution may be perturbed by reaction taking place on the interface. Substitution of eqn (7) in eqn (6) followed by elimination of x_{HL} gives such a complicated result that it cannot be used for mechanistic diagnosis. A better procedure is to obtain simpler expressions by assuming that one of the terms on the left-hand side of eqn (6) is dominant. The results for irreversible extraction and stripping are collected in table 1.

THE RATE-LIMITING STEPS

Inspection of the results in table 1 show that it is unlikely that transition state 2 is rate limiting. In the extraction direction, unless the $K_1 m_0$ term is significant, the reaction would be first order in m_0 throughout; on the other hand if that term is significant, then the flux would pass through a maximum as m_0 increased. Thus this expression cannot explain the limiting flux observed in fig. 2. In the stripping direction a similar argument applies to the variation of the flux with c_0. The reaction is first order in c_0 unless the last term in the bracket is significant, and, if it is, then the flux would pass through a maximum as c_0 increased. In fact the reaction is zero order in c_0.

The remaining expressions for extraction in table 1, E$\ddagger$1 and E$\ddagger$3, do contain terms that can explain the observed rate laws; each expression has a term that is first order in m_0 and first order in l_0 and a term, k_3, that gives a common limit at high concentrations of either Cu^{2+} or of HL. Similarly the remaining expressions for stripping, S$\ddagger$1 and S$\ddagger$3, both contain the required zero-order term k_{-1}. In table 2 we collect together various inequalities that are required for the different combinations of an extraction and a stripping expression; these are based on the dominance of the terms in eqn (6) and on the insignificance of the penultimate terms in E$\ddagger$3 and S$\ddagger$1. The fact that the zero-order limits observed in the extraction and stripping

Table 1. Expressions for the flux from eqn (6) and (7)

dominant term in eqn (6)	irreversible extraction $n/j =$	irreversible stripping $-n/j =$
$\ddagger 1$	$\dfrac{1}{K_L k_1 m_0 l_0} + \dfrac{1}{k_1 m_0} + \dfrac{c_0}{K_L k_1 K_3 m_0 l_0} + \dfrac{1}{k_3}$	$\dfrac{K_2 K_3 K_L l_0}{k_{-1} c_0} + \dfrac{K_2 K_3 (K_L l_0)^2}{k_{-1} c_0} + \dfrac{K_2 K_L l_0}{k_{-1}} + \dfrac{1}{k_{-1}}$
$\ddagger 2$	$\dfrac{1}{K_1 k_2 m_0} \left(\dfrac{1}{K_L l_0} + 1 + K_1 m_0 + \dfrac{c_0}{K_2 K_L l_0} \right)^2$	$\dfrac{K_3}{k_{-2} c_0} \left(1 + K_L l_0 + K_1 K_L l_0 m_0 + \dfrac{c_0}{K_3} \right)^2$
$\ddagger 3$	$\dfrac{1}{K_1 K_2 k_3 m_0 (K_L l_0)^2} + \dfrac{1}{K_L K_1 K_2 k_3 m_0 l_0} + \dfrac{1}{K_L K_2 k_3 l_0} + \dfrac{1}{k_3}$	$\dfrac{1}{k_{-3} c_0} + \dfrac{K_L l_0}{k_{-3} c_0} + \dfrac{K_L K_1 l_0 m_0}{k_{-3} c_0} + \dfrac{1}{k_{-1}}$

Table 2. Inequalities for combinations of extraction and stripping expressions

expressions	inequality	however
E$\ddagger$1 & S$\ddagger$1	$(k_{-1})_S \ll [k_{-1}/K_2 K_L l_0]_S \ll (k_3)_S \approx (k_3)_E$	$(k_{-1})_S \approx (k_3)_E$
E$\ddagger$3 & S$\ddagger$1	$[k_3 K_2 K_L l_0/k_{-1}]_E \ll [k_3 K_2 K_L l_0/k_{-1}]_S$	$[K_2/k_{-1}]_E \gg [K_2/k_{-1}]_S$
E$\ddagger$3 & S$\ddagger$3	$(k_3)_E \ll [k_3 K_2 K_L l_0]_E \ll (k_{-1})_E \ll (k_{-1})_S$	$(k_{-1})_S \approx (k_3)_E$

reactions are comparable means that at first sight the E$\ddagger$1, S$\ddagger$1 and E$\ddagger$3, S$\ddagger$3 combinations are unlikely. However, as we shall see below it may be that k_3 is not responsible for the zero-order limit in extraction. In that case it is possible to have the E$\ddagger$1, S$\ddagger$1 combination. The E$\ddagger$3, S$\ddagger$1 combination is ruled out because the change in acid concentration means that the inequality changes in the wrong direction. The reverse is true for the most likely combination, E$\ddagger$1, S$\ddagger$3, where the change in acid concentration alters the inequality in the correct direction. As one drives a sequence of irreversible steps harder the rate-limiting transition state will be found earlier in the sequence.

To evaluate the rate constants we need a value for n, the number of sites. From the geometry of the oxime ligand we estimate that

$$n = 0.5 \text{ nmol cm}^{-2}. \tag{8}$$

Using this value, in the stripping direction, from the data in fig. 5 we can find a value for k_{-1} in 1.5 mol dm^{-3} H$_2$SO$_4$:

$$k_{-1} = 60 \text{ s}^{-1}. \tag{9}$$

The data in fig. 6 show that this step is roughly first order in H$^+$. The inhibition of the reaction by HL is too small to be certain about which other term is responsible.

In the extraction direction there is no product inhibition so we can ignore the c_0 term in E$\ddagger$1 in table 1. Fig. 8 shows plots of the intercepts, j_0^{-1}, from the $j^{-1}/W^{-1/2}$ plots for series of experiments at constant bulk [Cu^{2+}] plotted against [HL]$^{-1}$.

Table 3. Analysis of data in fig. 8.

pH	$[Cu^{2+}]$ /mmol dm^{-3}	gradient /ks cm^{-1}	$K_L k_I$ /dm^3 mol^{-1} ms^{-1}	K_L /dm^3 mol^{-1}
4	10	1.6	2.5	12
4	100	1.4	1.6	15
3	10	8	2.8	8
3	100	ca. 1	—	—

Reasonable straight lines are obtained. Results for the gradients are collected in table 3. First of all, at pH 3 we find that as expected from the first term in E‡1 the gradient is inversely proportional to $[Cu^{2+}]$. Comparing the experiments at $[Cu^{2+}] = 10$ mmol dm^{-3} we find that increasing acid decreases the rate. In their respective acidity ranges both k_1 and k_{-1} vary with $[H^+]$. This means that there must be two parallel routes, and the simplest hypothesis is

$$\text{extraction:} \quad HL + Cu^{2+} \underset{k_{-1}}{\overset{k_1}{\rightleftarrows}} L^- + Cu^{2+} + H^+ \overset{k_{II}}{\rightarrow} CuL^+ + H^+$$

$$\text{stripping:} \quad H^+ + CuL^+ \rightleftarrows HCuL^{2+} \rightarrow HL + Cu^{2+}.$$

In the extraction direction one would then find that

$$1/k_I m_0 = 1/k_1 + [H^+] k_{-1} k_r k_{II} m_0. \tag{10}$$

Substitution of eqn (10) in E‡1 and ignoring the c_0 term gives

$$\frac{n}{j} = \left(1 + \frac{1}{K_L l_0}\right) \frac{1}{k_1} \left(1 + \frac{k_{-1}[H^+]}{k_{II} m_0}\right) + \frac{1}{k_3}. \tag{11}$$

In the results discussed above the H^+ term dominated giving a k_1 step that was first order in Cu^{2+} and inhibited by H^+. At large concentrations of Cu^{2+} (100 mmol dm^{-1}) and at pH 4, this term becomes small. The results in fig. 8 and table 3 show that the gradients are very similar. From the gradients we can conclude that the two terms are approximately equal at pH 4 when $m_0 = 10$ mmol dm^{-3}. Hence

$$k_{-1}/k_{II} \approx 100.$$

For reactions close to diffusion control, this seems a reasonable result. Using eqn (10) and (11) we can calculate the values of $K_L k_I$ given in table 3. Reasonably constant values are found.

Inspection of eqn (11) shows that the limiting reciprocal flux at high concentrations of Cu^{2+} and HL is given by the sum of $1/k_1$ and $1/k_3$. Unfortunately our data are not precise enough to make an unambiguous distinction between these two terms. In either case we find

$$k_1 \text{ or } k_3 \approx 10^2 \text{ s}^{-1}. \tag{12}$$

If the k_3 term is negligible then eqn (11) shows that we can find K_L from the intercept on the x axis. Results are given in table 3. Again reasonably constant values are found. Furthermore the value of k_1 in eqn (12) is close to the value to be expected for the dissociation of an acid of pK 9. In the bulk homogeneous phase the pK of Acorga P50 is 9.5.[9]

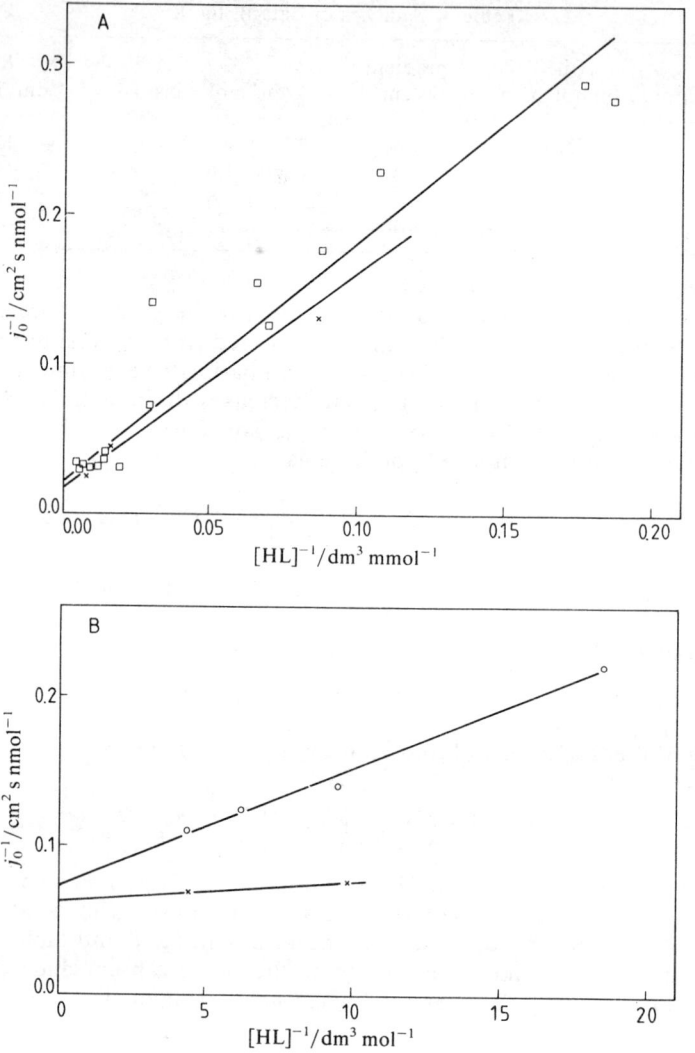

Fig. 8. Intercepts from plots, similar to those shown in fig. 3, plotted against $[HL]^{-1}$. In A the pH was 4.0, and in B it was 3.0. The concentrations of Cu^{2+} in mmol dm^{-3} were as follows: (A) □, 10; ×, 100; (B) ○, 10; ×, 100.

Although more work is required on the fine details of the mechanism, the salient features are clear. Transition-state one is the most rate limiting in both the extraction and stripping directions.

IMPLICATIONS FOR EFFICIENT SOLVENT EXTRACTION

First of all from the results in fig. 2 and table 3 we can see that when the oxime concentration is >150 mmol dm^{-3} the rate reaches a limit. This concentration is therefore the optimum oxime concentration, and is close to the concentration actually used in practice of 200 mmol dm^{-3}.[10] In the extraction direction the limiting rate as given by eqn (8) and (12) is *ca.* 50 nmol cm^{-2} s^{-1}. Under typical extraction conditions the concentration of Cu^{2+} is 0.1 mol dm^{-3},[10] and the radius of the droplets, r, is *ca.* 100 μm.[11] The transport limited flux, j_L, is given by

$$j_L = D[\text{Cu}^{2+}]/r \approx 100 \text{ nmol cm}^{-2} \text{ s}^{-1}.$$

Hence the system is close to being governed by the mass transfer of Cu^{2+} to the interface.

In the stripping direction we have found in 1.5 mol dm^{-3} H$_2$SO$_4$ that the zero-order rate as given by eqn (8) and (9) is *ca.* 30 nmol cm^{-2} s^{-1}. A similar argument suggests that under these conditions the system is again evenly balanced between kinetic and transport control. Increasing the acid concentration above 1.5 mol dm^{-3} would therefore lead to no significant increase in the rate of stripping since mass-transfer control would become dominant. Therefore this concentration of acid, which is the one used in practice, is probably the optimum concentration.

We thank the S.E.R.C. and I.C.I. for financial support. We are grateful to Dr R. F. Dalton for helpful conversations. We thank Mr M. J. Pritchard for making the rotating diffusion cells, Dr M. J. Lee for making the evaporated-ring electrodes and Mr L. R. Svanberg for assistance in pH-stat measurements.

[1] W. J. Albery, A. M. Couper, J. Hadgraft and C. Ryan, *J. Chem. Soc., Faraday Trans. 1*, 1974, **70**, 1124.
[2] W. J. Albery, J. F. Burke, E. B. Leffler and J. Hadgraft, *J. Chem. Soc., Faraday Trans. 1*, 1976, **72**, 1618.
[3] W. J. Albery and J. Hadgraft, *J. Pharm. Pharmacol.*, 1979, **31**, 65.
[4] W. J. Albery and P. R Fisk, *Hydrometallurgy*, 1981, F**5**, 15.
[5] V. G. Levich, *Physicochemical Hydrodynamics* (Prentice-Hall, New Jersey, 1962).
[6] W. J. Albery and M. L. Hitchman, *Ring–Disc Electrodes* (Clarendon, Oxford, 1971).
[7] W. J. Albery and S. Bruckenstein, *Trans. Faraday Soc.*, 1966, **62**, 1920.
[8] R. J. Whewell, M. A. Hughes and C. Hanson, in *ISEC'77* (Canadian Institute of Mining and Metallurgy, 1979) vol. 21, p. 185.
[9] J. S. Preston and R. J. Whewell, *J. Inorg. Nucl. Chem.*, 1977, **39**, 1675.
[10] R. F. Dalton, personal communication.
[11] J. Giles, C. Hanson and H. A. M. Ismail, in *Industrial and Laboratory Nitrations*, A.C.S. Symp. Ser. No. 22, ed. L. F. Albright and C. Hanson (A.C.S., Washington D.C., 1976), chap. 12.

Separation of Metal Ions by Ligand-accelerated Transfer through Liquid Surfactant Membranes

By Darsh T. Wasan* and Zhongmao M. Gu†

Illinois Institute of Technology, Chicago, Illinois 60616, U.S.A.

and Norman N. Li

UOP Inc., Des Plaines, Illinois 60016, U.S.A.

Received 5th December, 1983

The rate of extraction of heavy-metal ions is greatly accelerated by the presence of a ligand in the aqueous solution containing the metal ions. The ligand effect on interfacial mass-transfer rates has been defined by measuring the rate of extraction of cobalt by di-2-ethylhexyl phosphoric acid using sodium acetate as ligand in a modified Lewis cell. The effect of a surfactant such as polyamine on mass transfer at liquid/liquid interfaces has been investigated and is found to be quite significant. This interfacial resistance to metal extraction by surfactant-membrane processes must be taken into account when modelling such systems.

Since the discovery of liquid surfactant membranes for the separation of hydrocarbons over a decade ago,[1] this novel separation technique has been widely studied. It appears that liquid membranes may potentially become effective tools for the separation and purification of many substances. In recent years many authors have reported their studies on the recovery and enrichment of valuable heavy metals[2-8] and the removal of trace contaminants from waste water.[9,10] The formation of liquid surfactant membranes and the general separation process have been described elsewhere.[11]

For metal-ion extraction in the liquid-membrane system, the process can be facilitated by utilizing the mechanism of carrier-mediated transport.[12,13] In this type of facilitation, an ion-exchange reagent is incorporated in the membrane phase to carry the diffusing species across the membrane to the receiving phase. As an example, the extraction of cobalt is achieved according to the following chemical reactions:

extraction: $2HR + Co^{2+} \rightleftharpoons CoR_2 + 2H^+$ (1)
 org.   aq.   org.   aq.

stripping: $2H^+ + CoR_2 \rightleftharpoons Co^{2+} + 2HR$ (2)
 aq.   org.   aq.   org.

where HR represents the protonated form of a liquid exchange agent which is used as the carrier or 'transport facilitator'. In this system, extraction [eqn (1)] occurs at the membrane/external-aqueous-phase interface while stripping [eqn (2)] occurs at the membrane/internal-aqueous-phase interface. The cobalt is effectively concentrated in the encapsulated phase of the emulsion by the continuous permeation of hydrogen ions from the encapsulated phase to the external phase.

† Permanent address: Institute of Atomic Energy, Academia Sinica, Beijing, China.

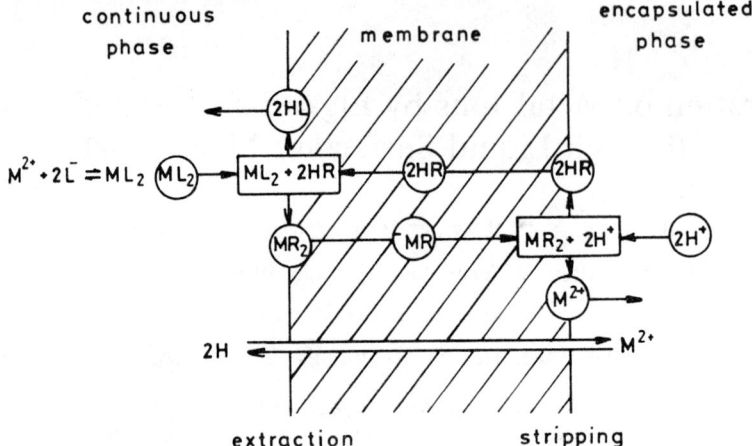

Fig. 1. Mechanism of ligand-accelerated liquid-membrane extraction. M = metal, L = ligand and MR = extractant.

A kinetic study has shown that the process of the extraction of cobalt by di-2-ethylhexyl phosphoric acid (D2EHPA) is slow.[14] This has also been shown by the published kinetic data for the extraction of cobalt by a liquid surfactant membrane.[15]

In our recent experiment on the liquid-membrane extraction of cobalt we found that the introduction of certain anionic ligands (such as acetate) to the aqueous solution containing Co^{2+} greatly accelerated the extraction rate.[16] This phenomenon coincides with the effect found by some other authors.[17-19]

The ligand effect combined with the previously discussed mechanism of carrier-mediated transport becomes the mechanism of ligand-accelerated liquid-membrane extraction, illustrated in fig. 1.

In a liquid-surfactant-membrane system, surfactant is usually incorporated into the membrane phase for its stability. Adsorption and orientation of surfactant molecules at the membrane/aqueous interface lead to the formation of dense, viscous interfacial films which often offer considerable resistance to metal-ion transfer through a liquid-surfactant-membrane system.

. In order to elucidate the effect of ligands on the liquid-surfactant-membrane extraction of heavy-metal ions, interfacial mass-transfer rates were determined for the extraction of cobalt(II) by di-2-ethylhexyl phosphoric acid using a stirred liquid–liquid contactor (Lewis cell). The effects of different ligands on the kinetics of the extraction and the influence of the surfactant on the interfacial resistance to mass transfer were examined in the present study.

EXPERIMENTAL

LIQUID-MEMBRANE EXPERIMENTS

The liquid-membrane phase was composed of 20–40 g dm^{-3} ECA 4360, 2–5% (v/v) D2EHPA and LOPS. ECA 4360 (a non-ionic polyamine, from Exxon) was added as a surfactant, D2EHPA (di-2-ethylhexyl phosphoric acid, from Sigma) was used as an extractant and LOPS (low-order paraffin solvent, from Exxon) was used as membrane solvent and had an average molecular weight of 180, a specific gravity of 0.799 and a viscosity of 2.6 cSt at 60 °F (ca. 289 K).

The internal-phase concentration of the liquid membrane was 50–200 g dm^{-3} H$_2$SO$_4$, which served as a stripping agent. The external aqueous phase of the liquid-membrane system was a CoSO$_4$ solution containing 1000 ppm Co^{2+}. The initial pH value was adjusted to 5.0.

The water-in-oil emulsion was prepared by mixing a 50 cm^3 solution of the internal aqueous phase with a 50 cm^3 solution of the oil membrane phase in a Waring blender for 2 min at ambient temperature. The prepared emulsion was examined microscopically and the internal droplets were found to be <1 μm in diameter. The emulsion (40 cm^3) was then added to a 400 cm^3 vessel containing 200 cm^3 of the CoSO$_4$ solution to be extracted. The system was stirred by a variable-speed mixer equipped with a marine-type impeller; the mixing speed was 200 r.p.m. Samples of the raffinate were taken periodically and analysed with a Beckman u.v. spectrophotometer for cobalt concentration.

During the experiments, the feed solution was preconditioned with different ligands, such as acetate, tartrate, salicylate, succinate and formate, to investigate their effects on the mass-transfer rate.

INTERFACIAL MASS-TRANSFER MEASUREMENTS

All interfacial kinetic studies were conducted in a Lewis cell. The Lewis cell consisted of a cylinder 10 cm in diameter and 8 cm high. The cylinder was constructed of two 4 cm long glass pipes of 10 cm i.d., which were clamped between two flat end plates. These and all other metal parts inside the cell were made of stainless steel. The two glass sections were separated by a circumferential baffle, which, together with the central baffle, divided the cell into two identical halves, each of 250 cm^3 volume, the interfacial area (the annular gap) being 27.3 cm^2.

The two stirrers of the two phases were driven by two variable-speed d.c. motors (from Boding Electric). The stirring-speed range was 0–300 r.p.m.

In order to obtain meaningful results, mass-transfer data for different systems should be compared for the same pH value. Therefore a pH controller (from Cole-Parmer) was used to control the pH of the aqueous phase.

In each run the 250 cm^3 organic phase contained LOPS as solvent and 5% (v/v) D2EHPA as extractant, and the 250 cm^3 aqueous phase was a CoCl$_2$ solution containing 500 ppm Co^{2+}. The pH value of the aqueous phase was maintained at 4.6 ± 0.1 during the course of experiment. The ligand effect was examined by adding different ligands at various concentrations to the aqueous phase.

The effect of surfactant on interface mass transfer was investigated using different concentrations of ECA 4360.

RESULTS AND DISCUSSION

Several ligands were tested and their effects on the liquid-membrane extraction kinetics of cobalt are shown in fig. 2. Among these ligands, acetate was found to be the most effective. With no ligand in the aqueous feed solution it took ca. 15 min for 80% cobalt recovery, while only 2 min were needed for 98% recovery in the presence of 0.1 mol dm^{-3} acetate in the continuous aqueous phase.

An examination of the ligand effect reveals that the selected ligands act as phase-transfer catalysts and accelerate the transfer of the metal ions from one phase to another.[17-19]

Some authors[18,19] have pointed out that some hexa-aquo–metal complexes are very inert kinetically, so that the extraction of metal ions from the aqueous phase to the organic phase is limited by the speed of the release of water molecules. However, the introduction of certain anionic ligands to the aqueous phase may accelerate the extraction rate significantly. It is assumed that the added ligand replaces the coordinated water molecules surrounding the metal ions to form a

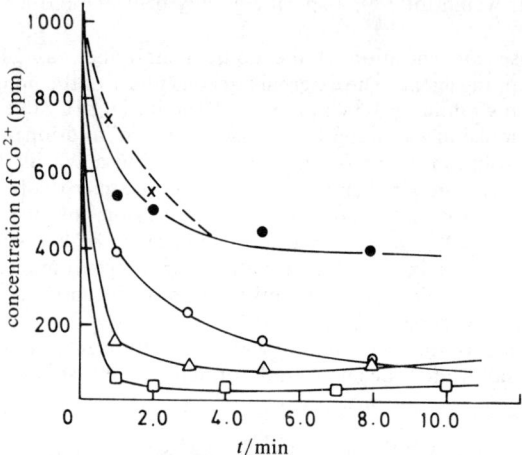

Fig. 2. Effect on the kinetics of different ligands in the external phase: ×, no ligand; □, acetate; △, succinate; ○, formate; ●, tartrate (all ligand concentrations 0.1 mol dm^{-3}).

thermodynamically less stable but a kinetically more labile complex with the metal ions

$$[Co(H_2O)_6]^{2+} + 2Ac^- \rightleftharpoons Co(H_2O)_4 \cdot (Ac^-)_2 + 2H_2O.$$

Such an intermediate ligand–metal complex reacts rapidly with an organic extractant such as D2EHPA and/or with the surfactant at the interface between the membrane phase and the aqueous phase to form either a binary complex

$$\underset{\text{aq.}}{Co(H_2O)_4 \cdot (Ac^-)_2} + \underset{\text{org.}}{2HR} \rightleftharpoons \underset{\substack{\text{binary}\\\text{complex}}}{CoR_2} + \underset{\text{aq.}}{2Ac^-} + \underset{\text{aq.}}{4H_2O}$$

or a ternary complex

$$CoR_2 + R'NH_2 \rightarrow \underset{\text{ternary complex}}{R'H_2N-Co-R_2}$$

where HR is D2EHPA and R'NH$_2$ is polyamine. The whole extraction process is accelerated in this manner.

The equilibrium extraction experiments on the distribution coefficients of Co^{II} as a function of the hydrogen-ion concentration with and without the addition of acetate to the aqueous solution as well as our n.m.r. and visible absorption spectral observations revealed that the ligand does not enter the organic-membrane phase.[16,20,21] Thus the thermodynamic equilibrium for extraction has not been changed by the addition of the ligand to the aqueous phase.

INTERFACIAL MASS-TRANSFER STUDY

Several organic ligands were tested to examine their effects on the kinetics of the extraction of cobalt by D2EHPA. In all the cases studied, 0.03 mol dm^{-3} of different ligands were added to the aqueous phase. The stirring speed of both phases was 150 r.p.m.

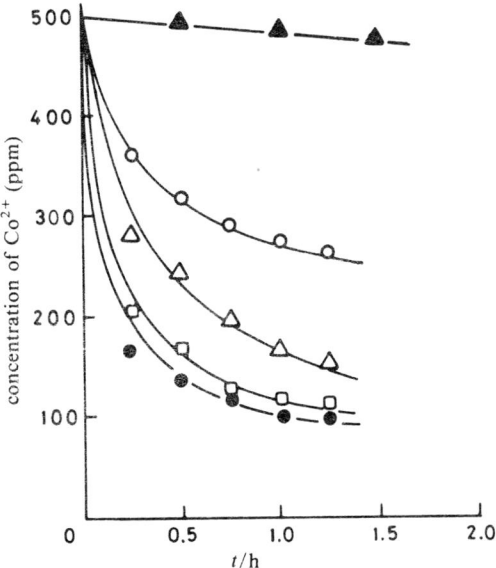

Fig. 3. Effect of various ligands (0.03 mol dm^{-3}) on the kinetics of cobalt extraction: ▲, no ligand; ○, salicylate; △, formate; □, succinate; ●, acetate.

Table 1. Mean value of overall mass-transfer coefficients, $\bar{K}_w$, and the interfacial resistance to mass transfer, $\bar{r}_i$, within the first hour of mass transfer for the extraction of cobalt by D2EHPA with the addition of 0.03 mol dm^{-3} of different ligands to the aqueous phase

ligand	$\bar{K}_w/10^{-3}$ cm s^{-1}	$\bar{r}_i/10^2$ s cm^{-3}
—	0.094	105.7
salicylate	1.64	5.0
formate	3.51	1.7
succinate	5.34	0.8
acetate	6.09	0.5

The kinetic curves for the extraction of cobalt(II) on the addition of different ligands to the aqueous phase are shown in fig. 3. The mean value of the overall mass-transfer coefficients within the first hour of extraction, $\bar{K}_w$, and the related interfacial resistances, $\bar{r}_i$, for different ligands in the aqueous phase were calculted and the values are given in table 1. The sequence of the ligand effect for the different ligands is: acetate > succinate > formate > salicylate. These data show that the ligand effect of the different ligands is in good agreement with that found in our liquid-membrane study of the extraction of cobalt(II) (fig. 2). Acetate is confirmed as being the best ligand for accelerating the extraction.

The effect of ligand concentration on the kinetics of the extraction of cobalt(II) by D2EHPA is shown in fig. 4. Here the average interfacial resistance, $\bar{r}_i$, calculated within the first hour of mass transfer is plotted as a function of the concentration of sodium acetate, which acts as the ligand.

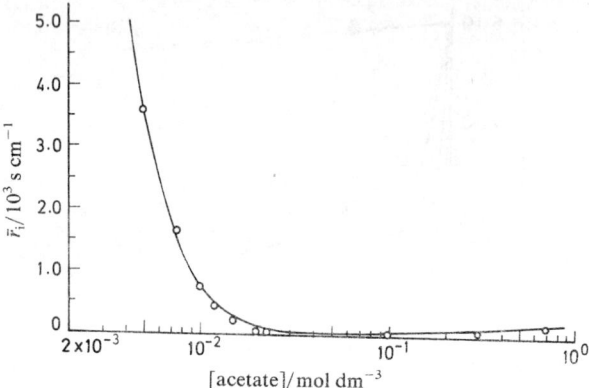

Fig. 4. Interfacial resistance as a function of sodium acetate concentration.

In the present system there is no surfactant other than D2EHPA (extractant) in the organic phase and acetate (ligand) in the aqueous phase. Both D2EHPA and acetate exhibit weak surface activity and their relative concentration at the water/oil interface is favourable for cobalt extraction. Therefore the estimated interfacial resistance should be attributed to the extraction reaction occurring at the interface.

In the absence of a ligand in the aqueous phase $\bar{r}_i$ is as high as 10.57×10^3 s cm^{-1}; it decreases to approximately zero after >0.025 mol dm^{-3} acetate is added to the aqueous phase, as shown in fig. 4. This result indicates that the ligand effect changes the slow interfacial chemical reaction to a very fast reaction.

From fig. 4 it can be seen that the ligand effect increases sharply when the concentration of acetate in the aqueous phase is <0.01 mol dm^{-3}. The ligand effect reaches a maximum (*i.e.* the minimum value of interfacial resistance) near 0.025 mol dm acetate and then remains constant over a wide range of acetate concentration.

Since the initial concentration of Co^{2+} in the aqueous phase is 500 ppm (*i.e.* 0.0083 mol dm^{-3}), the 0.025 mol dm^{-3} acetate in the aqueous phase is approximately three times as high as the molar concentration of Co^{2+} ions. This implies that the reaction for the extraction of cobalt(II) by D2EHPA at the water/oil interface is accelerated after 2 or 3 coordinated water molecules surrounding the Co^{2+} ions are replaced by acetate ions.

A further increase in sodium acetate concentration (>0.2 mol dm^{-3}) slightly decreases the ligand effect. This can be explained by the competitive extraction of sodium with cobalt.

The effect of surfactant on the interfacial mass transfer has been reported to be significant in many previous studies.[22-26] In our study of extraction by a liquid surfactant membrane, ECA4360, a non-ionic polyamine, was used as surfactant to stabilize the liquid membrane. The influence of this surfactant on cobalt extraction was also investigated.

With the presence of a surfactant in the organic phase, note that for the extraction process at the water/oil interface the interfacial resistance to mass transfer results from chemical kinetic resistance as well as from the interfacial barrier caused by surfactant orientation at the interface hindering the passage of solute across the interface. Also, it has been shown in fig. 4 that with the addition of >0.025 mol dm^{-3} acetate to the aqueous phase the interfacial resistance to mass transfer because of chemical reaction may be overcome by means of a ligand effect.

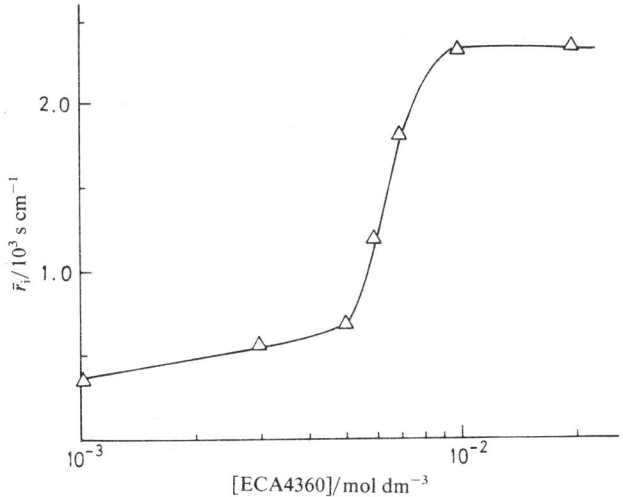

Fig. 5. Interfacial resistance as a function of surfactant concentration.

In our study of the effect of surfactant on interfacial mass transfer, we added 0.1 mol dm^{-3} acetate to the aqueous phase in each run. By doing this we could retard the interfacial resistance to mass transfer as the result of a surfactant layer at the interface by excluding the resistance of a chemical reaction. The interfacial resistance $\bar{r}_i$ due to ECA4360 is now plotted against the concentration of ECA4360, as shown in fig. 5. It is seen from fig. 5 that in the range of low ECA4360 concentration ($<10^{-3}$ mol dm^{-3}) $\bar{r}_i$ increases slowly with increasing ECA4360 concentration. When ECA4360 in the oil phase is $>5.0\times10^{-3}$ mol dm^{-3}, $\bar{r}_i$ rises sharply and then remains approximately constant. This can be interpreted as follows: In the range of low ECA4360 concentration ($<10^{-3}$ mol dm^{-3}), ECA4360 adsorbed at the interface is far from saturation. There exists no rigid surfactant film at the interface; therefore $\bar{r}_i$ in this region rises slowly with ECA4360 concentration. In the region from 5.0×10^{-3} to 1.0×10^{-2} mol dm^{-3} ECA4360, the interface appears to be packed with ECA4360 molecules, forming a dense, rigid interfacial barrier through which the solute must pass. Therefore, $\bar{r}_i$ rises sharply in this region.

Once a densely packed monolayer (or multilayer) of surfactant is formed at the interface, $\bar{r}_i$ reaches a maximum and any further increase in ECA4360 concentration does not significantly contribute to the rigidity of the interfacial film. A substantial constant value of $\bar{r}_i$ is maintained.

We find from fig. 5 that the interfacial resistance due to a surfactant such as polyamine is as high as 2500 s cm^{-1} at a surfactant concentration of ca. 0.01 mol dm^{-3}. This suggests that for liquid-membrane metal extraction, which is accelerated by the ligand effect, the major resistance to mass transfer is concentrated on the peripheral surfactant layers of the emulsion globules, i.e. the interfacial surfactant layer offers the predominant barrier to mass transfer for the liquid-surfactant-membrane system.

The formation of highly viscous interfacial films during extraction in the presence of surfactant was confirmed by the interfacial viscosity measurements recently made in our laboratory.[20] Based on this experimental finding, a model of diffusion-controlled mass transfer for the liquid-surfactant-membrane system, in which the rate of the extraction reaction is increased by means of the ligand effect and the major resistance to interfacial mass transfer is from the surfactant layer, has recently been developed by us and the details of this model are discussed elsewhere.[27]

This work was supported by an EPA grant awarded to the Industrial Waste Elimination Research Center at Illinois Institute of Technology.

[1] N. N. Li, *U.S. Patent* 3 410 794, 1968.
[2] E. L. Cussler and D. F. Evans, *J. Membr. Sci.*, 1980, **6**, 113.
[3] E. L. Cussler, *AIChE J.*, 1971, **17**, 405.
[4] R. M. Izatt, M. P. Biehl, J. D. Lamb and J. J. Christensen, *Sep. Sci. Technol.*, 1982, **17**, 1351.
[5] K. H. Lee, D. F. Evans and E. L. Cussler, *AIChE J.*, 1978, **24**, 860.
[6] T. P. Martin and G. H. Davies, *Hydrometallurgy*, 1978, **2**, 315.
[7] J. W. Frankenfeld, P. P. Cahn, and N. N. Li, *Sep. Sci. Technol.*, 1981, **16**, 4, 385.
[8] A. Hochhauser, and E. L. Cussler, *AIChE Symp. Ser.*, 1975, **71**, 136.
[9] N. N. Li and A. L. Shrier, in *Recent Developments in Separation Science* (CRC Press, Cleveland, OH, 1972), vol. 1, p. 163.
[10] T. Kitagawa, Y. Nichikawa, J. W. Frankenfeld and N. N. Li, *Environ. Sci. Technol.*, 1977, **11**, 602.
[11] R. P. Cahn and N. N. Li, *Sep. Sci.*, 1974, **9**, 505.
[12] E. S. Matulevicius, and N. N. Li, *Sep. Purif. Methods*, 1975, **4**, 73.
[13] N. N. Li, *J. Membr. Sci.*, 1978, **3**, 265.
[14] M. L. Brisk and W. J. McManamey, *J. Appl. Chem.*, 1969, **19**, 109.
[15] J. Strzelbicki and W. Charewicz, *Sep. Sci. Technol.*, 1978, **13**, 141.
[16] Z. Gu, R. M. Kurzeja, D. T. Wasan and N. N. Li, paper presented at the *Am. Inst. Chem. Engr. Meeting*, Los Angeles, 1982.
[17] P. R. Subbaraman, Sr. M. Cordes and H. Freiser, *Anal. Chem.*, 1967, **41**, 1878.
[18] H. L. Finston and Y. Inone, *J. Inorg. Nucl. Chem.*, 1967, **29**, 199.
[19] H. Eccles, G. J. Lawson and D. J. Rawlence, in *ISEC'77* (Canadian Institute of Mining and Metallurgy, 1979), vol. 21.
[20] Z. Gu, R. M. Kurzeja, D. T. Wasan and N. N. Li, paper presented at the *Am. Inst. Chem. Eng. Meeting*, Washington D.C., 1983.
[21] P. Becher, *Emulsion, Theory and Practice* (Reinhold, New York, 2nd edn, 1965), p. 15.
[22] S. Ross, E. S. Shen, P. Becher and H. J. Ranato, *J. Phys. Chem.*, 1959, **63**, 1681.
[23] H. Eccles, G. J. Lawson, D. J. Rawlence, in *ISEC'77* (Canadian Institute of Mining and Metallurgy, 1979), vol. 21, p. 203.
[24] A. H. Ghanem, W. I. Higuchi and A. P. Simonelli, *J. Pharm. Sci.*, 1969, **58**, 165.
[25] V. Surpuriya and W. I. Higuchi, *J. Pharm. Sci.*, 1972, **61**, 375.
[26] T. Yotsuyangi, W. I. Higuchi and A. H. Ghanem, *J. Pharm. Sci.*, 1973, **62**, 41.
[27] Z. M. Gu, H. F. Zhang, D. T. Wasan and N. N. Li, paper presented at the *Am. Inst. Chem. Eng. Meeting*, Denver, 1983.

Faraday Discuss. Chem. Soc., 1984, 77, 75-84

A General Model to Account for the Liquid/Liquid Kinetics of Extraction of Metals by Organic Acids

By Michael A. Hughes*

Schools of Chemical Engineering, University of Bradford, Bradford, West Yorkshire

and Vladimir Rod

Czechoslovak Academy of Sciences, Prague, Czechoslovakia

Received 9th November, 1983

A model based on two-film theory is developed for the case of the rate of extraction of a divalent metal from an aqueous acid phase with an organic acid HR held in a second immiscible solvent phase. The model involves a reaction zone of variable thickness so that the case of reaction at an interface of molecular dimensions can be accommodated as well as the case of reaction extending into the diffusion film on the aqueous side of the interface. Four parameters are used, one involving a partition coefficient for the acid, one involving diffusivities in the films together with the two-film mass-transfer coefficients. Rate data from three techniques used in laboratory liquid/liquid contacting are fitted by the model.

The important commercial liquid/liquid extraction systems for the metals copper, cobalt and nickel involve contacting an aqueous acid phase with some organic acid, HR, held in a diluent.[1] Typically, HR is the hydroxyoxime LIX64, SME529, P5000 for copper or di-2-ethylhexylphosphoric acid for cobalt and nickel. Several papers describe the kinetics of extraction in these systems, but a variety of concentration conditions, in both the aqueous and organic phases, have been employed together with different contacting techniques. The latter range from the single-drop experiment[2] through the constant-interface stirred cell[3] to the rotating diffusion cell.[4] Danesi[5] has shown how a combination of experimental technique with concentration conditions can result in true chemical-kinetic control or true mass-transfer control or mixed control.

Rod *et al.*[6,7] have proposed a model for mass transfer with a fast reversible reaction and product extraction and have applied it to the extraction of copper(II) by hydroxyoximes. It is now necessary to show how this model is generally applied and also how examples proposed by other workers are specific cases of the general model.

THE MODEL

The concentration profiles for this model are shown in fig. 1. The concentrations shown at the interface are arbitrary; in some cases a high interfacial concentration may exist if the extractant is surface active. It is taken that reaction must be at the interface and/or extending out into the aqueous diffusion film. Note that as $P_{HR} \to \infty$ and $P_{MR_2} \to \infty$ then the profiles for c_{HR} and c_{MR_2}, in the aqueous phase, are coincident at zero. Also, under these conditions, the c_H and c_M profiles in the aqueous phase become linear, as shown by the dotted lines in fig. 1. In fig. 1 λ_d is the diffusion-film thickness and λ_r is the reaction-zone thickness.

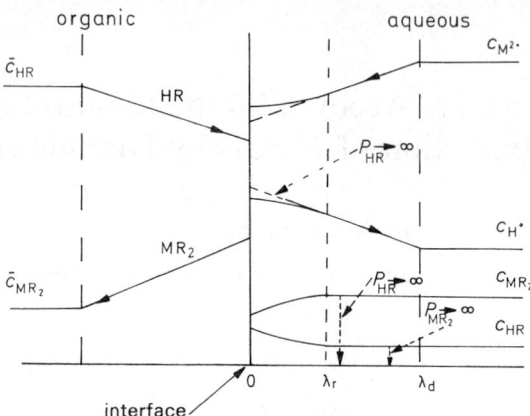

Fig. 1. Concentration gradients for the model.

For a divalent metal M^{2+}:

$$H^+ + R^- \rightleftharpoons HR \qquad K_{HR} = c_{HR}/c_{H^+}c_{R^-} \qquad (1)$$

$$M^{2+} + R^- \underset{}{\overset{k_R}{\rightleftharpoons}} MR^+ \qquad K_{MR} = c_{MR^+}/c_{M^{2+}}c_{R^-} \qquad (2)$$

$$MR^+ + R^- \underset{}{\overset{k_R^1}{\rightleftharpoons}} MR_2 \qquad K_{MR_2} = c_{MR_2}/c_{MR^+}c_{R^-} \qquad (3)$$

$$M^{2+} + 2HR \rightleftharpoons MR_2 + 2H^+ \qquad K_{EX} = \bar{c}_{MR_2}c_{H^+}^2/c_{M^{2+}}\bar{c}_{HR}^2. \qquad (4)$$

Reaction (2) or (3) may be rate controlling.

In the derivations, superscript bar will indicate the organic phase, subscript i will denote an interface condition and N will denote a flux. The film coefficients will be shown as k_o for organic and k_{aq} for water. D is a diffusivity within the films.

It can be shown that if reaction (2) is rate determining then:

(i) by Astarita's method:

$$N_{HR} = \sqrt{\frac{k_R D_{HR}}{P_{HR}^2 K_{HR}} \frac{c_{M,i}}{c_{H,i}}\left(1 - \frac{c_{H,i}^2 \bar{c}_{MR_2,i}}{K_{EX}\bar{c}_{HR,i}^2 c_{M,i}}\right)(\bar{c}_{HR,i}^2 - P_{HR}^2 c_{HR}^2)} \qquad (5)$$

(ii) for the organic phase:

$$\bar{c}_{HR,i} = \bar{c}_{HR} - N_{HR}/k_{HR,o} \qquad (6)$$

$$\bar{c}_{MR_2,i} = \bar{c}_{MR_2} + N_{HR}/2k_{MR_2,o} \qquad (7)$$

(iii) for the aqueous phase:

$$c_{HR} = \frac{1}{P_{HR}}\left[\frac{-P_{MR_2}D_{HR}c_H^2}{4P_{HR}D_{MR_2}K_{EX}c_M} \right.$$

$$\left. + \sqrt{\left(\frac{P_{MR_2}D_{HR}c_H^2}{4P_{HR}D_{MR_2}K_{EX}c_M}\right)^2 + \frac{c_H^2}{K_{EX}c_M}\left(\bar{c}_{MR_2,i} + \frac{D_{HR}P_{MR_2}\bar{c}_{HR,i}}{2D_{MR_2}P_{HR}}\right)}\right] \qquad (8)$$

$$c_{MR_2} = K_{EX}(c_{HR}^2 c_M/c_H^2)(P_{HR}^2/P_{MR_2}) \qquad (9)$$

$$c_{M,i} = c_M - 0.5 N_{HR}/k_{M,aq} + [0.5 D_{HR}/(D_H P_{HR})](\bar{c}_{HR,i} - c_{HR}P_{HR}) \qquad (10)$$

$$c_{H,i} = c_H + N_{HR}/k_{H,aq} - [0.5 D_{HR}/(D_H P_{HR})](\bar{c}_{HR,i} - c_{HR}P_{HR}). \qquad (11)$$

Table 1. Special cases of the general model

particular case	values of key parameters		
	k_R	P_{HR}	$\theta_1 = \dfrac{k_R D_{HR}}{P_{HR}^2 K_R}$
reaction in the film	finite	finite	finite
equilibrium reaction in the film	∞	finite	∞
reaction at the interface	∞	∞	finite
instantaneous reaction at the interface (Chapman's model)	∞	∞	∞

Some special cases may be noted which are summarised in table 1.

The model of Chapman et al.[8] is of particular interest but is only one special case, for under their assumed conditions then $k_R \to \infty$, $\theta = k_R D_{HR}/P_{HR}^2 K_{HR} \to \infty$, $P_{HR} \to \infty$, K_{EX} is finite, thus eqn (5) transforms to:

$$\frac{c_{H,i}^2 \bar{c}_{MR_2,i}}{K_{EX} \bar{c}_{HR,i}^2 c_{M,i}} - 1 = 0. \tag{12}$$

Eqn (6) and (7) remain as before.

Eqn (8) transforms to $\quad c_{HR} = 0.$ (13)

Eqn (9) transforms to $\quad c_{MR_2} = 0.$ (14)

Eqn (10) transforms to $\quad C_{M,i} = c_M - 0.5 N_{HR}/k_{M,aq}.$ (15)

Eqn (11) transforms to $\quad c_{H,i} = c_H + N_{HR}/k_{H,aq}.$ (16)

It may be noted here that if reaction (3) is the rate-determining step then eqn (5) becomes:

$$N_{HR} = \sqrt{\frac{k_R^1 D_{HR} K_{MR} c_{M,i}}{P_{HR}^3 K_{HR}^2 c_{H,i}} \left(1 - \frac{c_{H,i}^2 \bar{c}_{MR_2}}{K_{EX} c_{M,i} \bar{c}_{HR}^2}\right)(\bar{c}_{HRi}^3 - P_{HR}^3 c_{HR}^3)} \tag{17}$$

but the other equations remain as before.

IMPORTANT PARAMETERS OF THE MODEL

The most sensitive parameters in the model are:

$$\theta_1 = \frac{k_R D_{HR}}{P_{HR}^2 K_{HR}} \text{[eqn (5)]} \quad \text{or} \quad \theta_1' = \frac{k_R^1 D_{HR} K_{MR}}{P_{HR}^3 K_{HR}^2} \text{[eqn (17)]}$$

together with P_{HR}. Their values decide the location of the reaction and characteristics of the transfer process; table 2 illustrates this in a general way.

Table 2. Location of the reaction and characteristics of the transfer process

P_{HR}	θ_1 or θ'_1	particular case	description of process
finite	finite	reaction in the film	diffusion coupled with kinetics of reaction in the film
finite	$\to \infty$	equilibrium reaction in the film	diffusion coupled with equilibrium in the film
$\to \infty$	finite	reaction at the interface	diffusion and reaction kinetics at the surface
$\to \infty$	$\to \infty$	instantaneous reaction at the interface	diffusion and equilibrium at the surface

THE POSSIBILITY OF BULK PHASE REACTION

In order to consider the possibility of reaction in a bulk aqueous phase, the thickness of the reaction zone $\dot{\xi}$ must be considered. Now:

$$\dot{\xi} = \frac{\lambda_r}{\lambda_d} \approx \frac{3}{P_{HR}} \left(\frac{k_{HR} \bar{c}_{HR}}{N_{HR,i}} - \frac{k_{HR,aq}}{k_{HR,o}} \right) \qquad (18)$$

only if $\dot{\xi} > 1$ can reaction occur in the bulk phase. Suppose that the partition coefficient of the extractant was relatively low at ca. 100 and the mass-transfer coefficients relatively high at $k_{HR,aq} = k_{HR,o} \approx 10^{-4}$ m s^{-1} then with $\bar{c}_{HR} = 10^{-2}$ kmol m^{-3} eqn (18) gives $N_{HR,i} < 3 \times 10^{-9}$ kmol m^{-2} s^{-1} which is too low for practical purposes. Extractants of this nature, which react in the bulk, would be of no practical use.

MATHEMATICAL TREATMENT OF THE SIMULTANEOUS EQN (5)–(11)

The implicit function for N_{HR} is:

$$F[N_{HR}, c_j(K_{EX}, K_{HR}, P_{HR}, P_{MR_2}, k_R, D_j)(k_{aq}, k_o)] = 0 \qquad (19)$$

in which c_j is the bulk concentration and D_j the molecular diffusivity of species j. The function is therefore made up of: (1) flux, (2) concentration and (3) physical and hydrodynamic parameters. The physical parameters are regrouped to give:

$$(K_{EX}, \theta_1, \theta_2, D_j/D_{HR})$$

in which $\theta_1 = k_R D_{HR}/P_{HR}^2 K_{HR}$ and $\theta_2 = D_{HR} P_{MR_2}/D_{MR_2} P_{HR}$. The ratio D_j/D_{HR} refers to a transferring species j with diffusivity D_j, and this ratio is relatively easy to obtain from generalised correlations: in any case the flux N_{HR} is not very sensitive to this ratio. Thus in any problem K_{EX}, $D_{HR} P_{MR_2}/D_{MR_2} P_{HR}$ and D_j/D_{HR} are known or may be estimated for a given system and the data from the technique (be it single drop, stirred cell etc.) may be fitted by the model optimising the best value of θ_1, k_{aq} and k_o. It is k_{aq} and k_o, the hydrodynamic parameters, which change from one technique to another.

A computer program is written using the Runge–Kutta–Merson method for numerical integrations and the Marquardt method for the optimisation technique to give parameter estimation. In the case of HR $\equiv$ hydroxyoxime, the program

Table 3. Most important parameters of F [eqn (19)] dictated by position of rate and bulk concentrations

regions of rate	$\bar{c}_{HR} \gg c_{M^{2+}}$	$\bar{c}_{HR} \approx c_{M^{2+}}$	$c_{HR} \leqslant c_{M^{2+}}$
far from equilibrium, e.g. initial rates	k_{aq}	θ_1	k_o
near to equilibrium, e.g. in real contactors	K_{EX}, k_{aq}	K_{EX}, θ_1	K_{EX}

includes the chemical model of Whewell and Hughes[10] to calculate the thermodynamic concentrations, c_{HR} and c_{H^+}, $c_{M^{2+}}$.

A reasonably large number of points taken from kinetic experiments on a particular system can be fitted to the model. Alternatively, the values of constants making up certain parameters can be estimated or measured separately and the parameters can be inserted into the model to calculate fluxes which can be compared with experiment.

THE SENSITIVITY OF THE MODEL TO THE PARAMETERS

The sensitivity of the model to the parameters depends on whether the rate is measured near to equilibrium or far away from equilibrium together with the bulk concentration conditions; the most important parameters for the varying conditions are highlighted in each case in table 3.

The behaviour of the model can be demonstrated using assumed parameter values and selected concentrations for the aqueous metal ion, the aqueous proton and the 'free' organic ligand, HR.

In fig. 2 the flux (or extraction rate) is seen to depend upon the extractant concentration at variable values of the rate parameter θ_1. As the values of θ_1 increase, the rate approaches the maximum theoretical diffusion-controlled transfer curve A. This theoretical curve A would take on new positions in the plot when the $c_{M^{2+}}$, k_{aq} and k_o values are altered. Note that with increasing extractant concentration the extraction rate is approaching the region where the mass transfer of metal ions is the rate-controlling step. Only at very low values of θ_1 is chemical control possible, and for $\theta_1 < 10^{-6}$ (or near) the model is not very sensitive to this parameter.

The dependence of the extraction rate on the concentration of the metal ion is shown in fig. 3. As $c_{M^{2+}}$ increases the rate approaches the limit of $k_o c_{HR}$, and for a given $c_{M^{2+}}$ value the flux is higher as the θ_1 increases. Again chemical control becomes significant when θ_1 becomes very small, e.g. ca. 10^{-12} or less.

The importance of the influence of the reverse reaction is partly measured in the K_{EX} value, and for set values of all the parameters this influence on the rate is illustrated in fig. 4, where c_{HR} and K_{EX} are varied. As expected, for a given c_{HR} value the rate increases as K_{EX} increases, chemical control is forced upon the system when K_{EX} becomes very small and, in any case, as c_{HR} increases the diffusion-controlled limit at $2k_{aq}c_{M^{2+}}$ is approached.

The sensitivity of the flux to the parameters can be best summarised in fig. 5, in which a fractional change in flux produced by a fractional change in a parameter is plotted. The left-hand side of the graph represents excess metal in the aqueous phase and the right-hand side represents excess extractant in the organic phase. The sensitivity of the flux with respect to the mass-transfer coefficient k_{aq} increases

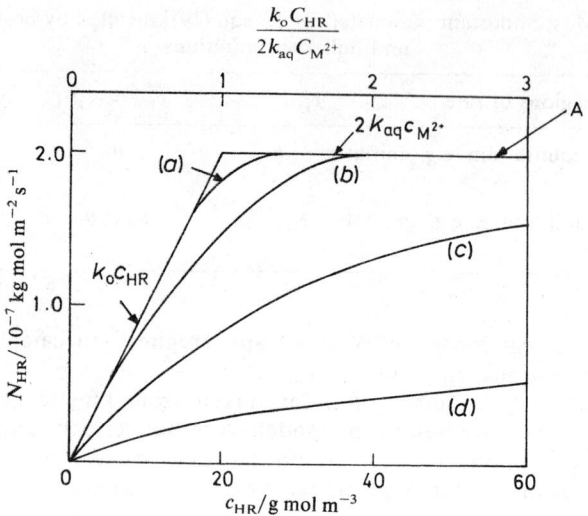

Fig. 2. Dependence of the flux of HR on the parameter θ_1 and the concentration of HR in the organic bulk phase. $\theta_1 = $ (a) 10^{-6}, (b) 10^{-8}, (c) 10^{-10} and (d) 10^{-12}.

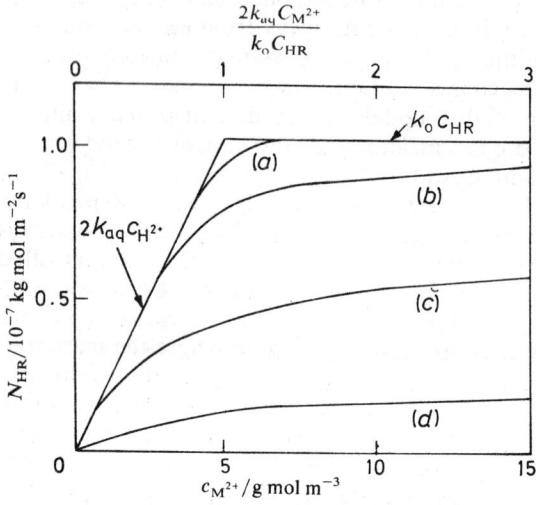

Fig. 3. Dependence of the flux of HR on the parameter θ_1 and the concentration of metal in the aqueous bulk phase. $\theta_1 = $ (a) 10^{-6}, (b) 10^{-8}, (c) 10^{-10} and (d) 10^{-12}.

with the extractant concentration and with the ratio $k_o c_{HR}/2k_{aq} c_{M^{2+}}$. On the other hand, the sensitivity with respect to k_o decreases with increasing c_{HR} and the ratio above.

A point on the graph at $c_{HR} = 20$ kg mol m^{-3}, corresponding to a value of 1.0 for $\beta = k_o c_{HR}/2k_{aq} c_{M^{2+}}$, is a stoichiometric point for the $c_{M^{2+}}$ value chosen for this calculation. It is now seen that the sensitivity of the flux to k_{aq} is high if $\beta \gg 1$ and its sensitivity to k_o is high if $\beta \ll 1$. The sensitivity of the flux to θ_1 is a maximum if $\beta \approx 1$. These observations demonstrate that in order to obtain good estimates of

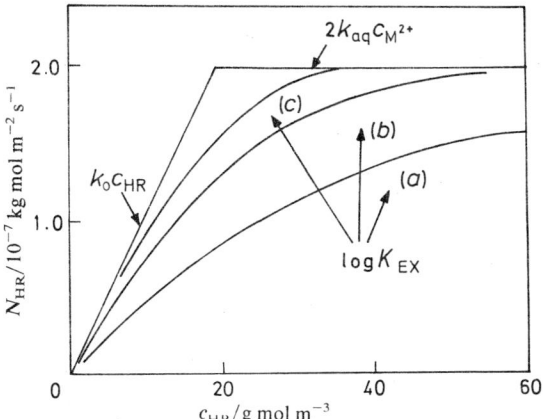

Fig. 4. Dependence of the flux of HR on the value of log K_{EX} and the concentration of HR in the organic bulk phase. log $K_{EX} = $ (a) 1.0, (b) 2.0 and (c) 3.0.

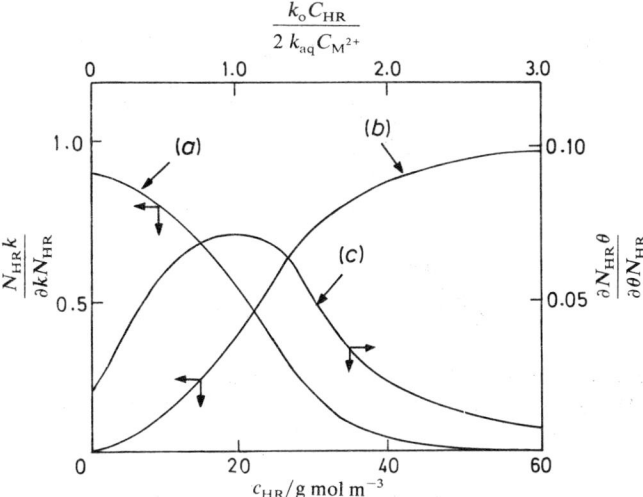

Fig. 5. Dependence of the rate of change in flux of HR on the fractional change in the parameters (a) k_o, (b) k_{aq} and (c) θ_1.

the three major parameters of the model then experimental data should be obtained from three different regions. Experiments in the region $c_{M^{2+}} \gg c_{HR}$ will provide good estimates of k_o and experiments in the region $c_{HR} \gg c_{M^{2+}}$ will provide good estimates of k_{aq}. The reaction rate parameter θ_1 is best estimated if the experimental conditions are such that

$$\frac{c_{HR}}{2c_{M^{2+}}} \approx \frac{k_{aq}}{k_o}.$$

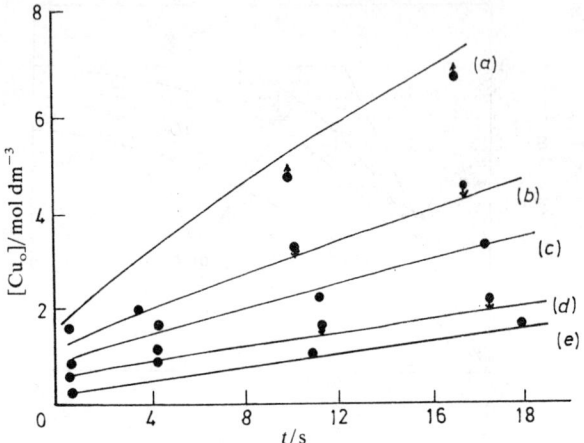

Fig. 6. Typical fit of the model (solid line) to experimental points from the rising-drop experiment. The oxime concentration is constant at 20 vol % HR and the aqueous copper concentration is 8 g dm^{-3}. [H$_2$SO$_4$]/g dm^{-3} = (a) 2, (b) 3.5, (c) 5, (d) 8 and (e) 12.

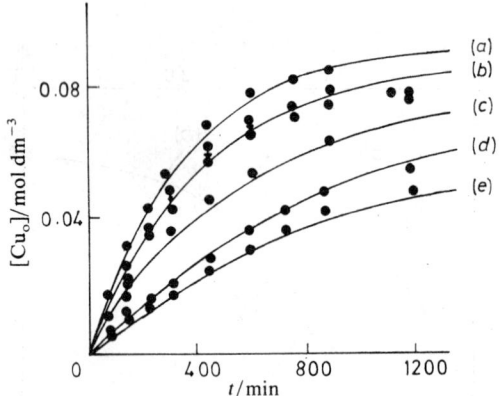

Fig. 7. Typical fit of the model (solid line) to experimental points from the gauze-cell experiment. The oxime concentration is constant at 20 vol % HR and the aqueous copper concentration is 8 g dm^{-3}. [H$_2$SO$_4$]/g dm^{-3} = (a) 1, (b) 2, (c) 4, (d) 6 and (e) 8.

APPLICATION OF THE MODEL TO DATA FOR THE LIX64N + CuSO$_4$ + H$_2$SO$_4$ SYSTEM

We have tested this model on the LIX64N + CuSO$_4$ + H$_2$SO$_4$ system using data from two entirely different techniques of contacting the two phases but using bulk concentrations in both phases which are of commercial interest. In particular, the pH range is commercially more realistic than that adopted by Albery and Fisk.[4]

The model was fitted to the initial rate data from the single-drop experiments,[3] see fig. 6.

Rate data may be obtained over more extensive times using a 'gauze cell',[9] where a fixed volume of organic phase is continually stirred at an interface, of known area, with an aqueous phase of constant composition continuously flowing through

Table 4. Parameter estimates for copper extraction systems

data	θ_1 /(m s^{-1})^2	k_o /μm s^{-1}	k_{aq} /μm s^{-1}	θ_2
Cu+LIX64N gauze cell	7.0×10^{-12}	6.7	3.3	11.0
Cu+LIX64N rising drops	7.0×10^{-12}	150	100	11.0
Cu+P5000 membrane cell (Albery and Fisk)	6.8×10^{-9}	10.4	21.8	1.0

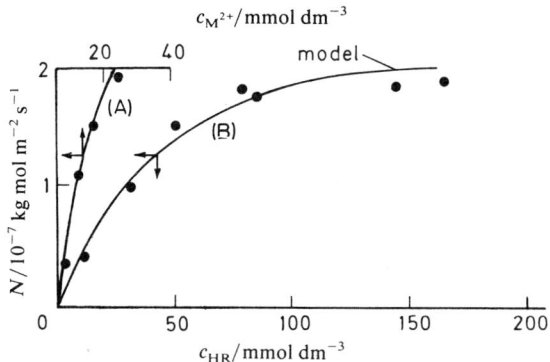

Fig. 8. Fit of the model to the data from the rotating diffusion cell. Data reported by Albery and Fisk.[4] Curve B, the concentration of aqueous copper is constant at 10 mmol dm^{-3} but HR is varied. Curve A, the concentration of HR is constant at 68 mmol dm^{-3} but aqueous copper is varied; the pH is 4.1.

the cell and in contact with the organic phase at that interface. Data from these experiments, which involve concentrations in the same range as those for the single drop, have been reported elsewhere.[9] The model fitted these data, see fig. 7.

The parameters are reported in table 4. In the gauze cell the rate of mass transfer is controlled by diffusion with chemical reaction. The parameters θ_1, k_o and k_{aq} are determined with confidence because the data are measured in concentration–time ranges which allow the model to be sensitive to all three parameters. The model is not sensitive to θ_2 in this range. It is probable that the relatively low k_o and k_{aq} values are caused by inefficient stirring near to the interface.

In the case of the rising-drop experiments the rate is mainly controlled by chemical reaction. Because these are initial rate data the range over which they are measured means that the rate is not so sensitive to k_o and k_{aq}, so these are not as well determined as in the case of the gauze-cell work.

Note that the value of θ_1 is 6.9×10^{-12} if it is determined independently for the drop data alone.

The model appears successful especially since it accounts for rates measured both at initial times and near to equilibrium.

We now turn to the experiments of Albery and Fisk,[4] who used a rotating diffusion cell to study the extraction of copper with P5000 from slightly acid media. The present model also accounts for their results, see fig. 8. In table 4 the value of θ_1 is greater than that found for the Cu+LIX64N system; this is to be expected since P5000 has a lower partition coefficient than LIX64N.[10] A theoretical value of $k_o = 11.4$ can be calculated for this system using the classical equation for a rotating disc, developed by Levich. So the value of 10.4 found by optimisation of the parameters is satisfactory. The relatively low value of k_o shows that there is a diffusion resistance but it is not possible to say if this is due to the membrane itself or some film on the organic side of the membrane.

In all the above cases the equation involving the addition of the first ligand [eqn (2) and thus (8)] gave the best fit.

[1] P. J. Bailes, C. Hanson and M. A. Hughes, *Chem. Eng.*, 1976, 86.
[2] R. J. Whewell, M. A. Hughes and C. Hanson, *J. Inorg. Nucl. Chem.*, 1975, **87**, 2323.
[3] C. A. Fleming, *Natl Inst. Metall., Repub. S. Afr.*, Rep. no. 1793, 1976.
[4] W. J. Albery and P. R. Fisk, in *Hydrometallurgy '81* (Soc. Chem. Ind., London, 1981), F5/1-F5/15.
[5] P. R. Danesi and R. Chiarizia, in *Critical Reviews in Analytical Chemistry*, ed. B. Campbell (CRC Press, Boca Raton, Florida, 1980), chap. 10, p. 1.
[6] V. Rod, *Chem. Eng. J.*, 1980, **20**, 131.
[7] V. Rod, L. Strnadova, V. Hančil and Z. Šir, *Chem. Eng. J.*, 1981, **21**, 187.
[8] W. Chapman, R. Caban and M. Tunison, *Am. Inst. Chem. Eng. Symp. Ser.*, 1975, **71**, 152.
[9] M. Bhaduri, C. Hanson, M. A. Hughes and R. J. Whewell, in *ISEC'83* (American Institute of Chemical Engineers, 1983), p. 293.
[10] R. J. Whewell, M. A. Hughes and C. Hanson, in *ISEC'77* (Canadian Institute of Mining and Metallurgy, 1979), vol. 21, p. 185.

The Concept of Interfacial Reactions for Mass Transfer in Liquid/Liquid Systems

BY WALTER NITSCH

Lehrstuhl für Technische Chemie I, Technische Universität München, Lichtenbergstraße 4, 8046 Garching, Federal Republic of Germany

Received 15th December, 1983

Interfacial reactions during reactive mass transfer between liquid phases are the central point of this contribution. In order to go from the general problem to identify the site of the reaction and the rate-controlling step, a kinetic treatment of interfacial reactions is proposed that considers the individual processes of all the species involved. This concept, together with a new method for the determination of individual transport coefficients, leads to an efficient concept of resistances and to the possibility of treating chemically coupled systems (coextraction). Finally, the pronounced effects of adsorption layers are discussed with respect to diffusional- and interfacial-controlled mass transfer.

The literature concerning the kinetics of liquid/liquid reactions has greatly increased in volume during the last decade,[1,2] indicating a pronounced interest in chemical-extraction systems. However, until now it has been difficult or impossible to nominate a system for which the kinetic features are known and accepted: The 20-year-long discussion concerning kinetics in dithizone systems[3-5] and the various disagreements about kinetics in Purex[6-8] are typical of the state of controversy.

We have been studying the fundamentals of liquid/liquid reactions for nearly twenty years, during which time we have identified the following problem areas which give rise to the present unsatisfactory state of knowledge of the subject: (1) a lack of agreement or certainty concerning suitable experimental methods to provide kinetic conclusions, (ii) 'overall treatments' of the measured results, (iii) uncontrolled effects of surfactants and (iv) incomparability of systems for methodical reasons. Our concepts and results concerning and surrounding the problem of interfacial reactions are summarized in the following sections.

LIQUID/LIQUID REACTIONS AS HETEROGENEOUS SYSTEMS

A water phase loaded with metal ions and in contact with a solvent phase loaded with an extracting agent leads to a typical liquid ion-exchange reaction:

$$M^{2+} + 2\overline{HX} \rightarrow \overline{MX_2} + 2H^+$$

which obviously contains (in the interference of a chemical reaction with transport processes) the characteristics of heterogeneous reactions.

In table 1 the kinetic features of various two-phase systems are compared. The differences in the nature of the transport processes are evident. However, in addition to their having small enthalpies and often well defined stoichiometric relations,[9] liquid/liquid reactions are characterized, first of all, by the appearance of gradients for reactants and products in both phases. It is in this aspect that the most pronounced familiarity is established with heterogeneous processes of membrane transport.

Table 1. Comparison of various two-phase heterogeneous systems

system	transport	important gradients	site of reaction
solid/fluid	pore diffusion	fluid phase	surface
gas/liquid	fluid dynamic + diffusion	liquid phase	liquid phase
liquid/liquid	fluid dynamic + diffusion	both phases	liquid 1, liquid 2 or interface

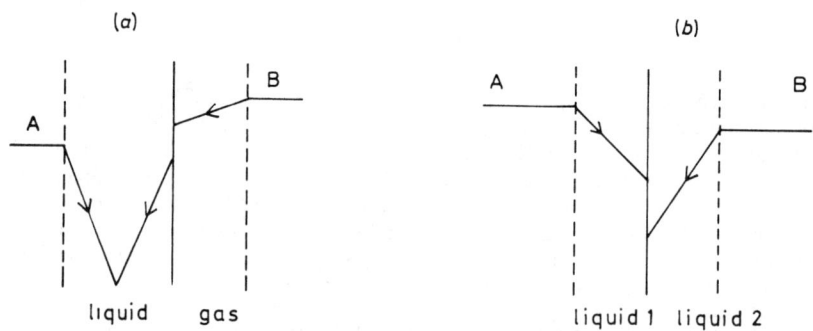

Fig. 1. (a) Scheme of concentration gradients for (a) an instantaneous irreversible reaction (gas/liquid) and (b) an interfacial reaction (liquid/liquid).

THE SITE OF REACTION

In relation to the fluid-dynamic nature of the transport processes, liquid/liquid systems are similar to liquid/gas systems, but if one considers the literature concerning reactions in these two types of system a decisive difference is evident. While gas/liquid reactions are always treated as occuring inside the liquid phase (homogeneous reactions),[10,11] for liquid/liquid reactions the concept of interfacial, i.e. heterogeneous, reactions is the usual mode.[1,2] This remarkable situation calls for an explanation.

Without a systematic consideration of mass transfer accompanied by chemical reaction the problem of homogeneous or interfacial reaction can be explained by looking at the two heterogeneous reactions

$$A_{liquid} + B_{gas} \rightarrow C_{liquid}$$

$$A_{liquid\,1} + B_{liquid\,2} \rightarrow C_{liquid\,2}.$$

A gas/liquid reaction exclusively inside the liquid phase is realized only in the case of the instantaneous irreversible reaction (Hattamodel)[10] between A and B_{gas}, because A is completely depleted at the plane of the reaction at a definite distance from the interface [fig. 1(a)]. In all other regimes the reactants A and B will meet together not only in the bulk but also at the interface, which means that interfacial reactions are not to be excluded. The degrees of conversion in the bulk phase and at the interface, however, will depend on the kinetic properties of the system.

The case of a pure interfacial reaction is established if reactant A is exclusively soluble in one phase and reactant B is exclusively soluble in the other, which means that the interface must be the only locale of reaction [fig. 1(b)]. Proceeding from

Table 2. Distribution data of typical gaseous reactants and extractants

system	gas/water	system	solvent/water
O_2/water	32	dithizone, CCl_4/water	1.5×10^5
CO_2/water	1, 2	dithizone, $CHCl_3$/water	10^4
H_2S/water	3.9×10^{-1}	LIX^a/paraffin/water	ca. 10^4
SO_2/water	2.5×10^{-2}	tributylphosphate/paraffin/water	3×10^3

^{a} 2-Hydroxy-5-t-octylacetophenone oxime.

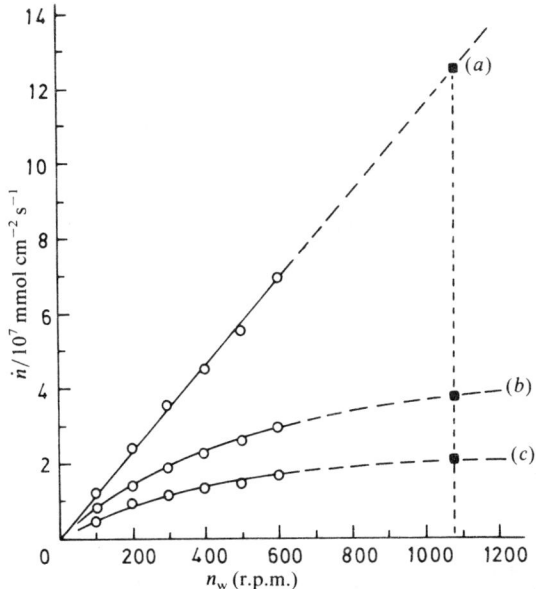

Fig. 2. Influence of forced convection on the initial flux $\dot{n}$ into CCl_4 for the complexation of zinc ions with dithizone. [Zn]/mmol cm^{-3}: (a) 10^{-2}, (b) 10^{-4} and (c) 5×10^{-5}. Filled symbols, free falling drops; open symbols, stirred cell. Initial concentration of HDz = 1.25×10^{-4} mol dm^{-3}, pH 5.

this extreme kinetic situation the transition to reactions in the bulk phase and their degree will depend at first on the distribution coefficients of B and/or A between both phases.

Table 2 shows a comparison of the distribution coefficients for typical gas-phase reactants B_{gas} with typical solvent-phase reactants B_{liquid}. The values of the coefficients demonstrate that the 'solubility' of typical extracting agents in the aqueous phase is extremely low compared with the gas-phase reactants, which means that interfacial reactions are more probable in liquid/liquid than in liquid/gas systems.

In a real kinetic situation, however, interfacial reactions are only to be identified if they are rate-controlling: (i) fluxes must be independent of the forced convection (plateau rates), (ii) specific fluxes must be independent of the area and (iii) plateau rates should be sensitive to surfactants.

The behaviour of the zinc/dithizone system[12] is an example of such an identified interfacial reaction (see fig. 2 and fig. 8). However, in the case that transport processes are rate-determining, which means that under the influence of forced convection identification of the site of reaction requires indirect methods.

THE RATE-CONTROLLING STEP

The basic problem in the kinetic treatment of individual liquid/liquid reactions is identification of the rate-controlling step (diffusion or reaction).

The most usual (and until now probably the only applicable) method for the identification of the rate-determining step is the measurement of mass transfer for different forced convections, as realized in different types of stirred cells.[13-16] In addition to the serious problem caused by impurities[17] brought about by the long time of phase contact, such stirred cells should be calibrated to ensure that the kinetic conclusions are valid.

For the stirred cell used in our investigations we know, from the so-called calibration measurements of 'diffusional' heat transfer and physical mass transfer, that the transfer rates or coefficients have a linear dependence on the stirring rate if the stirring ratio is kept constant.[18] Therefore we can be sure that the appearance of plateaus in the rate plots (see fig. 2) shows that chemical reactions are rate-determining. Without such a calibration the plateau rates may be caused by the fluid-dynamics of the type of stirred cell being used.

KINETIC TREATMENT OF INTERFACIAL REACTIONS AT LIQUID/LIQUID BOUNDARIES

Sinks (reactants) and sources (products) of concentrations located at the interface point to an interfacial reaction. For such kinetic situations the relevant set of equations corresponds to the scheme shown in fig. 3.

For each component one individual transport equation of the type

$$\dot{n}_i = \pm \beta_i (c_i - c_i^*) \tag{1}$$

may be applied, together with the rate equation of the interfacial reaction, derived from the analysis of the plateau rates

$$\dot{n} = \vec{k} \prod_i c_i^{*P_i} - \overleftarrow{k} \prod_i c_i^{*P_i}. \tag{2}$$

The individual fluxes $\dot{n}_i$ are coupled with stoichiometric mass balances.

For transport processes to be rate-determining, instead of the rate equation the equilibrium condition for the interfacial reaction must be applied:

$$K = \prod_i c_i^{*r_i}. \tag{3}$$

This set of equations [eqn (1) and (2) or (1) and (3)], characterized by the consideration of the individual fluxes, substitutes the usual descriptive overall equations[7,19,20] and proposes a physico-chemical treatment of interfacial reactions.

TRANSPORT PROCESSES

The general problem of the above-mentioned kinetic treatment is that the interfacial concentrations of the individual species c_i^* and the individual transport coefficients β_i are unknown quantities. Because interfacial concentrations c_i^* are

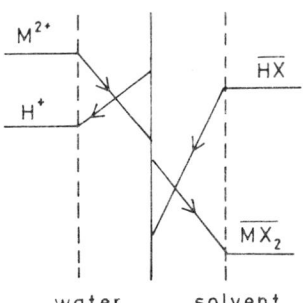

Fig. 3. Scheme of gradients for a liquid ion-exchange reaction at the interface.

not measurable, values for the individual transport coefficients are necessary to solve the set of equations.

In this context it is remarkable that in the numerous publications concerning the kinetics of liquid/liquid reactions at droplets very seldom has an attempt been made to use the relevant engineering literature[21,22] containing so-called correlations that would be suitable for calculating individual transport coefficients.

However, for mass transfer in stirred cells, especially for the prototype used in our work, efficient correlations for calculating β do not exist. Therefore we have developed a new approach to the individual coefficients.

The applied concept was originally elaborated for the system uranyl nitrate, nitric acid and tributylphosphate,[23] but it should be applicable in general for the case when transport processes are rate-determining. For the concentration region of transport-controlled mass transfer, the set of eqn (1) and (3) can be solved numerically with reasonable assumptions for the values of the different transport coefficients.[24] Thereby, with the equilibrium constant [eqn (3)] the interfacial concentrations c_i^* may be calculated.

For the particular concentration region in which, for one (or more) of the reactants or products, the condition $c_i^* \ll c_i$ remains established in spite of strong variations in the assumed coefficients, the corresponding transport coefficient is accessible from measurements using the simplified equation $\beta_i = \dot{n}_i/c_i$.

In table 3 are compared individual coefficients for different systems measured in the same stirred cell for equal stirring numbers in the water phase and with $Re_w = Re_s$. The values are close together, and are thus suitable to describe the time dependence of the concentrations in the respective systems, which shows that the evaluation method applied on the basis of eqn (1) and (3) is justified. A further proof concerns measurements of mass transfer at droplets in the system uranyl nitrate, nitric acid and tributylphosphate. Using the published correlations for the state of circulating droplets the calculated individual coefficients are in a good agreement with measured values,[29] which also confirms the method proposed above. In addition, with such physico-chemical treatments of diffusion-controlled interfacial reactions it is possible to prove a proposed kinetic configuration and to identify the effects of impurities and of interfacial instabilities, by comparing individual coefficients of different systems.

INTERFACIAL REACTIONS

As mentioned above, the rate equation for an interfacial reaction demands an analysis of plateau rates (see fig. 2). In such a manner, for the liquid ion-exchange

Table 3. Individual transport coefficients related to stirred cells of the same type at $Re_w = Re_s$ [a,b]

$\beta_w/10^3$ cm s^{-1}	solute	phases
8.5	I_2 [25]	H_2O/CCl_4
7.0	toluene[26]	H_2O/hexane
8.7	toluene[18]	H_2O/toluene
2.92	Cd^{27}	H_2O/Cl_4, HDz
5.0	UO_2^{2+} [23]	H_2O/hexane[c]

$\beta_0/10^3$ cm s^{-1}	solute	phases
5.7	HDz[28]	H_2O/CCl_4, $ChCl_3$
3.8	$ZnDz_2$ [12]	H_2O/CCl_4
1.78	Ko[23]	H_2O/hexane, T
6.84	Kx[23]	H_2O/hexane, T

[a] $n_w = 300$ min^{-1}. The effects of different stirrers are eliminated. [b] HDz = dithizone, Ko = $UO_2(NO_3)_2 \cdot 2T$, Kx = $HNO_3 \cdot T$, T = tributylphosphate. [c] $\Gamma = 1.1$ mol dm^{-3}.

of zinc and cadmium ions in water, in contact with a solvent phase loaded with dithizone, we found for conditions of unilateral equilibrium the rate equation[12]

$$\dot{n} = \vec{k}\frac{[Zn^{2+}]^* [HDz]^*}{[H^+]^*}. \qquad (4)$$

The simplest mechanism which agrees with this kinetic equation is the three-step consecutive reaction

$$HDz \underset{}{\overset{K_D^*}{\rightleftharpoons}} Dz^- + H^+ \quad \text{fast}$$

$$Zn^{2+} + Dz^- \overset{k}{\longrightarrow} ZnDz^+ \quad \text{slow}$$

$$ZnDz^+ + Dz^- \longrightarrow ZnDz_2 \quad \text{fast}$$

occuring at the interface, which means $\vec{k} = K_D^* k$. In order to compare the velocity constant k with the usual scale for the rate of homogeneous reactions it is necessary to make assumptions concerning the unknown interfacial properties: because the interfacial activity of the species involved (HDz, Me^{2+}, $MeDz_2$) is weak or negligible, equating the bulk concentration with the concentration in the adsorbed state seems to be justified. Note that inside the phases for the state of the plateau rates the concentration gradients are negligible. More critical is the value of the dissociation constant K_D^*. Following the suggestions of interfacial chemistry[30] the assumption $K_D = K_D^*$ should be justified and therefore the values for k_{homo} in table 4 are calculated with the 'water value' for K_D.

However, an attempt to understand the result $k_{CCl_4} \approx 100\, k_{CHCl_3}$ (see table 4) for both cations leads to the interpretation that K_D^* is dependent on the nature of the interface.[31] Nevertheless, recalculation into the scale of homogeneous reactions shows that the accessible interfacial reaction could be very fast. This aspect is of interest in the kinetics of permeation through biological membranes.

Table 4. Heterogeneous [eqn (4)] and derived homogeneous constants k_{homo} for measured interfacial reactions

cation	solvent	$\vec{k}/10^{-3}$ cm s^{-1}	k_{homo}/cm^3 s^{-1} mmol^{-1}
Cd^{2+}	CCl$_4$	2.94	1.47×10^9
Cd^{2+}	CHCl$_3$	0.03	1.5×10^7
Zn^{2+}	CCl$_4$	0.42	2.1×10^8
Zn^{2+}	CHCl$_3$	0.004	2×10^6

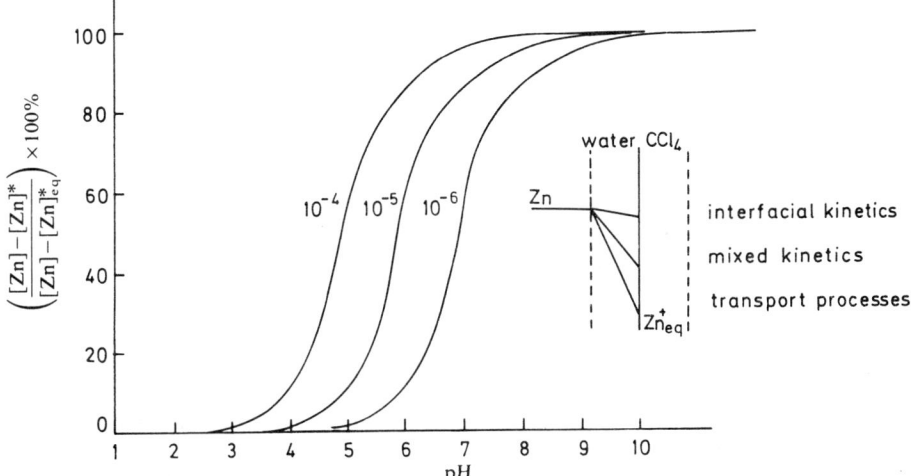

Fig. 4. Transport resistances for the zinc/dithizone system for different pH values. Parameter: concentrations [Zn]−[HDz] in mol dm^{-3}.

OVERLAPPING OF TRANSPORT AND REACTION

In principle, with the relevant set of equations and the corresponding kinetic parameters one may calculate fluxes and time dependences of concentrations. However, in order to survey the behaviour of a given system a suitable concept of resistances would be desirable, aiming at the separation of reaction and diffusion.

Such a concept is possible when the interfacial concentrations c_i^* can be calculated.[27] Using, for example, the metal-ion concentration in the dithizone system as the key component, the relation

$$I_T = \frac{[M^{2+}] - [M^{2+}]^*}{[M^{2+}] - [M^{2+}]^*_{\text{eq}}} \times 100\% \tag{5}$$

defines an actual measure of the transport resistance because the numerator corresponds to the actual gradient and the denominator is the gradient at equilibrium at the interface, i.e. a transport-controlled process.

The example for the zinc/dithizone system in fig. 4 shows the influence of pH and metal-ion concentration on the transport resistance I_T. In this calculation, the second term in eqn (2) for the back-reaction is derived with eqn (4) applying the kinetic-equilibrium condition.[12] In addition to the desirability of such a kinetic survey for a given system, these calculations are of interest because they suggest the possibility of realizing chemical control merely with a decrease in concentration (see fig. 2) or increase in pH.

COEXTRACTION

The two liquid/liquid reactions (bars indicate species in the solvent phase)

$$UO_2^{2+} + 2NO_3^- + 2\overline{T} \longrightarrow \overline{UO_2(NO_3)_2 2T}$$

$$H^+ + NO_3^- + \overline{T} \longrightarrow \overline{HNO_3T}$$

are competing for the extractant tributylphosphate, which means that the system is chemically coupled. The kinetic treatment of such a coextraction system is possible with the above-mentioned principles if one assumes that the interface is the site of both reactions.[23,29]

The decisive experimental results concerning this coextraction are the linear dependences of the fluxes on the stirring speed for all the participating species, which means that transport processes are rate-determining. Therefore the appropriate set of equations contains the equilibrium conditions at the interface for both reactions

$$K_H = \frac{[\overline{HNO_3T}]^*}{[H^+]^*[NO_3^-]^*[\overline{T}]^*}$$

$$K_U = \frac{[\overline{UO_2(NO_3)_2 2T}]^*}{[UO_2^{2+}]^*[NO_3^-]^{*2}[\overline{T}]^{*2}}$$

together with transport equations for all the individual fluxes by analogy with eqn (1). Applying the above-mentioned concept, it is possible to obtain the necessary individual coefficients for this system and therefore to calculate the time dependences of all the concentrations.

One outstanding result of these calculations is the strong coupling of both reactions which is evident at high uranium concentrations. The nitric acid concentration overshoots its equilibrium value in the course of mass transfer of uranium and nitric acid into the solvent phase very markedly.[23] This significant result of the numerical calculation is in agreement with the corresponding measurements (fig. 5), thus helping to confirm the proposed kinetic mechanism, and it shows that in the case of interfacial reactions chemically coupled systems can be treated.

A very important application of these kinetic results concerns the design of extraction columns. Until now the design of columns for reactive systems has been very empirical, because both important influences (kinetics and back-mixing) are unknown. Therefore knowledge of the detailed kinetics of a system represents a new approach to the calculation of concentration profiles, to the treatment of back-mixing effects and to the optimization of chemical separations.

ADSORPTION LAYERS

Decades ago, mass transfer at liquid/liquid boundaries was suggested as a model for biological permeation; at the present time mass transfer through monolayers between different liquid phases is of interest for liquid-membrane technology. In the context of this contribution it should be emphasized that the relations to such membrane processes are closely connected with the kinetics of mass transfer.

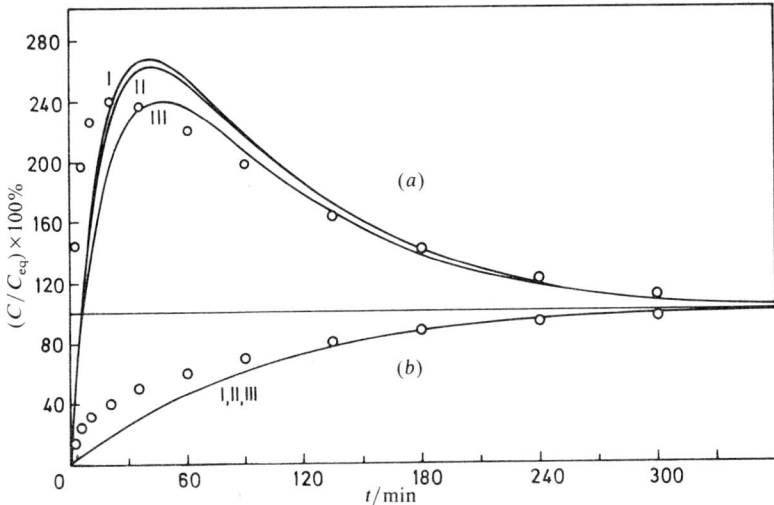

Fig. 5. Calculated (lines) and measured (open symbols) concentrations of (a) $\overline{\mathrm{HNO_3T}}$(Kx) and (b) $\overline{\mathrm{UO_2(NO_3)_2 2T}}$(Ko) for the mass transfer of $\mathrm{UO_2^{2+}}$ ($c = 0.5$ mol dm^{-3}) and $\mathrm{HNO_3}$ ($c = 2.4$ mol dm^{-3}) from water into hexane loaded with tributylphosphate [T] = 1.1 mol dm^{-3}. Stirred cell: $n_w = 300$ r.p.m. Initial deviations show interfacial instabilities; (I) (II) and (III) indicate different values for β_T.

In liquid/liquid permeation systems (water/adsorption-layer/solvent) the effects of surfactant layers are pronounced, but these effects are mostly of a fluid-dynamic nature. In this connection fig. 6 shows a very elaborate example of the influence of adsorption layers on liquid/liquid mass transfer for the case of transport limitation (stirred cell).[32]

Some peculiarities of fluid-dynamic behaviour in stirred cells should be emphasized. In the presence of surfactants, the state of the strongest depression of the rate (fig. 5) corresponds to rigid behaviour of the interface produced by a macroscopic gradient in the interfacial tension 'spread' over the whole interface. However, at a characteristic stirring number, connected with each individual surface coverage, the coefficient increases steeply, corresponding to a critical shear stress where the layer detaches from the edge of the interface (see sketch in fig. 7), leading to the appearance of an undisturbed fluid flow at the periphery of the interface.

In this fluid-dynamic region the true interfacial resistance is 'covered' by the rate-controlling diffusion step and therefore not recognizable. In spite of this fact knowledge about the fluid-dynamic mechanism with respect to the empirical behaviour of fluid-dynamic effects is important in order to identify true interfacial resistance.

Limited to the case that interfacial reactions influenced by the monolayer are rate-determining, the true interfacial resistance is accessible. The zinc/dithizone system is again a suitable example.[33] The typical behaviour of such a permeation system is markedly different compared with the fluid-dynamic effects: plateau rates indicate the reaction-controlling region, and the sensitivity of the plateau rates to the surfactant concentration indicates that an interfacial process is taking place (see fig. 8). In this context it should be emphasized that the separation of the fluid-dynamic and chemical effects of surfactants requires a measurement of mass transfer at variable forced convections.

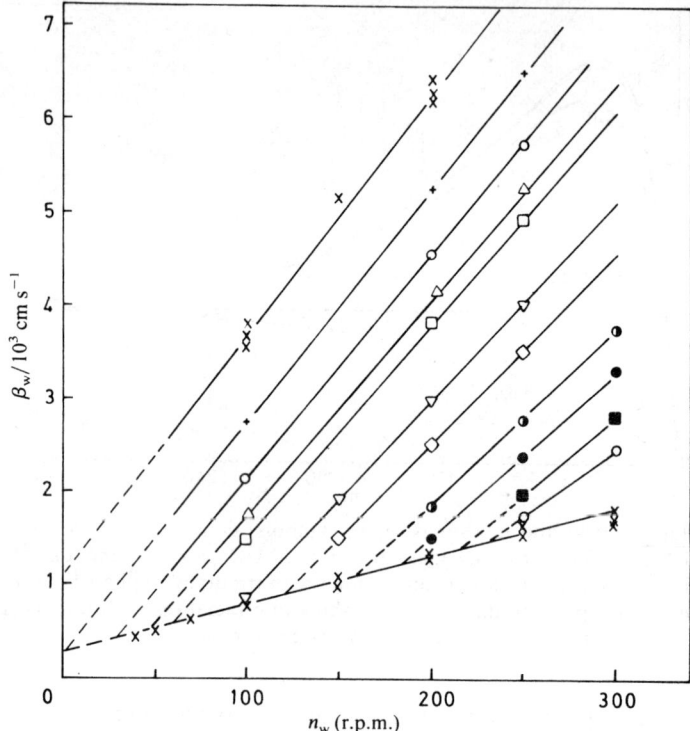

Fig. 6. Individual transport coefficient β_w for the transfer of toluene into an aqueous phase in the presence of insoluble poly(ethylenoxide) at the interface. Parameter: reciprocal coverage in Å^2 per monomeric unit. Highest line, pure interface; lowest line: rigid interface; intermediate lines as follows: ◆, 41.7; ■, 44.2; ●, 47.6; ◐, 51.5; ◇, 59.5; ▽, 66.8; □, 85.2; ○, 94.5; +, 114.0; ×, 168.4.

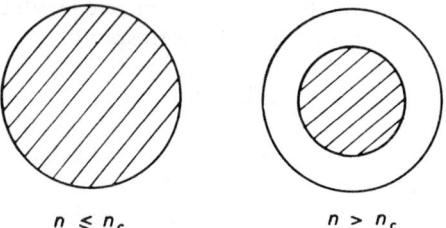

Fig. 7. Scheme showing the fluid-dynamic state of the interface below and above the critical stirring number.

The mechanism of such interfacial barriers has not been investigated until now. One interesting feature seems to be that in the case of an ionic interfacial reaction interfacial potentials are acting, because anionic layers decrease the rate of the interfacial reaction while cationic layers lead to its increase.[34]

The study of permeation through layers at the liquid/liquid boundary is in its infancy. To 'catalyse' its progress different chemically controlled systems occuring

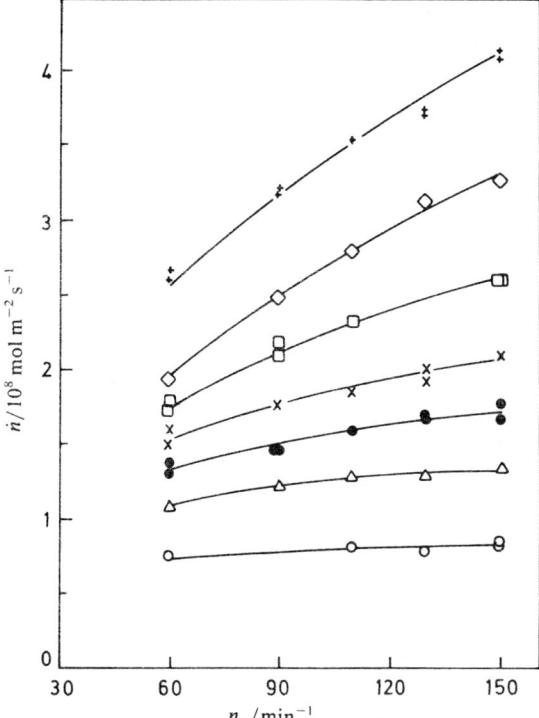

Fig. 8. Initial fluxes $\dot{n}$ for the complexation of zinc with dithizone ([HDz]= 1.25×10^{-4}, [Zn] = 5×10^{-5} mol dm^{-3}, pH 5) at the water/toluene interface in the presence of sodium dodecylsulphate (10^{-5} mol dm^{-3}) for various concentrations of sodium chloride, c/mol dm^{-3}: +, 0.01; ◇, 0.05; □, 0.1; ×, 0.15; ●, 0.2; △, 0.3; ○, 0.5.

at the interface are necessary in order to recognize and survey the different and decisive features of such processes.

CONCLUSIONS

Though a very small number of liquid/liquid reactions seem to have been thoroughly explained kinetically, it is the impression of the author that interfacial control in such systems will occur rather frequently.

For proceedings in this undeveloped field the introduction of an efficient and generally accepted type of stirred cell seems to be important so as to enable a comparison between different investigations to be made. Another catalysing feature should be the substitution of each overall treatment by an approach which involves individual fluxes. Finally, it is important to take account of the effects of surfactants, not only in the context of permeation but, more importantly, for the evaluation of experimental results. Without definite knowledge of a sufficiently pure interface, results are always doubtful.

---

[1] See the various Proceedings of the International Solvent Extraction Conferences (*Proc. ISEC*).
[2] P. R. Danesi and R. Chiarizia, *C.R.C. Crit. Rev. Anal. Chem.*, 1980, **10**, issue 1.
[3] C. B. Honaker and H. Freiser, *J. Phys. Chem.*, 1962, **66**, 127.

[4] H. Watari and H. Freiser, *J. Am. Chem. Soc.*, 1983, **105**, 191.
[5] W. Nitsch and K. Hillekamp, *Chem.-Ztg.*, 1972, **96**, 245.
[6] F. Baumgärtner and L. Finsterwalder, *J. Phys. Chem.*, 1970, **74**, 108.
[7] D. E. Horner, J. C. Mailen, D. W. Thiel, T. C. Scott and R. G. Yates, *Ind. Eng. Chem. Fundam.*, 1980, **19**, 103.
[8] M. F. Pushlenkov, N. N. Shchepetilnikov, G. I. Kuznetsov, F. D. Kasimov, A. L. Yasnovitskaya and G. N. Yakovlev, in *ISEC '74* (Society of Chemical Industry, London, 1974), vol. 1, p. 493.
[9] Y. Marcus and A. S. Kertes, in *Ion Exchange and Solvent Extraction of Metal Complexes* (Wiley, London, 1969).
[10] G. Astarita, in *Mass Transfer with Chemical Reactions* (Elsevier, Amsterdam, 1967).
[11] P. V. Danckwerts, in *Gas-Liquid Reactions* (McGraw-Hill, New York, 1970).
[12] W. Nitsch and B. Kruis, *J. Inorg. Nucl. Chem.*, 1978, **40**, 857.
[13] J. B. Lewis, *Chem. Eng. Sci.*, 1954, **3**, 248.
[14] J. Bulicka and J. Prochazka, *Chem. Eng. Sci.*, 1976, **37**, 137.
[15] H. Sawistowski and L. J. Austin, *Chem.-Ing.-Tech.*, 1967, **39**, 224.
[16] P. R. Danesi, R. Chiarizia and A. Saltelli, *J. Inorg. Nucl. Chem.*, 1976, **38**, 1687.
[17] W. Nitsch, M. Raab and R. Wiedholz, *Chem.-Ing.-Tech.*, 1973, **45**, 1026.
[18] W. Nitsch and J. Kähni, *Ger. Chem. Eng.*, 1980, **3**, 96.
[19] G. Petrich, in *ISEC '80* (Ass. des Ingénieurs sortis de l'Université de Liège, 1980), vol. 1, session 5A, pp. 80-42.
[20] J. A. Golding and V. N. Saleh, in *ISEC '80* (Ass. des Ingénieurs sortis de l'Université de Liège, 1980), vol. 1, session 2A, pp. 80-194.
[21] H. Brauer, in *Stoffaustausch einschließlich chemischer Reaktionen* (Verlag Sauländer, Aarau und Frankfurt, 1971).
[22] A. E. Handlos and T. Baron, *AIChE J.*, 1957, **7**, 127.
[23] W. Nitsch and A. van Schoor, *Chem. Eng. Sci.*, in press.
[24] A. van Schoor, *Doctoral Thesis* (Technical University of Munich, 1980).
[25] K. D. Heck, *Doctoral Thesis* (Technical University of Munich, 1974).
[26] K. Matt, *Diploma Thesis* (Technical University of Munich, 1980).
[27] A. Hoffmann, *Doctoral Thesis* (Technical University of Munich).
[28] A. v. Imhof, *Doctoral Thesis* (Technical University of Munich).
[28] W. Nitsch and U. Schuster, *Sep. Sci. Technol.*, in press.
[30] J. T. Davies and E. K. Rideal, in *Interfacial Phenomena* (Academic Press, New York, 2nd edn, 1963).
[31] W. Nitsch and O. Sillah, *Ber. Bunsenges. Phys. Chem.*, 1979, **83**, 1105.
[32] W. Kremnitz, *Doctoral Thesis* (Technical University of Munich, 1981).
[33] T. Michel, *Diploma Thesis* (Technical University of Munich, 1983).
[34] W. Nitsch and K. Roth, *Colloid Polym. Sci.*, 1978, **256**, 1182.

Facilitated Transport across Liquid/Liquid Interfaces and its Relevance to Drug Diffusion across Biological Membranes

BY NICHOLAS BARKER,[†] JONATHAN HADGRAFT* AND PAUL K. WOTTON

Department of Pharmacy, University of Nottingham, University Park, Nottingham NG7 2RD

Received 22nd November, 1983

The rotating diffusion cell has been used to study the facilitated transport of a dianionic dye, Resorcin Brown R, across a solid-supported liquid membrane. A pH gradient provides the chemical driving force for the co-transport mechanism. The carrier molecules incorporated into the liquid membrane were hydroxypropylamines and hydroxybutylamines. The transport mechanism was shown to be saturable. The amines also transported salicylate anions, and the proposed mechanism may be suitable for transporting anionic drug molecules across biological membranes. A further commercially available amine, Ethomeen S12, was also found to be capable of increasing the flux of salicylate across lipid membranes, and this compound may have potential in skin preparations.

In order to increase the efficiency of solvent extraction processes, solid-supported liquid membranes have been investigated recently.[1,2] Various workers have been able to utilise a pH gradient to facilitate the transport of ionic species from one aqueous compartment across a non-polar organic phase to a receptor aqueous phase against the ion concentration gradient. Either co- or counter-transport mechanisms are possible. In both cases a pH gradient is often the driving force, although inorganic anions have been used to facilitate cation flux in a co-transport scheme.[3] We have developed this principle in order to facilitate the transport of anionic drugs across the skin. In commercial solvent extraction systems there are not many constraints on the choice of the pH gradient, and often the difference may be from 2 to 12. However, in the case of skin and other biological membranes there are severe limitations and any proposed scheme must operate within narrow limits. Since the surface of the skin is slightly acidic, pH 4.2–5.6, and the lower layers of the skin are at physiological pH 7.3–7.4, these are the pH extremes under which the transport mechanism must operate.[4] However, we should be able to design a system which utilises this inbuilt pH gradient to facilitate the transfer of anionic drugs which normally do not penetrate the skin.

The potential of this technique has been assessed using a model system and the rotating diffusion cell has enabled us to simulate the lipids of the skin in a well defined *in vitro* model.[5] The epidermal barrier is simulated by using a Millipore filter impregnated with isopropyl myristate, a liquid representative of skin lipids.[6] The facilitated transfer scheme is represented in fig. 1. To effect transfer, a carrier molecule is introduced into the isopropyl myristate and at the lower pH becomes protonated at the interface. In this form it combines with the least hydrophilic anion present to form an ion pair. The ion pair partitions into the bulk lipid phase and diffuses down its concentration gradient to the opposite interface. In the

[†] Present address: Smith Kline & French Ltd, Welwyn Garden City, Hertfordshire AL7 1EY.

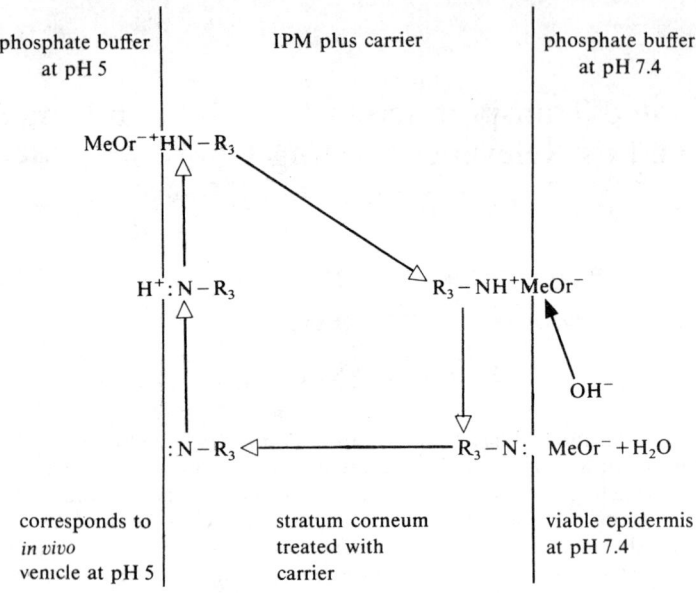

Fig. 1. Schematic diagram of the facilitated transport of MeOr.

interfacial region at the higher pH, the carrier molecule deprotonates and the anion is released. The carrier is then free to diffuse back to the other interface and repeat its role.

Various criteria will govern the relative efficiency of the transport. The ideal carrier should have a pK_a between 5 and 7.4, have a high partition coefficient and be water insoluble. Accumulation in the interfacial region will be improved if the carrier has slight surfactant properties; however, a compound with strong surface-active properties will cause problems owing to emulsion formation at the interface with concomitant interfacial breakdown. From the cosmetic and pharmaceutical standpoint the carrier must also be non-volatile, non-toxic, non-sensitising and inert to the different skin constituents. In previous studies we have shown that there are some substituted amine derivatives such as N-substituted bis(2-hydroxypropyl)amines which may fulfill the above criteria.[6] In previous work Methyl Orange was used as the model anion; in this study we have investigated ionised salicylate, since this is often used as a model drug, and Resorcin Brown R, a dianionic species. This dye was chosen since it has structural resemblances to a potential topical drug, disodium cromoglycate. The structures are presented in fig. 2.

EXPERIMENTAL

The mono-substituted amines were prepared by documented techniques. N,N-bis(2-hydroxypropyl)octadecylamine (1) was prepared by the method described by Boivin,[7] and N,N-bis(2-hydroxybutyl)hexadecylamine (2) and N,N-bis(2-hydroxybutyl)octadecylamine (3) were synthesised using the method of Perrault.[8] N-(2-hydroxypropyl)bis(octadecyl)amine (4) was prepared by refluxing 2-hydroxypropylamine with 1-bromoctadecane in chloroform with an excess of anhydrous potassium carbonate. Any quaternary ammonium compounds formed were removed by filtration and the product was distilled at reduced pressure. The

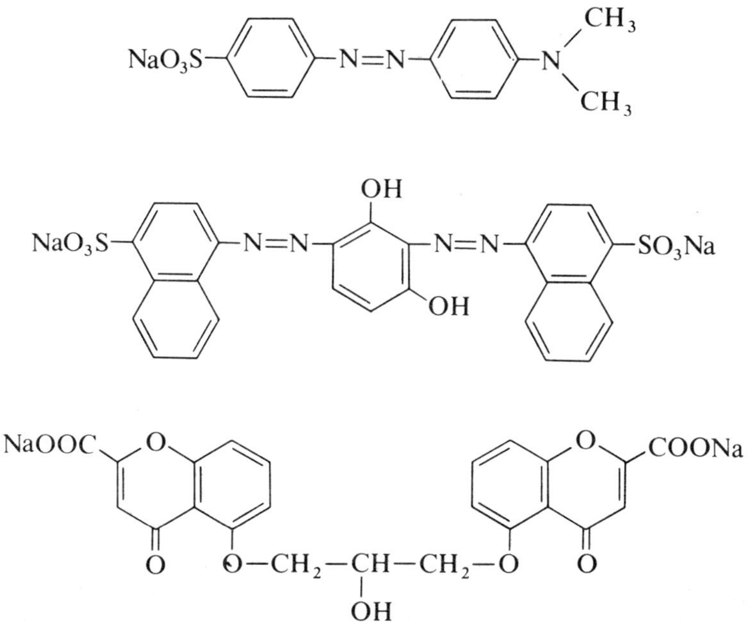

Fig. 2. Structures of (I) Methyl Orange (MeOr), (II) Resorcin Brown R (RBR) and (III) disodium cromoglycate.

Table 1. The carrier amines

amine	formula	boiling point/°C (pressure/mmHg)
N,N-bis(2-hydroxypropyl)octadecylamine (1)	$C_{18}H_{37}N(CH_2CH(OH)CH_3)_2$	211 (0.5)
N,N-bis(2-hydroxybutyl)hexadecylamine (2)	$C_{16}H_{33}N(CH_2CH(OH)C_2H_5)_2$	199 (0.4)
N,N-bis(2-hydroxybutyl)octadecylamine (3)	$C_{18}H_{37}N(CH_2CH(OH)C_2H_5)_2$	239 (0.2)
N-(2-hydroxypropyl)bis(octadecyl)amine (4)	$(CH_{18}H_{37})_2NCH_2CH(OH)CH_3$	221 (0.2)

purity of the amines was checked using i.r. and n.m.r. spectroscopies. The amines and their distillation temperatures are given in table 1.

Resorcin Brown R (RBR) was supplied by Hopkins and Williams, Methyl Orange (MeOr) and sodium salicylate by B.D.H.; all were recrystallised before use. Ethomeen S12 is a tertiary amine derived from soya-bean oil with an unsaturated C_{18} alkyl chain and was supplied by Akzo Chemie, U.K. Isopropyl myristate (IPM) was supplied by Croda chemicals and has a refractive index gradient of 1.4346 at 25 °C.

The rate of transfer of the solutes across a filter impregnated with IPM was studied using a rotating diffusion cell. This cell uses the hydrodynamics of the rotating-disc system to impose a known pattern of convective flow on either side of the filter. The rotation of the cell produces stagnant diffusion layers of known thickness on both sides of the filter. It is thus possible to

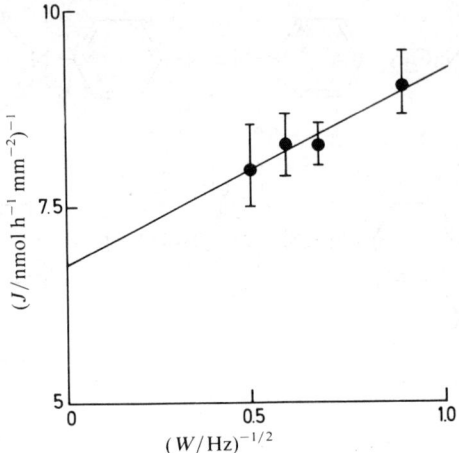

Fig. 3. Relationship between the inverse flux, J, and the square root of the rotation speed, W, for RBR with (2) at a concentration of 0.01 mol dm^{-3} as the carrier.

see whether diffusion across these layers is in any way rate-limiting. In all experiments 1 μm pore-size mixed cellulose ester filters were used which were impregnated with carrier solution in IPM by saturating the filter and carefully removing the excess solution with a tissue; in previous work this has been shown to be reproducible. The pH in both compartments was maintained using phosphate buffer. The rate of appearance of the solutes was monitored continuously using a flow-through cell in a spectrophotometer.

A typical experimental determination is shown in fig. 3, where the influence of rotation speed on the transfer rate of RBR using (2) as a carrier is illustrated. In all discussions following, the flux quoted is the effective flux at infinite rotation speed, i.e. there is no contribution from the stagnant diffusion layers. In the preliminary experiments with the carriers that were synthesised, equal concentrations of the solute were placed in the donor and receptor compartments. This was to ensure that any breakdown of the membrane integrity would be immediately apparent.

RESULTS AND DISCUSSION

In all the experiments with RBR the pH gradient was maintained using phosphate buffer, the ionic strength of which was kept constant. In previous studies we have observed that at high ionic strengths the transport rate is modified. This is possibly due to salting-out effects or because there is competition between the other anions and the dye anion for transfer sites at the interface. In all the studies with the synthesised amines the flux of RBR without the carrier present was negligible. The initial studies conducted with RBR compared the transfer rates achieved using the most efficient carrier described previously for Methyl Orange, compound (1). RBR was transported far less readily, which may be predicted if the carrier forms the 2:1 stoichiometric associate with the dianion. The results are given in fig. 4. The flux of RBR using this carrier was too low to be feasible for use in topical preparations and alternative amines were studied. (2) and (3) co-transported RBR against its concentration gradient far more readily than (1) (fig. 4). This result is difficult to rationalise as there are unlikely to be any significant pK_a or lipophilicity differences between the bis(2-hydroxypropyl)amines and the bis(2-hydroxybutyl)amines. The

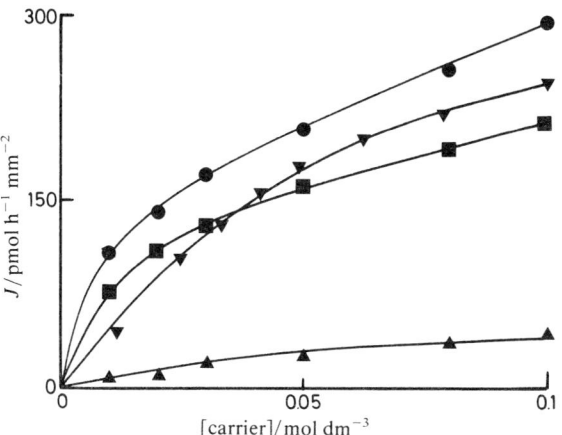

Fig. 4. Relationship between dye flux and carrier concentration: ▼, (1), MeOr; ▲, (1), RBR; ■, (2), RBR; ●, (3), RBR.

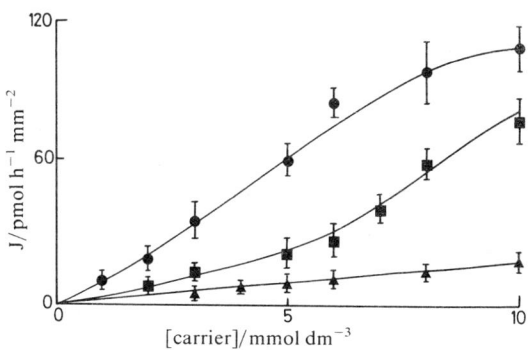

Fig. 5. Relationship between dye flux and carrier concentration: ●, (3), RBR; ■, (2), RBR; ▲, (4), RBR.

enhanced activity may possibly be attributed to a complex interfacial phenomenon related to the ion-association process. The transport rates at low carrier concentrations are shown in fig. 5. (4) also transported RBR at low concentrations, but its low solubility in IPM precluded its evaluation at higher concentrations. However, it was less efficient than its mono-substituted counterparts.

The influence of the RBR concentration was investigated (fig. 6); it appears that the coupled transport flux tends towards a limiting value, suggesting that a form of saturation carrier kinetics is occurring. A similar hypothesis has been tested by others.[9] The results may be analysed using a derivative of the Michaelis–Menten approach to enzyme kinetics. The classical constants V_{max} and K in the enzyme analysis now reflect the maximum attainable dye flux and an affinity constant between the RBR and the carrier. A double reciprocal plot for RBR with (3) as the carrier is shown in fig. 7.

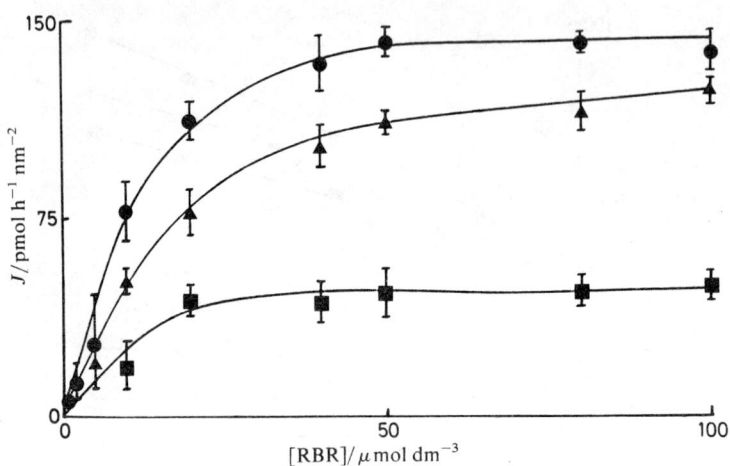

Fig. 6. Relationship between RBR flux and RBR concentration: ●, 0.01 mol dm^{-3} (3); ▲, 0.01 mol dm^{-3} (2); ■, 0.1 mol dm^{-3} (1).

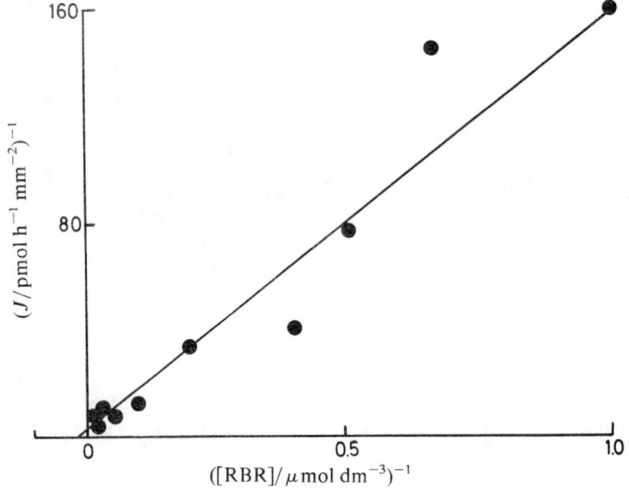

Fig. 7. Michaelis–Menten-type relationship; reciprocal RBR flux and reciprocal RBR concentration for 0.01 mol dm^{-3} (3).

One of the most commonly used model drugs is salicylic acid. Since the interfacial transfer kinetics have already been studied,[10] we investigated the facilitated transport of ionised salicylate. Fig. 8 shows that it is possible to facilitate the transport. However, salicylate is transported far less readily than RBR by the bis(2-hydroxybutyl)amines investigated. The cause of this reduction is probably the difference in the hydrophobicities of the two substrates and the stabilisation of the ion pair in the interfacial region. Compound (1), the less efficient carrier, transported the two anions equally, but on the basis of the stoichiometry required to produce a neutral ion pair we would expect salicylate to be transferred more efficiently.

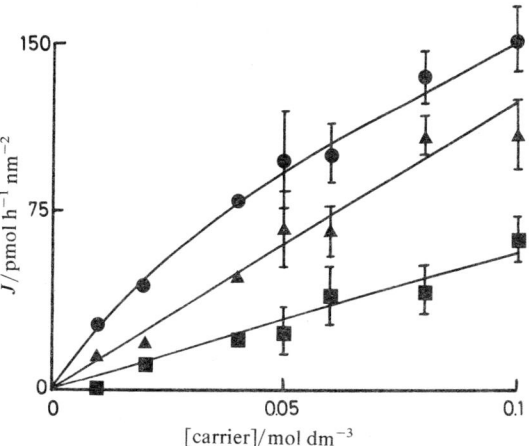

Fig. 8. Relationship between salicylate flux and carrier concentration: ●, (2); ▲, (3); ■, (1).

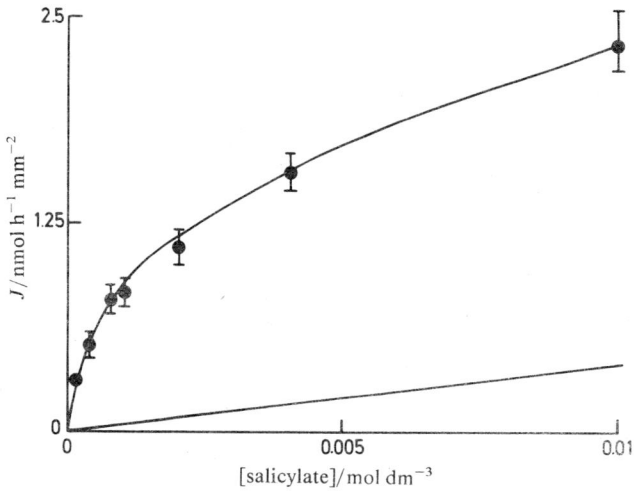

Fig. 9. Relationship between salicylate flux and salicylate concentration for 0.1 mol dm^{-3} Ethomeen S12. The lower linear relationship is diffusion without carrier transport and the upper curve is a combination of diffusion plus carrier transport.

In order to develop a facilitated transport scheme for commercial use it is necessary to use materials that are cosmetically and pharmaceutically acceptable. We have therefore investigated the ability of Ethomeen S12 to facilitate the transfer of ionised salicylate across an IPM-impregnated filter. In these experiments the substrate was not initially present in the receptor phase and we were not concentrating it against its concentration gradient as in the previous work (although we have shown that Ethomeen S12 is capable of achieving this). This enabled us to produce a better representation of the use *in vivo*.

To minimise the effect of transfer of un-ionised salicylate we used a pH gradient of 6.0–7.4 with the pH being maintained using a pH-stat. Diffusion in a system without carrier present occurred to a small extent as indicated in fig. 9. However, the flux obtained in the presence of 0.1 mol dm^{-3} Ethomeen S12 in IPM was at least an order of magnitude greater. The shape of the curve in fig. 9 has been reported previously for systems where diffusion and carrier transport occur simultaneously.[11] The results produced using this carrier and experimental conditions indicate that this type of phenomenon may be usefully utilised in a scheme to transfer anionic drugs across biological membranes. In preliminary work using a rabbit model we have shown that the penetration of ionised salicylate across intact skin can be enhanced by at least an order of magnitude using facilitated transfer and the amine carriers that we synthesised.

There are few reports in the literature where this type of mechanism has been used to advantage in the delivery of drugs. In a study of the percutaneous absorption of indomethacin [(1-p-chlorobenzolyl-5-methoxy-2-methylindol-3-yl) acetic acid] an enhanced penetration rate at pH 6.2 was attributed to the formation of an ion-pair complex.[12] This is possible since the pK_a of indomethacin is 5.2 and bis(2-hydroxypropyl)amine is present in the preparation. It is therefore possible that facilitated transport mechanisms may have been used fortuitously to deliver drugs across biological membranes.

We thank the S.E.R.C. for a studentship for N.B. and the S.E.R.C. and Fisons Pharmaceuticals for a CASE award to P.K.W.

[1] P. R. Danesi, E. P. Horowitz, G. F. Vandergrift and R. Chiariza, *Sep. Sci. Technol.*, 1981, **16**, 201.
[2] P. R. Danesi, E. P. Horowitz and P. Rickert, *Sep. Sci. Technol.*, 1982, **17**, 1183.
[3] F. Caracciolo, E. L. Cussler and D. Fennell-Evans, *AIChE J.*, 1975, **21**, 160.
[4] M. Katz, in *Design of Topical Drug Products: Pharmaceutics in Drug Design*, ed. E. J. Ariens (Academic Press, London, 1973), vol. IV, p. 97.
[5] W. J. Albery, J. F. Burke, E. B. Leffler and J. Hadgraft, *J. Chem. Soc., Faraday Trans. 1*, 1976, **72**, 1618.
[6] N. Barker and J. Hadgraft, *Int. J. Pharm.*, 1981, **8**, 193.
[7] J. L. Boivin, *Can. J. Chem.*, 1958, **36**, 1405.
[8] G. Perrault, *Can. J. Chem.*, 1967, **45**, 1063.
[9] T. Shimbo, M. Sugiura, N. Kamo and Y. Kobatake, *J. Membr. Sci.*, 1981, **9**, 1.
[10] R. H. Guy and J. Hadgraft, *J. Colloid Interface Sci.*, 1981, **81**, 69.
[11] K. D. Neame and T. G. Richards, *Elementary Kinetics of Membrane Carrier Transport* (Blackwell Scientific Publications, Oxford, 1972), p. 54.
[12] T. Inagi, T. Muramatsu, H. Nagai and H. Terada, *Chem. Pharm. Bull.*, 1981, **29**, 1708.

ns. Chem. Soc., 1984, 77, 105-113

Kinetics of Heterogeneous Nitration in Emulsions

BY JOHN E. CROOKS* AND JOHN M. CHISHOLM

Chemistry Department, King's College, University of London, Strand, London WC2R 2LS

Received 1st December, 1983

The two-phase nitration of aromatic hydrocarbons has been studied in a model system consisting of a solution of toluene in hexane dispersed as an emulsion in aqueous sulphuric acid, in the concentration range 63–76 wt%. Reaction half-lives were found to range from 400 to 0.4 s over this acid concentration range. The progress of the reaction was monitored by spectrophotometric determination of the nitrotoluenes produced; a stopped-flow apparatus was used for the faster runs. The experimental results are compared with those calculated from a theoretical model. The aqueous acid phase around each organic droplet is subdivided into a number of concentric spherical shells. The change in the amount of toluene in each shell owing to diffusion in and out and to reaction is calculated for a large number of short time intervals, during each one of which the concentration is assumed constant. Good agreement between theoretical and experimental results is obtained except for the fastest runs. For these, diffusion is faster than predicted for stagnant conditions, showing that stirring is occurring within 1 μm of the liquid/liquid interface.

The nitration of benzene is a reaction of great technical importance, being the first step in the production of aniline dyes. The nitration of benzene and its methyl derivatives has thus been the subject of much research. The reaction is between two immiscible liquids, the aromatic hydrocarbon and the nitrating mixture, which is a solution of nitric acid in aqueous sulphuric acid, of approximate composition 60 wt% acid.

Reaction occurs in the aqueous phase in a surface region closely adjacent to the interface with the organic phase.[1] The mechanism involves attack on the aromatic molecule by the nitronium ion, NO_2^+, which is insoluble in the organic phase, whereas the aromatic hydrocarbon is sparingly soluble (*ca.* 10^{-3} mol dm^{-3}) in the aqueous phase.

The reaction scheme may be represented by [2]

$$HNO_3 + H^+ \underset{k'_{-1}}{\overset{k'_1}{\rightleftharpoons}} NO_2^+ + H_2O$$

$$NO_2^+ + ArH \underset{k'_{-2}}{\overset{k'_2}{\rightleftharpoons}} ArHNO_2^+$$

$$ArHNO_2^+ \overset{k_3}{\longrightarrow} HArNO_2^+$$

$$HArNO_2^+ \longrightarrow H^+ + ArNO_2$$

where $[ArHNO_2]^+$ and $HArNO_2^+$ indicate reaction intermediates whose nature has been the subject of much research. It is fairly well established[3] that $HArNO_2^+$ has the Wheland structure. The nature of ArH NO_2^+ is less certain. It could be a simple encounter complex[4] (ArH, NO_2^+) or an *ipso* complex.[5] If $[NO_2^+]$ is assumed to be constant, i.e. the steady-state assumption is made, the overall rate of nitration is given by

$$\frac{-d[ArH]}{dt} = \frac{k_1' k_2' k_3' [ArH][HNO_3](H^+)}{k_{-1}'(H_2O)(k_{-2}' + k_3') + k_2' k_3'[ArH]} \quad (1)$$

where (H_2O) and (H^+) are the activities of H_2O and H^+, respectively. In concentrated aqueous sulphuric acid the activities of these species cannot be assumed to be equal to their concentration, not even as a first approximation. However, since (H_2O) and (H^+) may be considered to be constant during the course of the reaction, eqn (1) may be simplified by the substitutions

$$k_1 = k_1'(H^+) \quad (2)$$

$$k_2 = k_1'(H^+) k_2' k_3' / [k_{-1}'(H_2O)(k_3' + k_2')] \quad (3)$$

to give

$$\frac{-d[ArH]}{dt} = k_1 k_2 [HNO_3][ArH](k_1 + k_2[ArH]). \quad (4)$$

There are two limiting cases, depending on the relative magnitudes of k_1 and $k_2[ArH]$. In solutions of high concentration in sulphuric acid, (H_2O) is low and (H^+) is high, so that $k_2[ArH] \gg k_1$. First-order kinetics are observed, since eqn (4) reduces to

$$\frac{-d[ArH]}{dt} = k_1[HNO_3]. \quad (5)$$

In solutions of relatively low concentration of sulphuric acid, $k_2[ArH] \ll k_1$ and second-order kinetics are observed:

$$\frac{-d[ArH]}{dt} = k_2[HNO_3][ArH]. \quad (6)$$

For toluene, the transition from first-order to second-order kinetics occurs at ca. 78% sulphuric acid. In aqueous sulphuric acid of this composition, $k_2 \approx 1500$ dm^3 mol^{-1} s^{-1}, $[ArH] \approx 5 \times 10^{-4}$ mol dm^{-3} for a saturated solution and $k_1 \approx 0.6$ s^{-1}. Benzene, being less reactive than toluene, has k_2 values too low for first-order kinetics, and nitration of benzene is second-order for the whole range of sulphuric acid concentrations.

The nitration of toluene has been studied thoroughly[6] in homogeneous solution in aqueous sulphuric acid in the concentration range 52.5–80 wt% acid. However, the nitration of aromatic hydrocarbons in bulk is a two-phase process, so the interfacial kinetics must also be studied. Various theoretical models of stirred reactors have been produced, but these are semi-empirical and they introduce quantities which may only be evaluated from the kinetic measurements themselves, such as a mass-transfer coefficient[7] or the volume and surface area of an eddy in a rapidly stirred mixture.[8] The aim of this study is to produce a theory for which all the parameters may be evaluated by non-kinetic experiments and which may be tested by comparison with experimental results for a well defined system.

THEORETICAL DESCRIPTION

The organic phase is considered to consist of a set of spheres of uniform radius, stationary in a continuous aqueous medium. A small proportion of ArH and ArNO$_2$

is present in the aqueous phase but no NO_2^+ is present in the organic phase. Nitration occurs when the ArH molecules diffuse away from the surface of the spheres and meet NO_2^+ in the aqueous phase. The $ArNO_2$ produced diffuses back into the organic phase. An ArH molecule can only diffuse a short distance into the aqueous phase before reaction; the more concentrated the sulphuric acid, the shorter this distance is. The reaction zone thus consists of a spherical shell, across which [ArH] falls from its maximum value immediately adjacent to the organic phase to zero on the outside. The thickness of this shell depends on the relative magnitudes of the diffusion rate and the reaction rate. The differential equation describing the concentration gradient is extremely difficult to solve analytically but it lends itself to solution by numerical methods.

The organic sphere and its aqueous surroundings are split into a number of concentric spherical shells, of thickness δr, whose centre is the centre of the sphere. The interface separating the organic and aqueous phases is between the outermost shell of organic liquid and the innermost shell of aqueous liquid. The solute concentration, [ArH], is considered to be uniform across each shell at all times. The system as a whole has radius r_B; there are S elements of thickness δr_A containing the organic phase A, surrounded by a further S elements of thickness δr_B containing the aqueous phase B. The progress of the reaction is considered to take place in a series of short time intervals, each of duration δt. During each time interval the concentration change due to diffusion and reaction in each spherical shell is considered negligible in comparison with the concentration, [ArH], in each shell. This enables the amounts diffusing in and out across the boundaries of the shell to be calculated by simple linear equations, based on Fick's second law:

amount of solute crossing interface of unit area = D times the concentration gradient (7)

where D is the diffusion coefficient. For diffusion out from shell $(i-1)$ to shell i, the concentration gradient is taken as the difference in concentrations in these shells divided by δr. Diffusion out from shell i to shell $(i+1)$ also affects the concentration in shell i, unless the shell is shell S, the outermost in the model system. The change in concentration in shell i due to diffusion in and out, $\delta c_i(\text{diff})$, is given by

$$\delta c_i(\text{diff}) = \frac{3D\delta t(r_i - \delta r/2)^2(c_{i-1} + c_i) - (r_i + \delta r/2)^2(c_i - c_{i+1})}{(\delta r)^2[3r_i^2 + (\delta r/2)^2]}. \quad (8)$$

Eqn (8) is true for any shell i, except the outermost of phase B and also the shell l of phase B, the innermost, whose inner boundary is the interface with phase A. Since ArH is much more soluble in the organic phase A than in the aqueous phase B, free energy equal to the difference in solvation energy between the two solvents must be supplied to enable ArH to move out from phase A to phase B. This reduces the rate of diffusion out. The required free energy can be estimated from the partition coefficient, K, where

$$K = ([\text{ArH}] \text{ in phase A})/([\text{ArH}] \text{ in phase B}) \quad (9)$$

at equilibrium. A detailed study of the rate of flow of ArH molecules across the interface in both directions shows that the net rate of diffusion out is given by

$$\text{flux} = (D_{AB}/K\delta x)(c_{A,S} - Kc_{B,1}) \quad (10)$$

where $c_{A,S}$ and $c_{B,1}$ are the concentrations of ArH in the outermost shell of A and the innermost shell of B, respectively, and D_{AB} is the diffusion coefficient which would be observed if A and B had the same viscosity and K was unity. For shell 1 in phase B, eqn (10) is used to modify eqn (8). For all the shells in phase B the concentration is also changed by chemical reaction, by an amount δc_i (reaction) in the time interval δt

$$\delta c_i \text{ (reaction)} = c_{B,i}\{1 - \exp(-k_2[\text{HNO}_3]\,\delta t)\}. \tag{11}$$

This assumes that the sulphuric acid concentration is <78%, so that eqn (6) applies and HNO_3 is in excess over ArH. At the end of time δt, the flows are 'frozen' and all the concentration changes given by eqn (8), (10) and (11) are calculated. New values of c_i in phases A and B are calculated and substituted into eqn (8), (10) and (11) and the flows re-start and continue for another time interval δt. Calculations for a large number of steps, performed by a Fortran IV program run on the University of London's CDC 7600 computer, gave a complete picture of the conversion of ArH to ArNO_2. The computation was subject to computer time constraints which limited the number of spherical shells and the number of time intervals considered. If the subdivision was too coarse, it was no longer true that the concentration changes were negligible by comparison with the standing concentrations. This had the effect of sending the diffusion pattern into oscillation, c_{i+1} being alternately less and greater than c_i in successive time intervls. Experience showed that, to prevent this occurring, δt and δr should be chosen so that

$$D\delta t/(\delta r)^2 < 0.45. \tag{12}$$

Results from the theoretical model can be compared with experimental data from a system in which nitration is initiated by mixing an emulsion consisting of droplets of toluene dispersed in aqueous sulphuric acid with a solution of nitric acid in aqueous sulphuric acid of the same composition. The theoretical treatment must be extended to take account of the polydispersity of the emulsion; the spheres are of a range of radii. The droplet-size distribution is represented by five sets of spheres, each of a single radius, and the total amount of ArH reacted at any time is the sum of that reacted from each of the five sets of spheres, weighted according to the particle-size distribution.

In order to calculate the amount of toluene nitrated at any time, various data, independent of the experimental kinetic data, are required. The radius of the sphere of the toluene phase, r_A, were chosen for each of the five sets from photomicrographic measurements. The radius of the outer sphere of aqueous sulphuric acid, r_B, was calculated from the phase volume ratio, P, the ratio of the volumes of toluene and acid phases. This was kept at 0.0076 throughout. The value of r_B is hence 5.9 r_A. Values of D_A and D_B, the diffusion coefficients in the two phases, were calculated using the Stokes–Einstein equation

$$D = kT/6\pi\eta r. \tag{13}$$

Values of D for toluene in a solvent of known viscosity, namely hexane,[9] were used to obtain a value of r, 250 pm, then values of the viscosity of the aqueous sulphuric acid solutions[10] were used to calculate D_B. Values of δr and δt were chosen so that eqn (12) was valid. The total number of steps was chosen to allow time for nitration of almost all the toluene. Typically a few thousand steps were required. Values of k_2 were taken from the literature.[6] Fig. 1 shows some typical results of the calculation. Each circle shows the calculated concentration of toluene at the mid-point of each spherical shell in the aqueous acid phase. The number of circles on the concentration gradient curve shows the fineness of subdivision of r_B

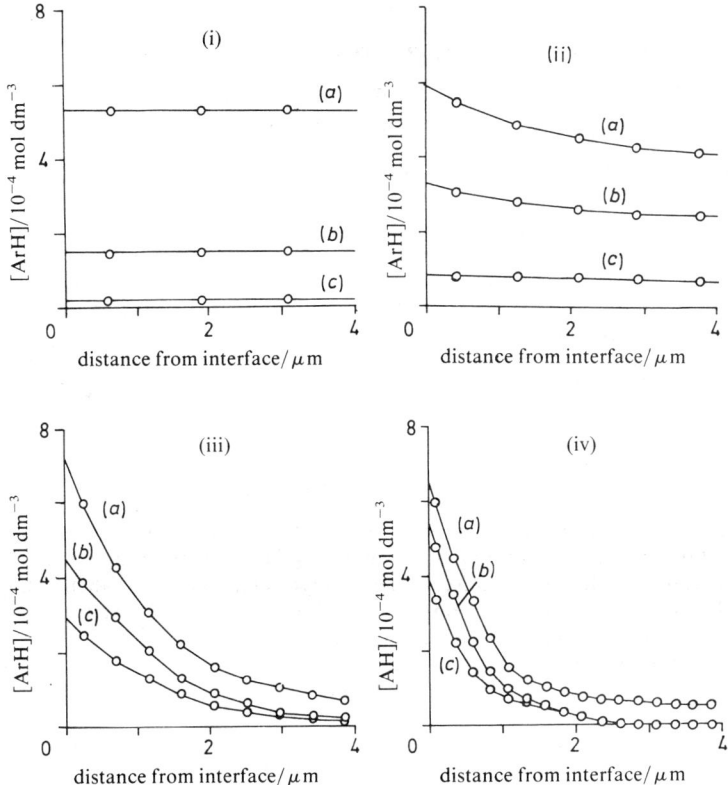

Fig. 1. Concentration of toluene in aqueous sulphuric acid at various times and distances from the interface, at 25 °C. (i) 63.2% Sulphuric acid, $r_A = 0.94\ \mu$m: (a) 20 s after mixing, (b) 400 s after mixing and (c) 980 s after mixing. (ii) 67.6% Sulphuric acid, $r_A = 1.63\ \mu$m: (a) 0.8 s after mixing, (b) 8.0 s after mixing and (c) 24.0 s after mxing. (iii) 71.9% Sulphuric acid, $r_A = 2.22\ \mu$m: (a) 0.1 s after mixing, (b) 2.5 s after mixing and (c) 4.9 s after mixing. (iv) 73.6% Sulphuric acid, $r_A = 1.21\ \mu$m: (a) 0.02 s after mixing, (b) 0.4 s after mixing and (c) 0.98 s after mixing.

required to obtain a reasonable representation of the reaction. The higher the acid concentration the faster the reaction and the larger the number, S, of subdivisions necessary.

For systems in which the reaction is much slower than diffusion, e.g. as for 63.2% acid in fig. 1, the observed rate of nitration depends only on k_2 and not on D or r_A. However, the observed rate constant is not k_2 since most of the toluene is held in the organic spheres and is not available for reaction until the toluene in the aqueous acid phase has reacted. If the ratio of the concentrations of toluene in the organic phase to that in the aqueous acid phase is K at all times then

$$k_{obs} = k_2/(KP+1) \tag{14}$$

where k_{obs} is related to the total number of moles of toluene in the whole system, n_{ArH}, by

$$dn_{ArH}/dt = k_{obs}[HNO_3]n_{ArH}. \tag{15}$$

EXPERIMENTAL

PREPARATION AND CHARACTERISATION OF EMULSIONS

The emulsifying agent used was a blend of Brij 30 and Brij 35, supplied by Sigma Chemicals and used without further purification. These surfactants are non-ionic polyoxyethylene lauryl ethers, $C_{12}H_{25}(OC_2H_4)_n OH$, for which $n = 4$ for Brij 30 and $n = 23$ for Brij 35. A blend of 43 wt% Brij 30 with 57 wt% Brij 35 had the correct HLB number, 14, for emulsifying hydrocarbon in aqueous sulphuric acid. In order to keep the droplet radius constant during the reaction, the organic phase was not pure toluene but a solution of toluene in hexane. To prepare an emulsion, 5 cm^3 of toluene and 2.15 g Brij 30 were dissolved in 10 cm^3 hexane; 1.14 g of Brij 35 were dissolved in 200 cm^3 aqueous sulphuric acid of known concentration and the hexane solution was dispersed in this in a tissue grinder using three strokes of the plunger. The emulsions produced were stable for >24 h.

The size distribution was obtained by photomicrography. Droplet sizes were measured from projections of the photomicrographs, which included a graticule scale. A total of at least 250 droplets was measured for each emulsion.

MEASUREMENT OF THE PARTITION COEFFICIENT, K

A solution of toluene in hexane was carefully added to an equal quantity of aqueous sulphuric acid of known concentration, in a separating funnel, with minimum agitation. The funnel was placed in a thermostat tank at the required temperature for 48 h. Samples were taken from the two layers and the concentration of toluene determined spectrophotometrically at 260 nm. If the layers were mixed by shaking, droplets of the organic phase, rich in toluene, were formed in the aqueous phase and could not be removed. The value of K so found, 1700 ± 150, was independent of sulphuric acid concentration in the range 61.2–76.1 wt%.

KINETIC MEASUREMENTS

To initiate the reaction, the emulsion was mixed with an equal volume of aqueous sulphuric acid containing 1 mol dm^{-3} HNO$_3$. The concentration of sulphuric acid in the reacting mixture took into account the water necessarily introduced by the addition of HNO$_3$ (66 wt%) to the stock aqueous sulphuric acid. For runs of duration >50 s, the increase in absorption at 300 nm from nitrotoluene formation was monitored using a conventional spectrophotometer (Perkin-Elmer Lambda 5). For faster runs, a stopped-flow spectrophotometer (Hi-Tech Scientific SF-3A) was used. The slower runs were made more complicated by the slow breaking-up of the emulsion. Although the emulsion in sulphuric acid was stable this was not so if nitric acid was also present. This had the effect of reducing the absorbance, as the emulsion droplets disappeared. A correction factor was obtained by repeating the runs with emulsions containing hexane only in the organic phase. In a typical run, in 60.4 wt% sulphuric acid, the absorbance at 370 nm for a 10 mm pathlength cell increased by 0.17 units for complete nitration, after 5700 s, even though the absorbance had been as high as 0.24 at 1500 s. The absorbance of a 10 mm cell containing the blank of emulsified hexane decreased by 0.25 units at 1500 s and by 0.65 units at 5700 s. The true absorbance increase for complete reaction was thus 0.82 units. Fortunately the theoretical progress of reactions proceeding this slowly is unaffected by droplet size, since the slow runs are kinetically controlled. Emulsion break-down was negligible for the fast runs, whose kinetics are partly diffusion controlled. The infinity values for the slow runs were calculated using the Swinbourne extrapolation,[11] which was permissible as the runs are first order.

RESULTS AND DISCUSSION

The experimental data for a representative range of a sulphuric acid concentration are shown in fig. 2, together with predicted results from various theoretical models.

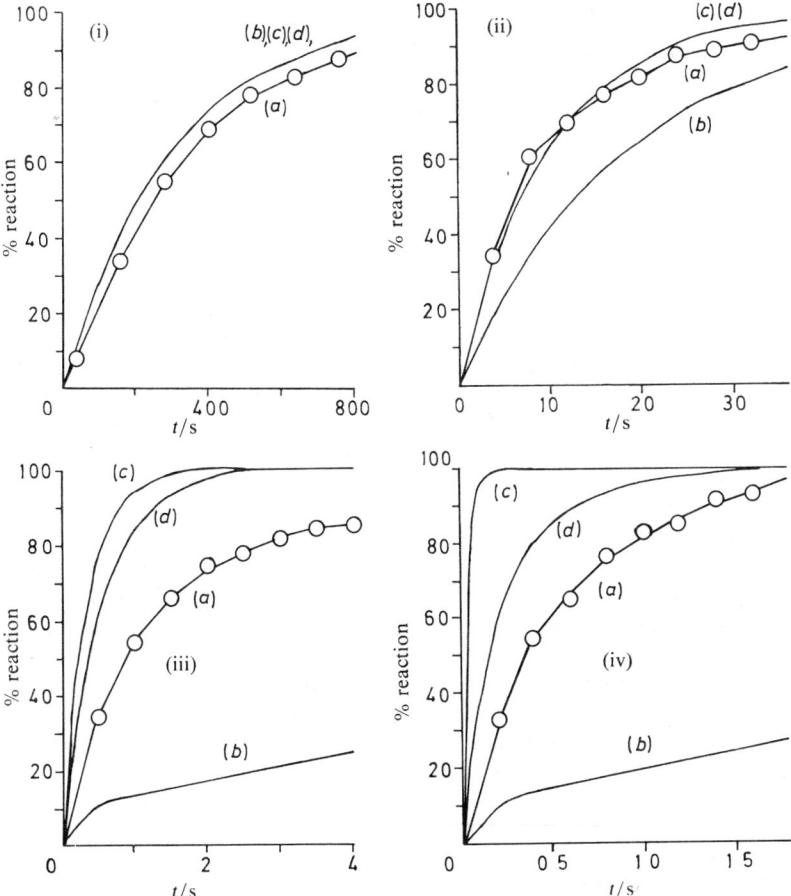

Fig. 2. Experimental and computed plots of % toluene nitrated as a function of time, at 25 °C. (a) Experimental data; (b) computed, assuming stagnant conditions; (c) computed, assuming fast diffusion; (d) computed, assuming stirred aqueous acid phase. (i) 63.2% Sulphuric acid; $a_g = 0.94\ \mu$m, $\sigma_g = 1.68\ \mu$m. (ii) 67.6% Sulphuric acid; $a_g = 1.63\ \mu$m, $\sigma_g = 1.59\ \mu$m. (iii) 71.9% Sulphuric acid; $a_g = 2.22\ \mu$m, $\sigma_g = 2.01\ \mu$m. (iv) 74.8% Sulphuric acid; $a_g = 2.00\ \mu$m, $\sigma_g = 1.99\ \mu$m.

For each acid concentration, curve (a) shows the experimental data. Curve (b) shows the curve produced by the calculation described above, using the photomicrographically determined droplet-size data. Observed values for the geometric mean radius, a_g, and geometric standard deviation, σ_g, are given for each system. These quantities are defined by

$$\ln a_g = \sum (n_i \ln a_i)/\sum n_i$$
$$\ln^2 \sigma_g = \sum (\ln a_i - n a_g)^2 / \sum n_i$$

where there are n_i particles of radius a_i. These statistical parameters are useful measures of the particle-size distribution, which to a good approximation is given by the log normal distribution[12]

$$n_i/\sum n_i = [\Delta a_i/(a_i 2\pi) \ln \sigma_g] \exp[-(\ln a_i - \ln a_g)^2/(2 \ln^2 \sigma_g)]$$

Table 1. Values of R_{obs} and K for various temperatures

$T/°C$	$k_{obs}/10^{-3} s^{-1}$	K
25.0	2.76	700
30.0	9.76	655
40.0	19.1	564
50.3	30.8	470
67.4	200	314

where there are n_i droplets in the range $(a_i - \Delta a_i/2)$ to $(a_i + \Delta a_i/2)$. Curve (c) shows the limiting behaviour if diffusion is so fast, or reaction so slow, that the equilibrium concentration of toluene is maintained in the aqueous acid phase at all times. In acid concentration of 63.2 wt%, the nitration rate is entirely kinetically controlled and independent of droplet size. The points for curve (c) were calculated using eqn (14). Unfortunately a discrepancy between theoretical and observed nitration rates was found even for this simple limiting case. This was ascribed to an incorrect value of the partition coefficient, K. If K was taken as 700 rather than 1700, the good agreement seen in fig. 2 is found. The discrepancy may be explained by the presence of surfactant in the solutions used for kinetic studies. The toluene is partially solubilised by the surfactant, so the toluene concentration in the aqueous phase is greater than for the surfactant-free solutions used in equilibrium studies. It was not possible to test this theory by measuring K directly for systems with added surfactant because the two phases could not be separated.

At high acid concentration where nitration is diffusion controlled, there is a large discrepancy between curves (a) and (b). This is attributed to a failure of the diffusion model. It has been assumed that the liquid for a few μm around each droplet could be regarded as stagnant, especially considering that the viscosity was around ten times that of water. However, the comparatively high nitration rates observed suggested that stirring was occurring. A representation of stirring is easily achieved in the mathematical model used in this study. At the end of each time interval δt, the total amount of toluene present in the aqueous phase, phase B, is calculated, and divided by the volume of phase B to obtain the concentration of toluene in each shell of phase B, which is used to calculate rates of change in the next time interval. Curve (d) shows the rate of nitration calculated by this technique. The observed rates lie between curves (b) and (d), showing that some stirring is occurring, but not right up to the interface.

The temperature variation of the first-order rate constant for the nitration in 64.4% sulphuric acid was used to calculate the activation energy parameters. Values of K were calculated by applying the temperature variation of K found for surfactant-free solutions (1700 at 25 °C, 970 at 57 °C) to the kinetically determined value of 700 at 25 °C. The results are shown in table 1. From these data a value of 66 ± 7 kJ mol^{-1} was calculated for $\Delta H^{\ddagger}$ and a value of -37 ± 5 J K^{-1} mol^{-1} was calculated for $\Delta S^{\ddagger}$.

We thank I.C.I. for the award of a CASE grant to J.M.C.

[1] J. Giles, C. Hanson and H. A. Ismail, *Industrial and Laboratory Nitrations*, A.C.S. Symp. Ser. No. 22, ed. L. F. Albright and C. Hanson (A.C.S., Washington D.C., 1976), chap. 12, p. 190.
[2] G. F. Sheats and A. N. Strachan, *Can. J. Chem.*, 1978, **56**, 1280.
[3] K. Schofield, *Nitration* (Cambridge University Press, Cambridge, 1980).

[4] C. P. Perrin, *J. Am. Chem. Soc.*, 1977, **99**, 5516.
[5] J. W. Barnet, R. B. Moodie, K. Schofield and J. B. Weston, *J. Chem. Soc., Perkin Trans. 2*, 1975, 648.
[6] R. B. Moodie, K. Schofield and P. G. Taylor, *J. Chem. Soc., Perkin Trans. 2*, 1979, 133.
[7] J. W. Chapman, P. R. Cox and A. N. Strachan, *Chem. Eng. Sci.*, 1974, 1247.
[8] F. Nabholz and P. Rys, *Helv. Chim. Acta*, 1977, **60**, 2937.
[9] Pin Chang and C. R. Wilkie, *J. Phys. Chem.*, 1955, **59**, 592.
[10] J. H. Ridd, *Adv. Phys. Org. Chem.*, 1978, **16**, 1.
[11] E. S. Swinbourne, *J. Chem. Soc.*, 1960, 2371.
[12] E. Shotton and S. S. Davies, *J. Pharm. Pharmacol.*, 1968, **20**, 430.

Faraday Discuss. Chem. Soc., 1984, 77, 115-126

Use of Microemulsions as Liquid Membranes

Improved Kinetics of Solute Transfer at Interfaces

By Christian Tondre* and Aristotelis Xenakis

Laboratoire de Chimie Physique Organique, ERA CNRS 222, Université de Nancy I, B.P. 239, 54506 Vandoeuvre-les-Nancy Cedex, France

Received 28th November, 1983

When the volume fraction occupied by the dispersed phase of a microemulsion is small, the microglobules making up this dispersed phase can be viewed as mobile carriers permitting the transport of substances which are either insoluble or very poorly soluble in the continuous phase. In this paper water-in-oil microemulsions composed of decane, water, tetraethyleneglycol dodecylether (TEGDE) and hexan-1-ol are used as liquid membranes and the microglobules are shown to transport alkali-metal picrates between two aqueous phases. The effect of changing the initial picrate concentration in the source compartment has been investigated and the resulting flux can be adequately described by a classical model of facilitated transport (fast transfer and 'chemical reaction' coupled with slow diffusion), as has been observed previously for the transport of lipophilic substances by oil-in-water microemulsions.

The presence of dicyclohexano-18-crown-6 (DC18C6) in the liquid membrane brings about an increase in the flux of alkali-metal picrates. At optimum conditions, the transport of K^+ picrate by the microemulsion alone is 5.1 times faster than with DC18C6 in pure decane, but it is 12.6 times faster when DC18C6 is added to the microemulsion. Although drastically reduced, selectivity for ion transport still exists in the latter situation.

Liquid membranes have long been used as models of biological membranes for studying the transport of different solutes (salts, metabolites, drugs *etc.*) either facilitated by a carrier or not.[1-9] Numerous applications have also been found in separation techniques which take advantage of their ability to perform selective permeations.[10-12]

The transport of substances in such experiments is usually facilitated by the incorporation of a carrier molecule in the liquid membrane. Macrocyclic compounds which are either naturally occurring (antibiotics)[2] or synthetic (crown ethers or cryptands for example)[4-8] have often been used for this purpose. On the other hand, the use of liposomes or vesicles as carriers for intracellular delivery of drugs constitutes an active area of research.[13,14]

We have developed experiments to demonstrate that the very small microemulsion droplets may behave like the mobile carrier molecules commonly used in studies of facilitated transport. In these experiments a microemulsion is used as a liquid membrane separating two liquid phases in thermodynamic equilibrium with it. We have previously examined the transport of lipophilic substances by oil-in-water microemulsion droplets,[15-17] but only very preliminary results have been given for the transport of hydrophilic solutes by water-in-oil microemulsions.[17] This paper will give a full account of the results obtained when water-in-oil microemulsion systems involving a non-ionic surfactant are used to transport alkali-metal picrates.

The role of carrier of the microemulsion droplets has implications regarding the liquid–liquid extraction of metal ions as it can be responsible for an improvement

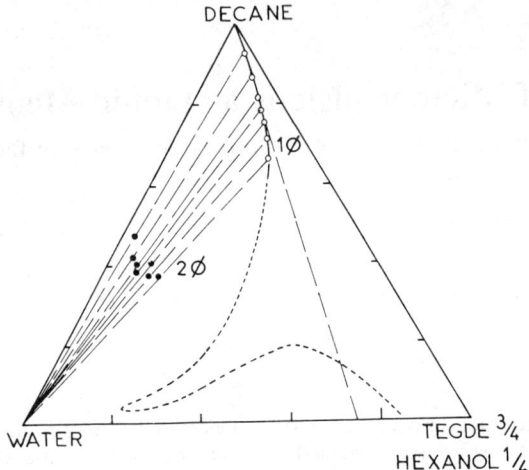

Fig. 1. Composition (filled circles) of the initial mixtures in the diphasic region of the pseudo-ternary diagram n-decane/water/tetraethyleneglycol dodecylether (3/4), hexan-1-ol (1/4). Phase separation occurs according to the tie-lines joining the water apex to the open circles indicating the composition of the water-in-oil microemulsions. The dashed line delimiting the monophasic domain was drawn approximately according to ref. (22).

in the kinetics of solute transfer at interfaces, as observed when surfactants are added to the extracting medium.[18–20] For this reason we have also investigated the transport of alkali-metal picrates by a crown ether, which can be considered as a model system, both when the membrane is a pure organic solvent and when the membrane is a water-in-oil microemulsion. Two alkali-metal cations with different stability constants[21] for complexation with the crown ether have been tested in order to compare the selectivity of ion transport in both media.

EXPERIMENTAL

CHOICE OF CHEMICAL COMPOUNDS

For these experiments, we required a water-in-oil microemulsion in thermodynamic equilibrium with an aqueous phase of composition as close as possible to pure water, *i.e.* a diphasic system of the so-called Winsor II type. In addition, the hydrophilic substance to be transported had to be able to be detected spectrophotometrically; this led us to choose the alkali-metal picrates. Because of the nature of the chosen hydrophilic solute we preferred to use a microemulsion system containing no salt, although most known diphasic systems of the Winsor II type require a salt in their formulation. This was intended to avoid an exchange of ions during the transport process. For this reason we sought a system involving a non-ionic surfactant and finally found a quaternary system meeting all the above requirements: decane/water/tetraethyleneglycol dodecylether (TEGDE)/hexan-1-ol. The monophasic region of the pseudo-ternary phase diagram obtained when keeping the ratio of TEGDE to hexanol constant (TEDGE/hexanol = 3) has previously been described by Friberg *et al.*[22] We found that the systems, the compositions of which are indicated in fig. 1, separate into two perfectly clear phases according to the tie-lines indicated. Unfortunately we could not find a simple ternary system having such a property: if there was no hexanol the system did not produce clear phases even after several weeks. It thus seems that it is easier to obtain Winsor II systems by adding an alcohol, as previously observed by Winsor himself in the case of ionic surfactants in the absence of salt.[23]

Table 1. Compositions (in wt%) of initial mixtures and microemulsion phases

	initial mixture				ϕ_S (water-in-oil microemulsion)					
system	H_2O	$C_{10}H_{22}$	$C_6H_{13}OH$	TEGDE	H_2O	$C_{10}H_{22}$	$C_6H_{13}OH$	TEGDE	d /g cm^{-3}	η /cp
I	50	47	0.75	2.25	1.83	92.28	1.47	4.42	0.742	1.06
II	53	42	1.25	3.75	3.40	86.32	2.57	7.71	0.753	1.30
III	53	40	1.75	5.25	4.80	81.03	3.54	10.63	0.762	1.55
IV	54	38	2	6	5.76	77.85	4.10	12.29	0.770	1.77
V	50	40	2.5	7.5	6.47	74.82	4.68	14.03	0.771	1.81
VI	52	37	2.75	8.25	8.05	70.96	5.27	15.82	0.781	2.37
VII	50	37	3.25	9.75	10.20	66.45	5.84	17.51	0.789	2.82

Our choice of dicyclohexano-18-crown-6 (DC18C6) for the experiments in the presence of a classical extractant was for the following reasons: (i) it has good solubility in decane and poor solubility in water, and will thus stay in the liquid membrane, and (ii) it complexes both K^+ and Na^+ ions with different stability constants,[21] which makes it convenient for testing the influence of the microemulsion on the selectivity of transport of alkali-metal ions.

ORIGIN OF CHEMICALS

Tetraethyleneglycol dodecylether (Nikko Chemicals, Japan), n-decane and dicyclohexano-18-crown-6 (Fluka purum) and hexam-1-ol (Fluka puriss.) were used as supplied. Potassium and sodium picrates were prepared from picric acid (Merck) according to the procedure described in ref. (24). The salts were recrystallized three times and the extinction coefficients measured in tetrahydrofuran were in agreement with previously published values.[4]

CHARACTERIZATION OF BIPHASIC SYSTEMS

Biphasic systems with the compositions shown in fig. 1 and table 1 were prepared by weighing the components directly in a separating funnel which was then placed in a thermostatted bath regulated at 20 °C (±0.2 °C) until separation into two clear phases was achieved (this took 3 days to 3 weeks, depending on the system). The compositions of the initial mixtures were chosen so as to give approximately equal volumes of the two separated phases.

Analysis of these phases was performed by the Karl Fisher method for the water content and by gas-phase chromatography for decane, hexan-1-ol and TEGDE, using columns 1 m long and $\frac{1}{8}$ in. in diameter containing 3% SE30 on chromosorb WAW 80/100 mesh with a linear variation of temperature from 50 to 280 °C. The composition of the superior phase ϕ_S is given in table 1. The inferior phase is essentially water (no other component could be detected by gas-phase chromatography). For this reason the representative points of both phases can be shown in the same pseudo-ternary diagram (fig. 1).

Densities and viscosities of ϕ_S (table 1) were measured with a digital Anton Paar DMA 10 densimeter and with an Ubbelohde-type viscometer (Schott–Geräte with automatic timing), respectively.

TRANSPORT EXPERIMENTS

The setup used for the transport experiments is similar to that previously described when investigating oil-in-water microemulsions,[16,17] except that the shape of the cell used in the reverse of the previous one. As shown in fig. 2, it resembles an inverted U tube with the two arms filled with the aqueous phase and the microemulsion phase at the top. The volumes were 14.8, 18.6 and 35 cm^3 for the source (S), receiving (R) and membrane (M) compartments,

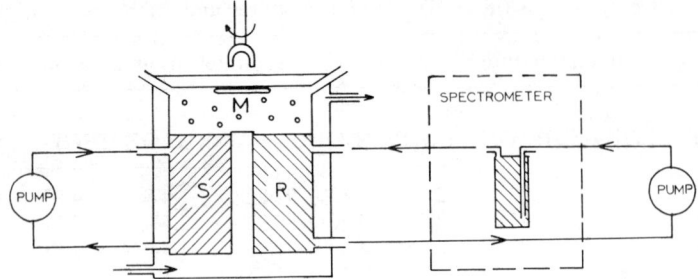

Fig. 2. Schematic diagram of the apparatus used for the transport experiments.

respectively. The light absorption of the receiving phase was measured continuously at 357 nm for K^+ picrate and at 351 nm for Na^+ picrate, thus allowing us to calculate from a standard curve the number of transported molecules at any time (Beer's law was found to be valid in the concentration range used).

Peristaltic pumps (Masterflux) were used to ensure fast homogenization in both branches of the transport cell.[9] The microemulsion compartment was stirred using a magnetic stirrer whose rotation speed was regulated at 120 r.p.m. The whole cell was thermostatted at 20 °C.

RESULTS AND DISCUSSION

TRANSPORT OF PICRATES BY PURE MICROEMULSIONS

The transport of K^+ or Na^+ picrates by the microemulsion systems referred to as I to VII in table 1 was first studied without adding a classical extractant to the membrane. Fig. 3 shows a plot of the number of moles of K^+ picrate transported against time for different initial concentrations of picrate in the source compartment, using system V. The curves look very much like those obtained for the transport of pyrene by oil-in-water microemulsions, with a time lag attributed to the time required for the picrate to reach an equilibrium concentration inside the membrane. The flux of picrate, calculated from the slopes of the straight lines observed when a steady state is established, is shown in fig. 4 as a function of the initial picrate concentration.

Blank experiments carried out with pure decane instead of the microemulsion in compartment M did not reveal any transport of picrates after 24 h.

We chose a picrate concentration of 2×10^{-3} mol dm^{-3} to study the influence of the composition of the microemulsion on the transfer rate of K^+ and Na^+ picrates. As can be seen in table 1, the composition of the microemulsion was varied by changing the amount of amphiphile compounds in the initial mixture: the larger the amount of amphiphiles, the larger the amount of water incorporated in the microemulsion phase. Fig. 5 shows the variation of the flux of K^+ and Na^+ picrates when increasing the volume fraction of water. The flux increases linearly up to a volume fraction of 0.05 and then decreases. Only a very small difference is observed between K^+ and Na^+ picrates, with the latter always giving a slightly lower value for the flux (table 2).

The maximum observed in the variation of the flux was unexpected because when the corresponding experiments had been conducted with oil-in-water microemulsions (increasing percentage of oil), instead of a decrease, a dramatic increase of the flux of pyrene occurred for a certain volume fraction of oil.[16,17] Such a

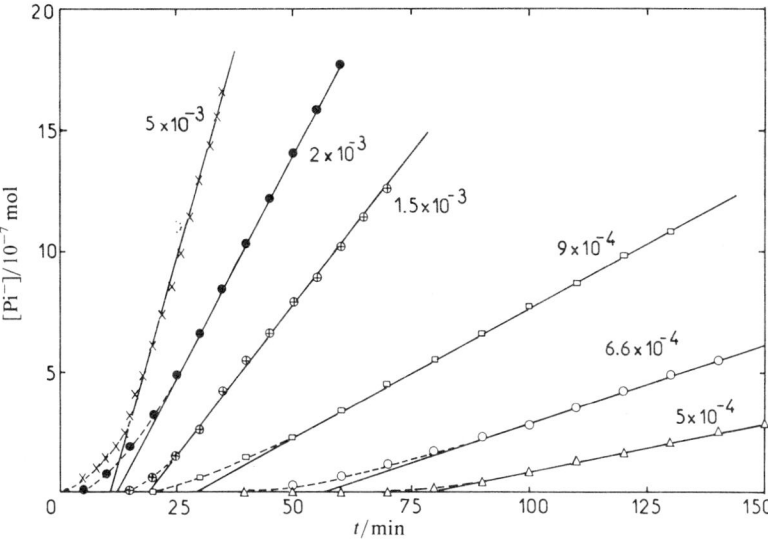

Fig. 3. Plots of the number of moles of K^+ picrate crossing the second interface against time in microemulsion system V. The numbers indicate the initial picrate concentration in the source compartment S (mol dm^{-3}). Cross-section of interface = 3.14 cm^2.

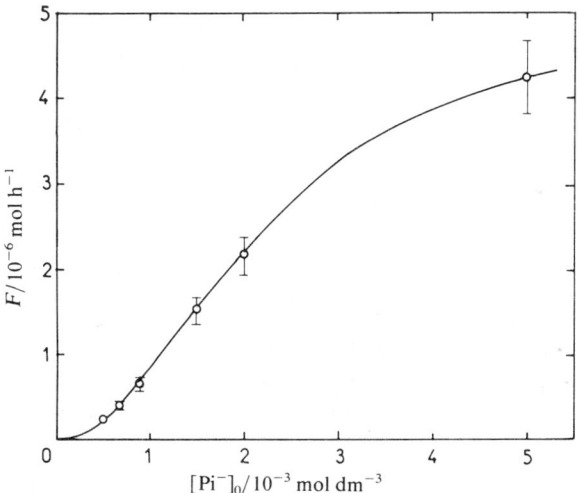

Fig. 4. Plot of the flux of K^+ picrate against the initial picrate concentration in the source compartment S. The curve is calculated from eqn (9) with the values of parameters given in fig. 6.

phenomenon was attributed to the percolation threshold of oil droplets. In the present situation a structural change of the microemulsion is probably responsible for the maximum observed. A possible interpretation is as follows: when the amount of water in the microemulsion is <5%, there is just enough water to hydrate the ethylene oxide groups constituting the hydrophilic part of the surfactant (we have between 2 and 2.4 water molecules per ethylene oxide group, in agreement with

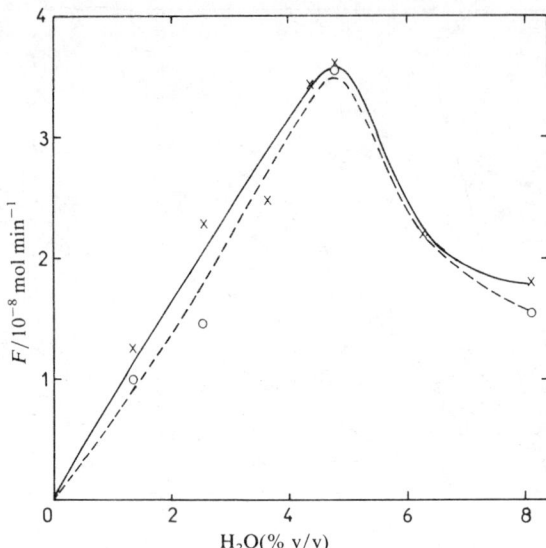

Fig. 5. Plots of the flux of K^+ picrate ($\times$) and Na^+ picrate ($\bigcirc$) against the percentage of water (v/v) in the microemulsion phase. Initial picrate concentration $= 2 \times 10^{-3}$ mol dm^{-3}.

previous estimations[25,26]); at >5% water there is enough water to start forming water pools (the sudden increase in the ratio of H_2O to TEGDE can be seen in fig. 1: it occurs when the representative points of the microemulsion phases depart from a straight line originating from the oil apex). In the first situation, only agregates of TEGDE and hexanol would exist, as previously postulated in comparable systems in the absence of alcohol.[25] Real water droplets (with unknown shape) with a palissade layer of amphiphile molecules would occur in the second situation. ^{13}C n.m.r. relaxation time measurements currently in progress should enable us to say whether the mobility of the hydrophilic chains is in agreement with such a model.[27] In this hypothesis it is clear that the equilibrium constant characterizing the interaction (solubilization or complexation) of the alkali-metal picrate with the 'carrier' will be very different if this carrier is an aggregate or a water droplet (in the latter case the ethylene oxide groups lining the inside of the droplet could act as a kind of crown ether towards the alkali-metal cations and make the releasing process more difficult). We will not attempt to speculate further on the maximum observed in fig. 5, but will merely try to interpret the results obtained in the linear part of the figure, i.e. where we can assume that there is only one sort of carrier (the exact structure of which does not matter). According to fig. 1 and table 1, the ratio between water and the amphiphile molecules is constant for all the systems containing <5% (v/v) water. Increasing the amount of amphiphile molecules (or water) can be considered as equivalent to increasing the concentration of carrier, which results in a linear increase of the flux.

In the previous case of oil-in-water microemulsions we tested different possible mechanisms in order to determine which step was rate controlling: transfer of solute across the interface, solubilization inside the droplet or diffusion through the non-stirred layers. The results were consistent with a model in which the diffusion of the droplet having solubilized the probe is much slower than transfer and

Table 2. Fluxes of transported picrates as a function of the nature of the liquid membrane and the presence or absence of a classical extractant

nature of liquid membrane	water (% v/v)	solute transported	flux^a/mol h^{-1} without DC18C6	flux^a/mol h^{-1} with DC18C6 (10^{-2} mol dm^{-3})
pure decane	~0	KPi	—	4.21×10^{-7}
		NaPi	—	8.6×10^{-9}
microemulsion I	1.36	KPi	0.74×10^{-6}	2.42×10^{-6}
		NaPi	0.61×10^{-6}	0.85×10^{-6}
microemulsion II	2.55	KPi	1.37×10^{-6}	
		NaPi	0.87×10^{-6}	
microemulsion III	3.63	KPi	1.48×10^{-6}	4.74×10^{-6}
		NaPi	—	2.28×10^{-6}
microemulsion IV	4.41	KPi	2.06×10^{-6}	
		NaPi	—	
microemulsion V	4.98	KPi	2.15×10^{-6}	5.30×10^{-6}
		NaPi	2.11×10^{-6}	2.80×10^{-6}
microemulsion VI	6.27	KPi	1.33×10^{-6}	
		NaPi	—	
microemulsion VII	8.11	KPi	1.06×10^{-6}	2.61×10^{-6}
		NaPi	0.94×10^{-6}	1.56×10^{-6}

^{a} Cross-section of interface = 3.14 cm^2.

solubilization reactions. Similar treatment permits a satisfying description of the present results.

The different steps involved in the transport process can be written as follows:

interface 1: $\quad Me_S^+ + Pi_S^- \rightleftharpoons Me^+, Pi^-_{\text{continuous phase of M}}$ (1)

membrane:
$\begin{cases} Me^+, Pi^-_{\text{continuous phase of M}} + (C) \rightleftharpoons (C, Me^+, Pi^-) & (2) \\ \text{diffusion of } (C, Me^+, Pi^-) \text{ across the non-stirred layers} & (3) \\ (C, Me^+, Pi^-) \rightleftharpoons Me^+, Pi^-_{\text{continuous phase of M}} + (C) & (4) \end{cases}$

interface 2: $\quad Me^+, Pi^-_{\text{continuous phase of M}} \rightleftharpoons Me_R^+ + Pi_R^-$ (5)

where the subscripts M, S and R refer to membrane, source and receiving compartments, respectively, Me^+ is the alkali–metal ion, Pi^- is the picrate ion and Me^+, Pi^- is the ion pair, and (C) and (C, Me^+, Pi^-) are the 'carrier' and the carrier–solute complex, respectively. Steps (1) and (5) correspond to the formation of the ion pair through the interfaces characterized by an equilibrium constant k. Steps (2) and (4) are governed by the equilibrium constant K for the solubilization (or complexation) of the metal picrate in (with) the carrier. Step (3) is characterized by a diffusion coefficient D, the thickness of the diffusion layer being $L = 2l$, where l is the non-stirred layer on the organic side of both interfaces (as previously observed with oil-in-water microemulsions, changing the rotation speed of the magnetic stirrer influences the transfer rate, but changing the rotation speed of the peristaltic pumps does not affect the result).

Solving the continuity equations for all the species present in the membrane with the assumptions that (i) the steady-state approximation is valid, (ii) the equilibrium characterizing steps (2) and (4) is always established and (iii) the unfacilitated transport of picrate is negligible, and with the following boundary conditions:

$$[\text{Me}^+, \text{Pi}^-]_M = k[\text{Me}^+]_1[\text{Pi}^-]_1 = k[\text{Pi}^-]_1^2 \quad \text{on interface 1} \tag{6}$$

$$[\text{Me}^+, \text{Pi}^-]_M = k[\text{Me}^+]_2[\text{Pi}^-]_2 = k[\text{Pi}^-]_2^2 \quad \text{on interface 2} \tag{7}$$

where $[\text{Pi}^-]_1$ and $[\text{Pi}^-]_2$ are the concentrations outside the membrane but right beside it in compartments S and R, respectively, at the beginning of the steady state, leads to the following equation for the flux:

$$F = \frac{DKk[(C)]}{L}\left(\frac{[\text{Pi}^-]_1^2}{1+Kk[\text{Pi}^-]_1^2} - \frac{[\text{Pi}^-]_2^2}{1+Kk[\text{Pi}_i^-]_2^2}\right). \tag{8}$$

As the concentration $[\text{Pi}^-]_2$ is practically zero at the beginning of the steady state, the equation reduces to the first term:

$$F = \frac{DKk[(C)]}{L}\frac{[\text{Pi}^-]_1^2}{1+Kk[\text{Pi}^-]_1^2}. \tag{9}$$

The essential difference from the previously developed treatment comes from the ion-pair formation equilibrium, which introduces a squared term in the concentration dependence of the flux.[4-6]

If the model is suitable for describing the experimental results, a plot of $1/F$ against $1/[\text{Pi}^-]_1^2$ should give a straight line with intercept $L/D[(C)]$ and slope $L/DKk[(C)]$. The concentration $[\text{Pi}^-]_1$ is not known and must be calculated first. It can be shown to be given by

$$[\text{Pi}^-]_1 = [\text{Pi}^-]_0[1 + V_M/k'(V_S + V_R)]^{-1} \tag{10}$$

where $[\text{Pi}^-]_0$ is the initial picrate concentration in the source compartment, k' is the partition coefficient of the picrate ion between the aqueous phase and the microemulsion phase as a whole (k' is a number without dimensions, not to be confused with k) and the V are the volumes occupied by the different phases. k' was experimentally measured in system V and found to be equal to 0.84, so the complete correcting factor is 0.44.

Fig. 6 shows that very good agreement exists between the theoretical prediction according to the proposed model and the experimental data. Combining the values of the intercept and slope enables us to determine the value of the product $Kk = (0.98 \pm 0.2) \times 10^6$ dm^6 mol^{-2}. It is difficult to discuss this value on a quantitative basis. Unfortunately we have no estimate of the equilibrium constant k, which could in principle be obtained from the solubility of picrate in water and in decane if one assumes that the continuous phase of the microemulsion is pure decane (it may also contain hexanol and TEGDE): we failed to detect any solubility in decane by measuring the absorption of an aqueous solution of picrate before and after shaking with decane. On the other hand, the only results we know concerning the dynamics of solubilization of a picric acid probe in reversed micelles are consistent with an equilibrium constant of 2.3×10^6 dm^3 mol^{-1}, *i.e.* of the same order of magnitude as

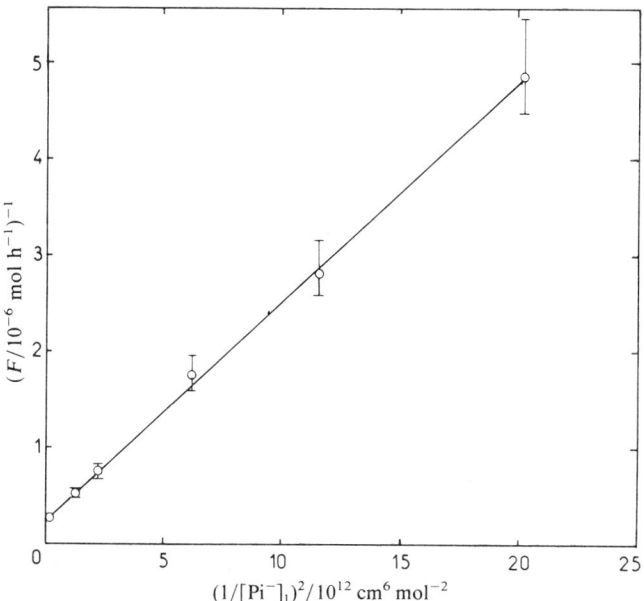

Fig. 6. Plot of the reciprocal flux of K^+ picrate against the square of the reciprocal concentration of picrate at steady state in the source compartment S. Intercept = $L/D(C)$ = 0.225×10^{10} mol^{-1} cm^2 s; slope = $L/DKk(C)$ = 2.29×10^{-3} mol cm^{-4} s.

the above value. This result was obtained by Tamura and Schelly[28] in the system Aerosol OT + benzene + water.

TRANSPORT OF PICRATES BY MICROEMULSIONS CONTAINING A CLASSICAL EXTRACTANT

When dicyclohexano-18-crown-6 is added to the microemulsion being used as a liquid membrane, an increase in the flux of transported picrates is observed for both Na^+ and K^+ salts. The results obtained with the different microemulsion systems containing 10^{-2} mol dm^{-3} DC18C6 are shown in fig. 7 and table 2, in which are also given for comparison the fluxes obtained when the liquid membrane is either pure decane containing 10^{-2} mol dm^{-3} DC18C6 or the microemulsion alone. All the experiments in which a microemulsion is involved show a maximum in the flux for the same volume fraction of water. At this maximum (system V) the transport of K^+ picrate by the microemulsion alone is 5.1 times faster than with DC18C6 in pure decane, but it is 12.6 times faster when DC18C6 is added to the microemulsion.

For the Na^+ picrate, the corresponding values are, respectively, *ca.* 245 and 325 because of the very weak flux obtained with DC18C6 in pure decane (the flux was so small that it was difficult to determine it accurately, which explains why the preceding values are approximate).

A comparison of the results obtained for K^+ and Na^+ cations shows that, although drastically reduced, some selectivity for ion transport still exists in the presence of the microemulsion. This is of importance when considering the use of microemulsions for liquid–liquid extraction of metal ions.

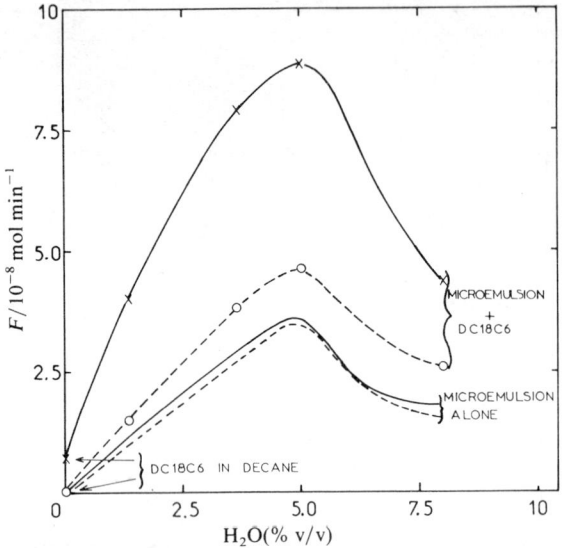

Fig. 7. Plots of the flux of K^+ picrate ($\times$, ———) and Na^+ picrate ($\bigcirc$, - - -) against the percentage of water (v/v) in the liquid membrane containing (or not containing) dicyclohexano-18-crown-6. Initial picrate concentration = 2×10^{-3} mol dm^{-3}; concentration of DC18C6 = 10^{-2} mol dm^{-3}.

Note also that when the system contains both the classical extractant and the microemulsion globules, the resulting flux of K^+ as well as Na^+ picrates is not equal to the sum of the fluxes obtained with the individual carriers.

The accelerating effects of microemulsions on the kinetics of the extraction of metal ions by a classic extractant have been reported by Fourre and Bauer,[18] who used a two-compartment cell of the type described by Lewis[29] and Allen[30] to study the extraction of gallium by Kelex 100. They suggested that the reason for the improvement in the extraction kinetics could be found in the fact that the microglobules of the microemulsion act as a relay between the aqueous phase and the extractant contained in the organic phase. According to this explanation the following simplified mechanism can be proposed for the present results:

$$Pi^-_{(\text{aqueous phase S})} \rightleftharpoons Pi^-_{(\text{microemulsion 'carrier'})} \rightleftharpoons Pi^-_{(\text{aqueous phase R})}$$
$$\searrow \quad \updownarrow \quad \swarrow$$
$$Pi^-_{(\text{crown ether carrier})}.$$

If there were no coupling between the two carrier complexes a simple additive effect would be expected on a first approximation. The coupling between the two species has the result of reducing the effective membrane thickness L and thus it improves the transfer rate according to equations similar to eqn (9). The detailed mechanism can be regarded as follows: the diffusion of DC18C6 in the non-stirred layer is certainly much faster than the diffusion of the microemulsion globule, so when the empty macrocycle diffuses back towards interface 1 it will meet microemulsion globules carrying a picrate ion, and if exchange occurs the driving force will pull it towards interface 2 before it has reached interface 1. The effective layer is thus decreased for both carrier species.

CONCLUSIONS

We have attempted to give a clear demonstration of the carrier properties of water-in-oil microemulsion globules by studying the transport of hydrophilic solutes (alkali-metal picrates) through an oil continuous phase. The behaviour observed is very similar to that previously reported for the transport of lipophilic substances by oil-in-water microemulsions. A quantitative interpretation is nevertheless more difficult because, contrary to the case of oil-in-water microemulsions, very little is known concerning the structure of the microemulsion systems used here, which involve a non-ionic surfactant.

This work also tries to clarify the mechanisms by which microemulsions can improve the kinetics of the liquid–liquid extraction of metal ions. It has been shown that transfer through a liquid membrane can help to test for the optimum conditions, such as the optimum water content of the organic phase. Nevertheless, when there is enough water to form water pools, the non-ionic detergent molecules present in the microemulsion system probably act as a strong chelating agent, and other systems will have to be found in order to investigate further the role of microemulsions in the mechanism of ion-extracting processes.

We thank J. L. Fringant and J. L. Vasseur for their technical assistance in the building of the experimental apparatus.

[1] H. L. Rosano, P. Duby and J. H. Schulman, *J. Phys. Chem.*, 1961, **65**, 1704.
[2] R. Ashton and L. K. Steinrauf, *J. Mol. Biol.*, 1970, **49**, 547.
[3] W. I. Higuchi, A. H. Ghanem and A. B. Bikhazi, *Fed. Proc., Fed. Am. Soc. Exp. Biol.*, 1970, **29**, 1327.
[4] K. H. Wong, K. Yagi and J. Smid, *J. Membr. Biol.*, 1974, **18**, 379.
[5] J. D. Lamb, J. J. Christensen, S. R. Izatt, K. Bedke, M. S. Astin and R. M. Izatt, *J. Am. Chem. Soc.*, 1980, **102**, 3399.
[6] C. F. Reusch and E. L. Cussler, *AIChE J.*, 1973, **19**, 736.
[7] Y. Kobuke, K. Hanji, K. Horiguchi, M. Asada, Y. Nakayama and J. Furukawa, *J. Am. Chem. Soc.*, 1976, **98**, 7414.
[8] M. Kirch and J. M. Lehn, *Angew. Chem., Int. Ed. Engl.*, 1975, **14**, 555.
[9] E. Pefferkorn and R. Varoqui, *J. Colloid Interface Sci.*, 1975, **52**, 89.
[10] N. N. Li and A. L. Shrier, in *Recent Developments in Separation Science*, ed. N. N. Li (CRC Press, Cleveland, 1972), vol. I, p. 163.
[11] N. N. Li, *Ind. Eng. Chem., Process Des. Dev.*, 1971, **10**, 215.
[12] N. N. Li, *AIChE J.*, 1971, **17**, 459.
[13] J. N. Weinstein, *Pure Appl. Chem.*, 1981, **53**, 2241.
[14] J. H. Fendler and A. Romero, *Life Sci.*, 1977, **20**, 1109.
[15] C. Tondre and A. Xenakis, *Colloid Polym. Sci.*, 1982, **260**, 232.
[16] A. Xenakis and C. Tondre, *J. Phys. Chem.*, 1983, **87**, 4737.
[17] C. Tondre and A. Xenakis, in *Surfactants in Solution*, ed. K. L. Mittal (Plenum, New York, 1983), vol. 3, p. 1881.
[18] P. Fourre and D. Bauer, *C.R. Acad. Sci., Ser. B*, 1981, **292**, 1077.
[19] P. Fourre, D. Bauer and J. Lemerle, *Anal. Chem.*, 1983, **55**, 662.
[20] J. Komornicki, *Thesis* (Université de Paris, 1981).
[21] J. Lamb, R. M. Izatt, J. J. Christensen and D. Eatough, in *Coordination Chemistry of Macrocyclic Compounds*, ed. G. A. Melson (Plenum Press, New York, 1979).
[22] S. Friberg, I. Lapczynska and G. Gilbert, *J. Colloid Interface Sci.*, 1976, **56**, 19.
[23] P. A. Winsor, *Trans. Faraday Soc.*, 1948, **44**, 376.
[24] M. Coplan and R. Fuoss, *J. Phys. Chem.*, 1964, **68**, 1177.
[25] G. Mathis, J. C. Boubel, J-J. Delpuech, J. C. Ravey and M. Buzier, in *Magnetic Resonance in Colloid and Interface Science*, ed. J. P. Freissard and H. A. Resing (D. Reidel, Dordrecht, 1980), p. 597.

[26] M. Buzier, *Thesis* (Université de Nancy, 1979).
[27] C. Tondre, A. Xenakis, A. Robert and G. Serratrice, to be published.
[28] K. Tamura and Z. A. Schelly, *J. Am. Chem. Soc.*, 1981, **103**, 1018.
[29] J. B. Lewis, *Chem. Eng. Sci.*, 1954, **3**, 260.
[30] K. A. Allen, *J. Phys. Chem.*, 1960, **64**, 667.

Solute Transport and Perturbation at Liquid/Liquid Interfaces

BY RICHARD H. GUY,* ROBERT S. HINZ AND MICHAEL AMANTEA

School of Pharmacy, University of California, San Francisco, California 94143, U.S.A.

Received 23rd November, 1983

Solute transfer kinetics across aqueous-solution/organic-liquid interfaces have been measured by two techniques. First, a new method based upon the capillary-tube procedure for self-diffusion has been developed. A short capillary containing radio-labelled solute dissolved in one phase is immersed in a large stirred volume of the second liquid phase, into which the movement of marker is followed. Demonstration and validation of the approach has been performed with a number of systems including the transport of salicylic acid (SA) at a water/isopropyl myristate (IPM) interface. In the second procedure, a rotating diffusion cell has been used to determine whether the rate and energetics of the transfer process are altered by the presence of dissolved anaesthetic alcohols. SA and methyl nicotinate (MN) transport across an aqueous-solution/IPM interface has been studied with various concentrations of different alcohols dissolved in the aqueous phase. The effects of alcohol are solute dependent: SA is not affected by any of the anaesthetics studied whereas 100 mmol dm^{-3} ethanol retards MN transfer by a factor of 2. For MN, furthermore, the favourable entropic contribution to the free energy of activation for interfacial transfer (which is observed in the absence of ethanol) is negated by the presence of 100 mmol dm^{-3} ethanol. This change may reflect a destabilization of interfacially ordered water molecules by the alcohol and appears similar to the effect of poly(ethylene glycol) on the same system.

The objectives of the work described in this paper were two-fold: (1) to develop a new procedure for the determination of solute transfer kinetics at liquid/liquid interfaces and (2) to investigate, using the physical chemistry of interfacial transport, the effect of anaesthetic alcohols at a model biomembrane interface.

The study of interfacial transport at liquid/liquid boundaries is important for its own sake, in terms of contributing to the understanding of a complex solution chemistry problem, and has relevance to numerous situations in industrial and biological processes (*e.g.* separation technology and membrane transport). Precise and reliable measurement of interfacial transport kinetics, however, is difficult and prone to artefact. The development of an alternative experimental approach is therefore both fundamentally useful and of value as a means with which to verify independently data from a single different technique. We describe here a novel application of the capillary-tube procedure for self-diffusion.[1] A capillary containing a radio-tagged solute dissolved, for example, in an organic liquid is immersed into a large stirred volume of aqueous buffer. Transport of the solute is followed by measuring the radioactivity lost from the capillary as a function of time. The solution to Fick's second law of diffusion with the appropriate boundary condition for the interface at the mouth of the capillary enables the phase-transfer coefficient of the solute to be evaluated from the data.

The correlation of anaesthetic potency with lipid solubility among the alcohols and other structurally unrelated small molecules has resulted in their biological effect being generally interpreted as a non-specific action (*e.g.* increasing fluidity) in the hydrophobic regions of membranes[2] rather than at discrete receptor sites.[3] However, the functional significance of anaesthetic-induced membrane fluidization has not been demonstrated absolutely. Furthermore, although alcohols do change the interfacial interactions between water and both surfactant and phospholipid monolayers[4-6] spread on an aqueous solution, the possibility that at least some of these molecules' effects on membranes could be due to interfacial events had, until recently, received little consideration.

In 1976 Eyring et al.[7] proposed that interfacial water structure and its alteration by anaesthetic agents may determine the state of membrane activity. This work has been developed[8] and expanded by showing (a) the dehydration of micelle surfaces by anaesthetics,[9,10] (b) a decreased tendency for inhalation anaesthetics to penetrate cell membranes[11] and (c) a growing body of theoretical and experimental evidence for the interfacial activity of these molecules.[12-15] In the study reported here we have chosen to use the kinetics and thermodynamics of solute transfer at a model biomembrane interface as a means to probe the interfacial influence of anaesthetic alcohols. The membrane surface is simulated with an aqueous-solution/isopropyl myristate liquid/liquid boundary, and the movement of two solute molecules has been considered. Experiments have been performed with a previously described rotating diffusion cell,[16] which has been used extensively in recent years for the measurement of interfacial transport kinetics.[16-24]

EXPERIMENTAL

CAPILLARY TECHNIQUE

A schematic diagram of the initial experimental configuration most often studied in this work is shown in fig. 1. A small glass capillary was first filled carefully with a solution of radio-labelled solute in isopropyl myristate (IPM) ensuring that the liquid surface was flush with the open end of the capillary. Over-filled capillaries were discarded. Generally, the capillaries used were 1.1–1.2 cm long with an internal diameter of 0.84 mm (corresponding to a volume of ca. 5 mm^3). Most experiments have been performed with salicylic acid (SA) (^{14}C-labelled in the carboxy position) as the solute. The specific activity was 56.5 mCi mmol^{-1}, which was equivalent to ca. 23 000 ±800 counts min^{-1} per 5 mm^3 of IPM solution. The initial SA molar concentration in IPM was ca. 40 μmol dm^{-3}.

The filled capillary was placed in a simple wire support and lowered into an aqueous receptor phase contained within a scintillation vial. The vial and contents were pre-positioned in a water bath held within a glass jacket through which thermostatted water at 25 °C was pumped. To allow the capillary to attain thermal equilibrium it was not fully submerged immediately but allowed sufficient time to reach 25 °C by being immersed to ca. 90% of its length into the receptor-phase. The receptor-phase volume was 2 cm^3 (i.e. ca. 400 times the entrapped volume in the capillary) and was stirred thoroughly by a small magnetic bar spinning at 800 r.p.m. We have found our results to be independent of stirring speed between 400 and 1000 rpm. Below 400 rpm it appeared that solution was not moved adequately across the mouth of the capillary and transport is impeded. Above 1000 rpm there was vortexing of the receptor phase and the possibility of convective removal of the liquid within the capillary.

At time zero the capillary was fully immersed into the receptor phase to a depth of ca. 0.5 cm below the surface and transport commenced. The duration of the experiment depends upon the system but was generally between 1 and 10 min, at the end of which the capillary was quickly and carefully removed from the receptor phase. We had found by experiment that the circumspect method of capillary submersion and retrieval did not induce a perturbation resulting in a burst of activity from the capillary. We therefore believe that the solute, which had exited from the entrapped phase, had done so according to the transport equation described below. The receptor phase was now mixed with an appropriate volume of scintillation fluid and the percentage of initially entrapped radioactivity, which had been transported

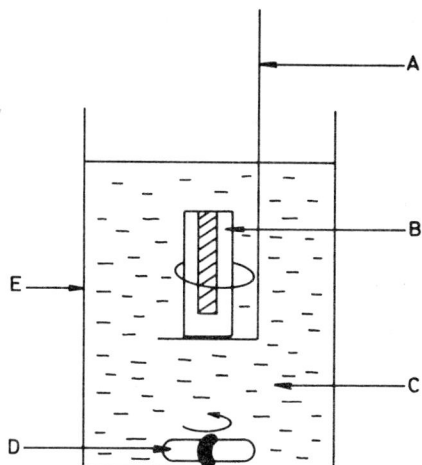

Fig. 1. Diagram of the most frequently considered experimental configuration for the capillary technique. For the sake of clarity, the thermostatting part of the apparatus has been omitted. A, wire support for capillary; B, glass capillary containing radio-labelled solute dissolved in IPM; C, aqueous receptor phase into which solute movement is followed; D, magnetic stirrer bar; E, scintillation vial.

Table 1. Initial experimental configurations of the systems studied with the capillary technique

system	solute[a]	capillary-entrapped phase	receptor phase
C1	salicylic acid	isopropyl myristate	water
C2	salicylic acid	isopropyl myristate	20 mmol dm^{-3} pentan-1-ol in water
C3	salicylic acid	60 mmol dm^{-3} pentan-1-ol in isopropyl myristate	20 mmol dm^{-3} pentan-1-ol in water
C4	salicylic acid	isopropyl myristate	0.15 mmol dm^{-3} sodium dodecyl sulphate in water
C5	acetic acid[b]	10 mmol dm^{-3} hydrochloric acid (aq)	isopropyl myristate
C6	acetic acid[b]	10 mmol dm^{-3} hydrochloric acid (aq)	dodecane

[a] Solutes were ^{14}C-labelled. [b] Specific activity of acetic acid was 56 mCi mmol^{-1}, equivalent to ca. 19 000 counts min^{-1} per 5 mm^3 of 10 mmol dm^{-3} HCl solution.

out of the capillary, was determined. The other information required for the interpretation of the data consists of (i) the capillary dimensions, which were known or measured, and (ii) the solute diffusion coefficient in the capillary-entrapped phase, the values of which (for the systems reported here) have been previously determined.

All chemicals used (for these and the rotating-diffusion-cell studies) were purchased commercially and aqueous solutions were prepared with water distilled from an all-glass apparatus. The initial experimental configurations studied are given in table 1.

ROTATING-DIFFUSION-CELL TECHNIQUE

The rotating-diffusion-cell (r.d.c.) procedure for the measurement of interfacial transfer kinetics has been described in detail.[16] In the systems reported in this paper (see table 2)

Table 2. Initial configurations of the systems studied by the rotating-diffusion-cell procedure

system	inner compartment[a]	filter	outer compartment[b]	T/°C
R1	5 mmol dm^{-3} salicylic acid (SA) in 10 mmol dm^{-3} HCl (aq)	IPM[c]	10 mmol dm^{-3} HCl (aq)	20, 25, 30, 37
R2	5 mmol dm^{-3} SA in 10 mmol dm^{-3} HCl + 100 mmol dm^{-3} ethanol (aq)	IPM[d]	10 mmol dm^{-3} HCl + 100 mmol dm^{-3} ethanol (aq)	20, 25, 37
R3	5 mmol dm^{-3} SA in 10 mmol dm^{-3} HCl + 20 mmol dm^{-3} pentan-1-ol (aq)	IPM[d]	10 mmol dm^{-3} HCl + 20 mmol dm^{-3} pentan-1-ol (aq)	20, 25, 30, 37
R4	5 mmol dm^{-3} SA in 10 mmol dm^{-3} HCl + 10 mmol dm^{-3} benzyl alcohol (aq)	IPM[d]	10 mmol dm^{-3} HCl + 10 mmol dm^{-3} benzyl alcohol (aq)	20, 25, 30, 37
R5	100 mmol dm^{-3} methyl nicotinate (MN) in water	IPM	water	20, 25, 30, 37
R6	100 mmol dm^{-3} MN in 100 mmol dm^{-3} ethanol (aq)	IPM[d]	100 mmol dm^{-3} ethanol (aq)	20, 25, 37

[a] Volume 40 cm^3; [b] volume 250 cm^3; [c] isopropyl myristate; [d] IPM was pre-equilibrated with the appropriate alcohol before impregnation into the r.d.c. filter.

the inner and outer compartments of the r.d.c. contained aqueous phases, while the rotating filter was impregnated with IPM. The choice of the organic phase reflects its use in previous work to model the characteristics of a biomembrane.[16,17,22,23,25-27] Liquid/liquid interfaces were thereby established on both sides of the spinning disc. The filters used were Millipore GS type 0.22 μm pore size. The porosity and thickness of these membranes were 0.75 and 150 μm, respectively, values that have been independently verified by an electrochemical technique.[28]

Solute flux in the r.d.c. was followed by periodic spectrophotometric assay of the outer aqueous phase. Experiments were conducted at three or four temperatures between 20 and 37 °C. Data analysis for this procedure requires knowledge of the solutes' diffusion coefficients in the aqueous and organic phases and the partition coefficients of the substrates. Diffusion coefficients in the control systems (no alcohol) were known.[16,22,23,29] In the presence of alcohol, the change in solute diffusion coefficient was assessed from the alteration induced in solvent viscosity. These relatively minor effects were determined using an Ostwald viscometer. Partition coefficients were found by shaking an aqueous solution of substrate with an equal volume of IPM (in the presence and absence of alcohol) for 48 h and then assaying for the solute spectrophotometrically. Diffusion and partition coefficients are reported in table 3.

THEORY

CAPILLARY EXPERIMENTS

Fig. 2 shows schematically the capillary system and defines the physicochemical and geometric parameters. The diagram assumes that an organic phase is contained within the capillary and that the receptor phase is aqueous (although this need not necessarily be the case).

Table 3. Diffusion[a] and partition coefficients

	R1–R4		R5 and R6		K^b					
$T/°C$	D_o /10^{-9} m^2 s^{-1}	D_a /10^{-9} m^2 s^{-1}	D_o /10^{-9} m^2 s^{-1}	D_a /10^{-9} m^2 s^{-1}	R1	R2	R3	R4	R5	R6
20	0.37	0.83	0.37	0.81	36.6	36.9	45.5	37.6	2.08	2.17
25	0.42	0.96	0.41	0.88	33.2	33.1	41.2	33.7	2.22	2.27
30	0.49	1.16	—	—	30.5	30.6	37.0	31.0	2.38	—
37	0.58	1.37	0.51	1.20	26.0	25.7	30.5	26.2	2.56	2.50

[a] Diffusion coefficients for the control systems (R1 and R5) have been reported in ref. (16), (22), (23) and (29). Within experimental error (±3%), addition of any of the alcohols at the level used did not cause solvent viscosity to change. Hence D_a and D_o values for R2–R4 and R6 are the same as the respective controls (R1 and R5). [b] Each value is the mean of at least 3 determinations. Standard deviations were not greater than 2%.

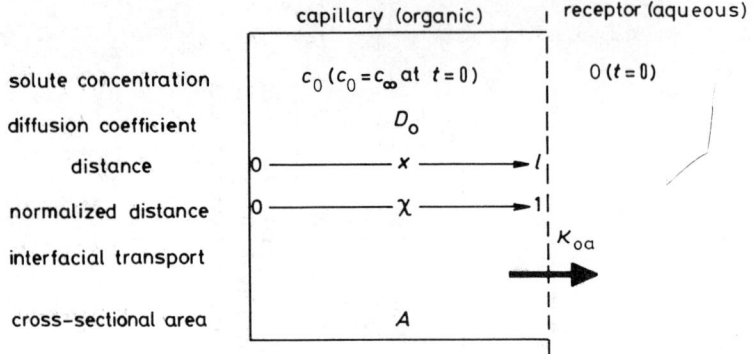

Fig. 2. Schematic representation of the capillary system and definitions of the relevant physicochemical and geometric parameters.

To describe the loss of solute from the capillary as a function of time requires that we solve Fick's second law of diffusion

$$\partial c_0/\partial t = D_o(\partial^2 c_0/\partial x^2) \tag{1}$$

with the appropriate boundary conditions. The mathematics are clarified by the use of the normalized variables

$$u = c_0/c_\infty \tag{2}$$
$$\tau = D_o t/l^2 \tag{3}$$
$$\chi = x/l \tag{4}$$
$$\kappa = k_{oa} l/D_o \tag{5}$$

such that eqn (1) becomes

$$\partial u/\partial \tau = \partial^2 u/\partial \chi^2. \tag{6}$$

The boundary conditions, in terms of the normalized variables, are

(a) for $0 < \chi < 1$, at $\tau = 0$, $u = 1$ (7)

(b) at $\chi = 0$, $(\partial u/\partial \chi)_0 = 0$ (8)

(c) at $\chi = 1$, $(\partial u/\partial \chi)_1 = -\kappa u_{\chi=1}$. (9)

The boundary condition given by eqn (7) shows that the initial solute concentration in the capillary is c_∞. Eqn (8) indicates that there is a finite supply of solute in the capillary (*i.e.* there is no replenishment at $\chi = 0$). Lastly, eqn (9) describes the flux of solute across the aqueous/organic interface at the open end of the capillary. The heterogeneous–interfacial-transfer rate constant for this process (k_{oa}) is related to the corresponding parameter for transport in the opposite direction (k_{ao}) by the solute's bulk organic/aqueous partition coefficient ($K = k_{ao}/k_{oa}$).

Eqn (6) is solved by the method of Laplace transformation following a similar procedure to that described for the calculation of drug release from controlled-delivery systems.[30] The transform of eqn (6) with boundary condition eqn (7) is

$$s\bar{u} - 1 = \partial^2 \bar{u}/\partial \chi^2 \tag{10}$$

the general solution of which is

$$\bar{u} = C_1 \cosh s^{1/2}\chi + C_2 \sinh s^{1/2}\chi + s^{-1} \tag{11}$$

where C_1 and C_2 are constants.

The cumulative amount (M_t) of material which has transported out of the capillary after a time t is

$$M_t = Alc_\infty \int_0^\tau -\kappa u_1 \, d\tau \tag{12}$$

and Alc_∞ is simply M_∞, the total amount of solute in the capillary at $t = 0$. Hence to evaluate M_t requires that we find u_1. This is achieved by using eqn (11) and the boundary conditions given by eqn (8) and (9) to yield, on substitution into eqn (12),

$$M_t = M_\infty \mathscr{L}^{-1}\{\kappa \tanh s^{1/2}/[s^{3/2}(s^{1/2} \tanh s^{1/2} + \kappa)]\} \tag{13}$$

where the integration with respect to τ has been achieved by division by the Laplace variable s.

A simple inversion of eqn (13) cannot be found. At short times ($\tau \ll 1, s \gg 1$), however, we can approximate the hyperbolic term

$$\tanh s^{1/2} \to 1$$

and eqn (13) reduces to

$$M_t/M_\infty = \mathscr{L}^{-1}\{\kappa/[s^{3/2}(s^{1/2} + \kappa)]\} \tag{14}$$

which on inversion gives[30]

$$M_t/M_\infty = 2(\tau/\pi)^{1/2} + \kappa^{-1}[\exp(\kappa^2\tau) \, \text{erfc}(\kappa\tau^{1/2}) - 1]. \tag{15}$$

If $\kappa \approx 1$, then for $\tau \ll 1$ eqn (14) can be further approximated by

$$M_t/M_\infty = \mathscr{L}^{-1}\{\kappa/s^2\} = \kappa\tau \tag{16}$$

i.e. the release rate of solute from the capillary becomes a zero-order process.

If no interfacial barrier exists at the interface at $x = l$ ($\kappa \to \infty$) then the solution to the diffusion equation is the classic ('burst') $t^{1/2}$ function:[30]

$$M_t/M_\infty = 2(\tau/\pi)^{1/2}. \tag{17}$$

Eqn (15) is plotted in fig. 3 for various κ values and is compared with the $t^{1/2}$ equation, eqn (17). For the smallest κ values eqn (16) provides a reasonable description of the release process.

At long times ($\tau \gg 1, s \ll 1$; $\tanh s^{1/2} \to s^{1/2}$) eqn (13) can again be inverted to give

$$M_t/M_\infty = 1 - \exp(-\kappa\tau). \tag{18}$$

However, Hadgraft[30] has shown that this expression is likely to be of value experimentally only if $\kappa = 0.1$ or less. The kinetics in our systems fall outside this range and hence attention is focussed upon the short-time approximation and solution.

R.D.C. EXPERIMENTS

Flux measurements in the r.d.c. yield an overall solute transport coefficient (P) (bulk inner aqueous phase → bulk outer aqueous phase) given by[16]

$$P^{-1} = 2(0.643 \, \nu^{1/6} D_a^{-2/3} W^{-1/2}) + h/\alpha K D_o + 2/\alpha k_{ao}. \tag{19}$$

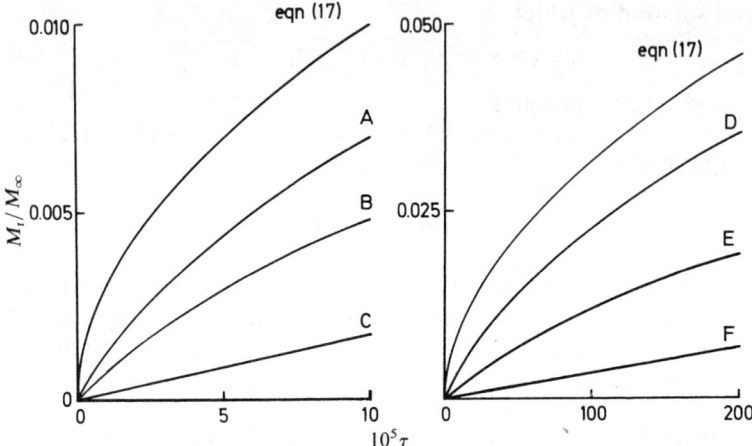

Fig. 3. Solute transport from the capillary as a function of time ($\tau \ll 1$). Predicted profiles [eqn (15)] for different interfacial barriers are compared with the release expected when no resistance to phase transfer exists [eqn (17)]. Each profile corresponds to a different value of κ: (A) 250, (B) 100, (C) 25, (D) 75, (E) 20, and (F) 5.

The first term describes substrate movement through aqueous diffusion layers on either side of the rotating filter (ν is the kinematic viscosity of aqueous phase and W the cell rotation speed). $h/\alpha KD_o$ is the barrier to diffusion through the IPM-impregnated filter of thickness h and porosity α. The third contribution to P^{-1} is the interfacial transfer resistance.

Experimentally we measure P^{-1} as a function of $W^{-1/2}$ and force the theoretical slope through the data to obtain the intercept terms in which k_{ao} is the only unknown. Since the partition coefficient (K) is determined independently, interfacial-transfer rate constants may be obtained for both aqueous → organic and organic → aqueous processes.

RESULTS AND DISCUSSION

The results from the capillary-technique experiments are summarized in table 4. The values of κ were deduced from the experimental data using eqn (15). Interfacial transfer coefficients could then be obtained by application of eqn (5) with the values of the capillary length and either D_o (systems C1–C4) or D_a (systems C5 and C6). Rate constants spanning two orders of magnitude are reported. For the systems and times studied, between 0.3 and 2.5% of the capillary entrapped counts reach the receptor phase. Transport kinetics determined using the r.d.c. procedure are given in table 5. The manner in which the raw r.d.c. data are handled to obtain these parameters has been discussed.[16–21] In table 6 the kinetics and partition coefficients are analysed in the normal way[16,18,20,21] to yield the thermodynamic parameters describing the phase-transfer processes.

The capillary-derived data suggest that the technique has potential for the measurement of interfacial transfer kinetics. Fig. 3 shows that an upper bound on the rapidity of the measurable range exists but also implies that, in our studies, we have considered systems for which transport is slower (in some cases much slower) than this theoretical maximum. Most of the systems investigated have been previously or concurrently studied by the r.d.c. technique. In most cases the capillary

Table 4. Capillary technique:[a] experimental results

system	experiment duration/s	$10^4 \tau^b$	$10^2 M_t/M_\infty{}^c$	κ^b	$k_{oa}/10^{-6}$ m s^{-1}	$k_{ao}/10^{-6}$ m s^{-1}
C1	200	8.40	1.40 ± 0.04	27	1.14[d]	—
C2	200	8.40	1.40 ± 0.12	27	1.14[e]	—
C3	75	3.15	1.41 ± 0.11	125	5.27[e]	—
	150	6.30	2.13 ± 0.34	110	4.62[e]	—
	200	8.40	2.45 ± 0.22	95	3.97[e]	—
C4	200	8.40	1.17 ± 0.10	20	0.84	—
C5	200	24	0.39	1.7	—	0.204[f]
	500	60	0.76 ± 0.16	1.4	—	0.168[f]
C6	1500	180	0.33 ± 0.04	0.2	—	0.024[g]

[a] $T = 25$ °C for all systems. [b] τ and κ are defined in eqn (3) and (5), respectively. Diffusion coefficients for the various systems (except C5 and C6) are given in table 3. For C5 and C6, $\tau = D_a t/l^2$ and $\kappa = k_{ao} l/D_a$ because the capillary holds the aqueous phase. [c] Except for system C5 at 200 s, all values of M_t/M_∞ are the mean (± standard deviation) of at least 5 determinations. [d] r.d.c. value found in this work was 1.54×10^{-6} m s^{-1}. [e] R.d.c. value found in this work was 1.68×10^{-6} m s^{-1}. [f] Ref. (16) gives $k_{oa} = 0.28 \times 10^{-6}$ m s^{-1} for this system; $D_a = 1.2 \times 10^{-5}$ cm^2 s^{-1}.[16] [g] Ref. (18) gives $k_{oa} = 0.086 \times 10^{-6}$ m s^{-1} for this system; $D_a = 1.2 \times 10^{-5}$ cm^2 s^{-1}.[16]

Table 5. Interfacial transfer kinetics determined with the rotating diffusion cell

	$k_{ao}/10^{-6}$ m s^{-1}				$k_{oa}/10^{-6}$ m s^{-1}			
T/°C:	20	25	30	37	20	25	30	37
R1	51	51	56	63	1.39	1.54	1.84	2.42
R2	56	62	—	69	1.52	1.87	—	2.68
R3	52	57	64	65	1.14	1.38	1.73	2.13
R4	52	51	57	61	1.38	1.51	1.84	2.33
R5	17	24	38	86	7.9	11	16	34
R6	9.1	12	—	22	4.2	5.1	—	8.8

Table 6. Thermodynamic parameters[a] (at 298 K) of phase transfer and partitioning for the systems studied with the rotating diffusion cell[b]

		R1	R2	R3	R4	R5	R6
k_{ao}	$\Delta G^\ddagger$	35.9	35.4	35.6	35.9	37.8	39.5
	$\Delta H^\ddagger$	9.9	8.9	10.3	8.0	69.8	39.1
	$\Delta S^\ddagger$	−87	−89	−85	−92	107	−0.1
k_{oa}	$\Delta G^\ddagger$	44.6	44.1	44.8	44.6	39.7	41.6
	$\Delta H^\ddagger$	25.0	24.8	28.2	24.0	61.0	33.2
	$\Delta S^\ddagger$	−66	−65	−56	−69	71	−28
K	ΔG	−8.7	−8.7	−9.2	−8.7	−2.0	−2.0
	ΔH	−15	−16	−18	−16	9.3	6.3
	ΔS	−21	−25	−30	−25	38	28

[a] ΔG and ΔH in kJ mol^{-1}, ΔS in J mol K^{-1}. [b] Deduction of the thermodynamic parameters from the kinetic constants has been described in detail elsewhere [e.g. ref. (16), (18) and (21)].

procedure yields slightly lower values. Relatively speaking, the two methods do show reasonable agreement; for example, both techniques find that the acetic acid systems (C5 and C6) have the slowest kinetics and that the transport into IPM from water is the more rapid of the two. The reasons why the capillary measurements are consistently lower, though, remain unclear. It is noted that systems C1 and C2 and systems R1 and R3 give identical k_{oa} values and that the agreement here between procedures is good. Specifically loading the IPM phase with pentan-1-ol in C3, however, facilitates the transport process and yields a rate constant faster than that found with the r.d.c. Again, the reason for this discrepancy is not yet understood. In system C4 we attempted to populate the interface with a surfactant to hinder solute transport. A 26% reduction in k_{oa} was observed and the difference is significant. The concentration of sodium dodecyl sulphate used is below the critical micelle concentration (c.m.c.). We found that above the c.m.c. the surfactant was able to solubilize some of the IPM (and dissolved radioactivity) out of the capillary, producing an inflated value of M_t/M_∞ in excess of that predicted by eqn (17).

For the r.d.c. experiments alcohol concentrations were chosen on the basis of previous investigations considering anaesthetic–membrane interactions.[31–36] It is first observed that the interfacial transport of salicylic acid (systems R1–R4) is not affected by ethanol, pentan-1-ol and benzyl alcohol. In these systems the activation free-energy barrier to transport is primarily entropic (table 6), $\Delta S^\ddagger$ being large and negative in every case. This observation has been reported for a number of alkanoic acids traversing the dodecane/water interface[18] and appears to be the effect of the carboxylic acid group. For methyl nicotinate the introduction of 100 mmol dm^{-3} ethanol does perturb the transfer behaviour. $\Delta G^\ddagger$ for this solute in the absence of alcohol is enthalpic in origin and a positive $\Delta S^\ddagger$ contribution of some magnitude exists (the size of $\Delta S^\ddagger$ in the system has been reported to be slightly higher in earlier work[20,23]). The $\Delta S^\ddagger$ value has been attributed in part to the disruption of a structured interfacial configuration[19–21,23] by the transporting solute. The implication of the results of the ethanol system (R6) is that the alcohol destabilizes this orientation (presumably of ordered water molecules) such that $\Delta S^\ddagger$ decreases significantly. The observation is consistent with the hypotheses and results of Ueda et al.[8–15] and very similar to recent studies in our laboratory in which the effect of poly(ethylene glycol) (PEG) on the same system was considered.[37] In the latter work PEG 400 concentrations of 10, 25 and 40% v/v reduced the $\Delta S^\ddagger$ terms from ca. 100 J mol^{-1} K^{-1} to zero.

Thus we have described a new procedure for studying liquid/liquid transfer and we have probed the action of various agents upon the interfacial region. Differences are identified between the procedures and between the effect of different perturbants on two different solute molecules. Because of the importance of interfacially related events to many biomembrane phenomena, it appears that further work is required in this area before an adequate physicochemical understanding is attained.

We thank Drs F. C. Szoka and C. A. Hunt for advice and comments and the U.S. National Institutes of Health (AA-05781-01) and the Donors of the Petroleum Research Fund administered by the American Chemical Society (PRF-13860-G5) for financial support. M. A. is a recipient of a President's Undergraduate Fellowship from the University of California, San Francisco. The expert typing of Andrea Mazel is gratefully acknowledged.

[1] R. A. Robinson and R. H. Stokes, in *Electrolyte Solutions* (Butterworths, London, 2nd edn, 1959), chap. 10, pp. 261–264.
[2] K. W. Miller, *Anesthesiology*, 1977, **46**, 2.

3. N. P. Franks and W. R. Lieb, *Nature (London)*, 1982, **300**, 487.
4. F. A. Vilallonga, E. R. Garett and J. S. Hunt, *J. Pharm. Sci.*, 1977, **66**, 1229.
5. D. A. Cadenhead and J. Osonka, *J. Colloid Interface Sci.*, 1970, **33**, 188.
6. H. L. Booij and W. Dijkshoorn, *Acta Physiol. Pharmacol. Neerl.*, 1950, **1**, 631.
7. H. Eyring, J. W. Woodbury and J. S. D'Arrigo, *Anesthesiology*, 1976, **38**, 415.
8. I. Ueda, H. Kamaya and H. Eyring, *Proc. Natl Acad. Sci. USA*, 1976, **73**, 481.
9. S. Kaneshina, I. Ueda, H. Kamaya and H. Eyring, *Biochim. Biophys. Acta*, 1980, **603**, 237.
10. S. Kaneshina, H. Kamaya and I. Ueda, *J. Colloid Interface Sci.*, 1981, **83**, 589.
11. A. Shibata, Y. Suezaki, H. Kamaya and I. Ueda, *Biochim. Biophys. Acta*, 1981, **646**, 126.
12. A. Shibata, H. Kamaya and I. Ueda, *J. Colloid Interface Sci.*, 1982, **90**, 487.
13. S. Kaneshina, H. Kamaya and I. Ueda, *Biochim. Biophys. Acta*, 1982, **685**, 307.
14. S. Kaneshina, H. Kamaya and I. Ueda, *J. Colloid Interface Sci.*, 1983, **93**, 215.
15. Y. Suezaki, S. Kaneshina and I. Ueda, *J. Colloid Interface Sci.*, 1983, **93**, 225.
16. W. J. Albery, J. F. Burke, E. B. Leffler and J. Hadgraft, *J. Chem. Soc., Faraday Trans. 1*, 1976, **72**, 1618.
17. W. J. Albery and J. Hadgraft, *J. Pharm. Pharmacol.*, 1979, **31**, 65.
18. N. H. Sagert, M. J. Quinn and R. S. Dixon, *Can. J. Chem.*, 1981, **59**, 1096.
19. R. H. Guy, T. R. Aquino and D. H. Honda, *J. Phys. Chem.*, 1982, **86**, 280.
20. R. H. Guy, D. H. Honda and T. R. Aquino, *J. Colloid Interface Sci.*, 1982, **87**, 107.
21. R. H. Guy, T. R. Aquino and D. H. Honda, *J. Phys. Chem.*, 1982, **86**, 2861.
22. M. Ahmed, J. Hadgraft and I. W. Kellaway, *Int. J. Pharmaceut.*, 1982, **12**, 219.
23. R. Fleming, R. H. Guy and J. Hadgraft, *J. Pharm. Sci.*, 1983, **72**, 142.
24. M. Ahmed, J. Hadgraft and I. W. Kellaway, *Int. J. Pharmaceut.*, 1983, **13**, 227.
25. B. J. Poulsen, E. Young, V. Coquilla and M. Katz, *J. Pharm. Sci.*, 1968, **57**, 928.
26. N. A. Armstrong, K. C. James and K. C. Wong, *J. Pharm. Pharmacol.*, 1981, **31**, 627.
27. N. Barker and J. Hadgraft, *Int. J. Pharmaceut.*, 1981, **8**, 193.
28. W. J. Albery and P. R. Fisk, in *Hydrometallurgy '81* (Soc. Chem. Ind., London, 1981), FS/1–FS/15.
29. A. D. Cadman, R. Fleming and R. H. Guy, *J. Pharm. Pharmacol.*, 1981, **33**, 121.
30. J. Hadgraft, *Int. J. Pharmaceut.*, 1979, **2**, 177.
31. J. C. Metcalfe, P. Seeman and A. S. V. Burgen, *Mol. Pharmacol.*, 1968, **4**, 87.
32. P. Seeman, *Pharmacol. Rev.*, 1972, **24**, 583.
33. J. H. Chin and D. B. Goldstein, *Mol. Pharmacol.*, 1977, **13**, 435.
34. J. H. Chin and D. B. Goldstein, *Science*, 1977, **196**, 684.
35. D. A. Johnson, N. M. Lee, R. Cooke and H. H. Loh, *Mol. Pharmacol.*, 1979, **15**, 739.
36. E. S. Rowe, *Biochemistry*, 1983, **22**, 3299.
37. R. H. Guy and F. C. Szoka, *J. Membr. Biol.*, submitted for publication.

GENERAL DISCUSSION

Prof. P. Meares (*University of Aberdeen*) said: The introduction of the rotating diffusion cell has been a major advance in the study of interfacial kinetics of reactions involving two liquid phases. It is necessary, however, to be satisfied that the postulated hydrodynamic conditions are fully met in practice. Three questions arise in this connection. The surface area available for transport of solutes is a function of the shape of the menisci at the open ends of the pores and so will be a function of the interfacial tensions, which will not, in general, be equal on opposite sides of the membrane. The theory of Levich deals with a solid disc rotating in a liquid. It has to be considered whether the shear forces between the liquid in the disc and in the bulk phases are sufficient to induce circulatory motion within the pores which could make a convective contribution to the total mass flow, *i.e.* reduce the effective diffusional length of the pores. Thirdly, the possibility of interfacial turbulence at the pore mouths, driven by the disequilibrium between the liquids in contact, has to be borne in mind.

Prof. W. J. Albery (*Imperial College, London*) said: Prof. Meares has frightened us all by drawing attention to the problems created by interfacial turbulence. What experimental methods are there for determining if interfacial turbulence is present or not?

Dr. M. A. Hughes (*University of Bradford*) replied: Interfacial turbulence is a common phenomenon observed in liquid–liquid extraction systems. It is especially common when high concentrations of solutes in either phase are present.

Simple direct observation is possible using a microscope,[1] but a more sophisticated technique is that of Schlieren photography. This system is described in ref. (2) and (3).

[1] M. A. Hughes, *Int. J. Hydrometallurgy*, 1978, **3**, 85.
[2] A. Orell and J. Westwater, *AIChE J.*, 1962, 351.
[3] H. R. C. Pratt, in *Handbook of Solvent Extraction*, ed. H. I. Baird, C. H. Nanson and T. C. Lo (Wiley, New York, 1982), chap. 3.

Dr. M. A. Hughes (*University of Bradford*) said: My first remark concerns the site of the reaction proposed in this work. It is suggested in Prof. Albery's paper that according to other workers [ref. (8)] 'The reaction takes place entirely in the aqueous phase.'. Actually, these workers did not say that: instead they suggest a thin reaction zone on the aqueous side of the interface where the interface is taken to mean a unimolecular layer of extractant/diluent molecules. Liquid/liquid interfaces are likely to be more diffuse than the air/liquid interface; thus the previous workers have considered 'interface zones' rather than 'unimolecular layers' as the site of reaction.

If the site of reaction was a unimolecular layer, as suggested by the present authors, it might be expected that these commercial processes would be much more sensitive to surface-active impurities; this is not the general experience.

(1) I have two questions for Prof. Albery. The model proposed in this paper does not appear to have the protonated ligand as a species. There is some evidence

from interfacial tension and acid-base titration measurements to show that oximes do protonate. However, the extent of this protonation relates to the structure of the oxime.[1]

Would the authors comment on this aspect of the mechanism, which would be important in the stripping process.

(2) There appears to be a fair amount of scatter in the points recorded in certain of the graphs. In particular, could the authors say (1) if the lines drawn through the points in fig. 3 are theoretical Levich lines or estimates of the best straight line through the points, (2) if the lines drawn through the points in fig. 5 could also be drawn as curves passing through the (0, 0) and (3) if the line drawn through the points in fig. 6 should not intercept the vertical axis, otherwise it implies that copper could be stripped under zero acid conditions.

I believe the technique described here is a useful contribution to the study of liquid-liquid extraction processes for metals. The other techniques, used to study the rate of mass transfer, are all complicated by hydrodynamics which are difficult to account for.

[1] T. A. B. Al-Diwan, M. A. Hughes and R. J. Whewell, *J. Inorg. Nucl. Chem.*, 1977, **39**, 1419; J. S. Preston and R. J. Whewell, *J. Inorg. Nucl. Chem.*, 1977, **39**, 1675.

Dr. B. H. Robinson (*University of Kent*) said: I have three questions for Prof. Albery.

(1) Have you studied the extraction of any other reactive metal ions in addition to Cu^{2+}(aq), *e.g.* Mg^{2+} or Co^{2+}? Would a correlation be expected between transfer rates (j_0 values) and the corresponding rate constants for complexation in bulk aqueous solution?

(2) Have you considered varying the alkyl chain length of the ligand, which should enable you to distinguish between k_1 and k_3?

(3) Have the authors considered adding a kinetically inert charged surfactant, *e.g.* $C_8H_{17}SO_4$, to the system? This should serve to concentrate the Cu^{2+} at the interface and facilitate the complexation rate. H .ve you carried out any experiments which support this?

Dr. A. Steinchen (*Université Libre de Bruxelles, Belgium*) added: I do not understand what is the actual surface of reaction. Indeed, in Millipore pores the surface of the meniscus may vary under the experimental conditions (pH and surfactant concentration), as may the interfacial tension.

Prof. A. Sanfeld (*Université Libre de Bruxelles, Belgium*) remarked: When dealing with surface kinetics one needs to know details of the evolution of interfacial tension with the concentrations of reacting species.

Prof. M. M. Kreevoy (*University of Minnesota*) said: Djugumović, Škerlak and I[1] have determined the profile of copper concentration in working membranes similar to those discussed by Albery and coworkers. As shown in fig. 1 of this comment, there is measurable accumulation of copper in a 5-layer, 0.3 mm thick, Gore-Tex[2]-supported membrane transferring Cu^{II} from an acetic acid–acetate buffer of pH 4, to 2 mol dm^{-3} H_2SO_4. The active carrier was 2-oximinobenzoyl-4-nonylphenol (LIX-65N[3]) [0.11 mol dm^{-3} in decahydronaphthlene (97% *cis*)]. Our apparatus was a modified dialysis cell[4] in which the hydrodynamic resistances have been previously determined as a function of pumping rate.[4] Experiments were interrupted at various

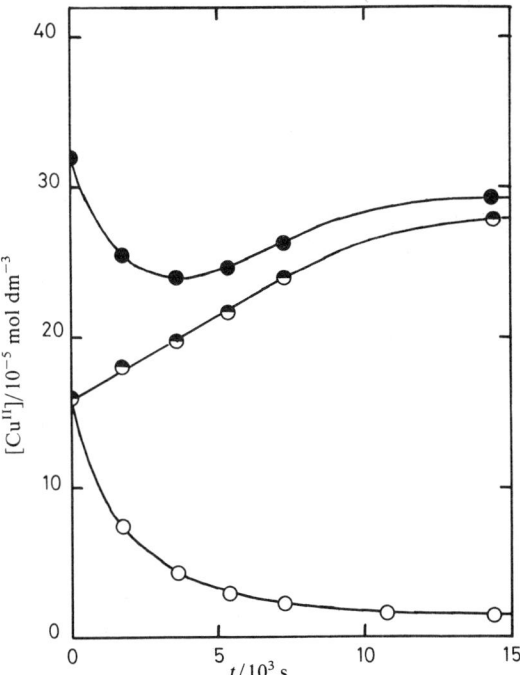

Fig. 1. Cu^{II} concentration in the feed (○), in the strip solution (◐), and the sum of the two (●) plotted as a function of time.

times, the membrane layers separated, and the Cu^{II} concentration determined in each layer. Several of the resulting profiles are shown in fig. 2.

From the slopes of these profiles it is evident that the internal resistance never becomes negligible in our experiments. Between 30 and 90 min the Cu^{II} content of the membrane is relatively constant, as shown by fig. 1, and the Cu^{II} profile within the membrane is also relatively invariant, so the flux of Cu^{II} out of the feed, into the strippant, and across the membrane should be about the same. From fig. 1 this common flux, J, is ca. 1×10^{-10} mol cm^{-2} s^{-1}. Within the membrane J is given by Fick's first law, eqn (1). A concentration near the interface with the feed is indicated by subscript f, i and a concentration near the interface with the strippant is indicated by subscript s, i:

$$J = \frac{([Cu]_{f,i} - [Cu]_{s,i}) \times D'}{10^3 l}. \qquad (1)$$

Since all the other quantities in eqn (3) are known, D', the apparent diffusion coefficient of the LIX-65N complex with Cu^{II}, can be calculated. A value of 5×10^{-7} cm^2 s^{-1} is obtained.

A value of D' can also be estimated from our previous value of 1.9×10^{-7} cm^2 s^{-1} for the migration of didodecylammonium picrate through a Gore-Tex-supported membrane with phenyl ether as solvent at 25 °C.[4] At that temperature phenyl ether has a viscosity of ca. 4 cP[4] while the viscosity of cis-decalin at 40 °C is 1.9 cP.[5] Assuming that D' is inversely proportional to the viscosity of the solvent leads to

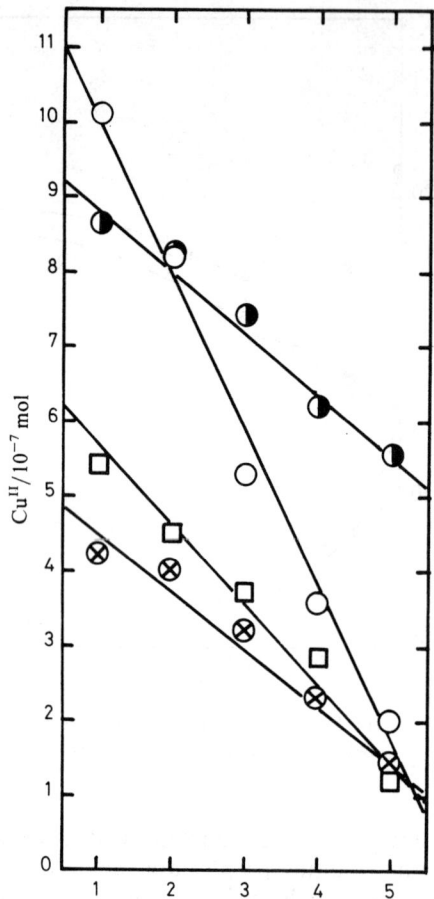

Fig. 2. Cu^{II} concentration profiles in 5-layered membranes at various times. The time of interruption, in minutes, is as follows: ○, 15; □, 15+30; ◐, 60 and ⊗, 120. Profiles similar to the 60 min profile were also obtained at 30 and 90 min. The profile marked 15+30 was obtained from an experiment interrupted after 15 min, but the membrane layers were not separated for a further 30 min, to evaluate the possible distortion introduced by the operational delay in the other experiments (ca. 2 min).

a value of 4.0×10^{-7} cm^2 s^{-1} for the present D'. Considering the uncertainties involved in the various estimates and assumptions, the agreement is respectable.

Our internal resistances are higher than Albery's, and our D' value lower by about a factor of 10. However, most of the difference can be accounted for by physical differences between the two systems. Our Gore-Tex support has a void volume only around 0.5,[4] while Albery's membranes are supported on Millipore filters, which have void volumes close to unity. The tortuousity of the Gore-Tex-supported membranes is also probably greater than that of those supported on Millipore filters. Finally, the viscosity of heptane, Albery's solvent, is only 0.4 cP at 25 °C.[6]

In order to compare the Cu^{II} concentrations close to the interfaces, obtainable from fig. 2, with those which would be at equilibrium with the feed and strip solutions, $K_{o/a}$, given by

$$K_{o/a} = \frac{[\overline{R_2Cu}][H^+]^2}{[\overline{RH}]^2[Cu^{2+}]}. \tag{2}$$

was evaluated in separator-funnel experiments. RH is LIX-65N and R_2Cu is its copper complex. Brackets indicate concentrations and the over-bar indicates a concentration in the organic phase. Activities were assumed to be proportional to concentrations. Complexing of Cu^{2+} with SO_4^{2-} and with acetate, and dimerization of LIX-65N were taken into account. A value of 2 was obtained for $K_{o/a}$.

Using this value of $K_{o/a}$, the concentrations of R_2Cu which would be in equilibrium with the feed and the strip solutions were calculated for the Cu^{II} concentrations prevailing at the time the experiments were interrupted for Cu^{II} profile determination. On the feed side, the Cu^{II} concentrations obtained by extrapolation to the interface ranged from 0.23 to 0.43 of the calculated values. However, on the strip side, the observed values exceeded the calculated values by factors between 10^4 and 10^5. The latter require a substantial interfacial barrier at the stripping interface, in agreement with the conclusions of Albery and coworkers. The results for the feeding interface, however, imply that there is little or no interfacial barrier there, contrary to the conclusions of Albery and coworkers. This discrepancy may be due to the efficiency of acetate ion,[7] present in our feeds but absent from Albery's, in catalysing the interfacial transport of metal ions.

Fig. 1 shows that the flux of Cu^{II} into the strip solution is nearly constant for over 120 min, while fig. 2 shows that the concentration of R_2Cu near the stripping interface varies considerably over this period. These observations support the contention of Albery and coworkers that rate of transport across the stripping interface is independent of $[R_2Cu]$.

[1] S. Djugumović, M. M. Kreevoy and T. Škerlak, *J. Phys. Chem.*, to be published.
[2] Gore-Tex is a registered trademark belonging to W. L. Gore Associates, Inc.
[3] LIX is a registered trademark belonging to Henkel-America, Inc.
[4] L. A. Ulrick, K. D. Lokkesmoe and M. M. Kreevoy, *J. Phys. Chem.*, 1982, **49**, 3651.
[5] F. Nauwelaers, L. Hellemans and A. Persoons, *J. Phys. Chem.*, 1976, **80**, 767.
[6] *Lange's Handbook of Chemistry*, ed. T. A. Dean (McGraw-Hill, New York, 12th edn, 1979), pp. 10-109.
[7] D. T. Wasan, Z. M. Gu and N. N. Li, *Faraday Discuss. Chem. Soc.*, 1984, **77**, 67.

Dr. R. D. Noble (*N.B.S., Boulder, Colorado*) said: I would like to make the following comments.

(1) Stability theory shows that there would not be convective cells present in supported liquid membranes. The wavelength of the instability is much larger than the membrane pore size.

(2) Concern over the size of the reaction zone at the interface depends on: (*a*) how one models the system and (*b*) what one wishes to calculate (flux or chemical reaction mechanisms).

Can Prof. Albery estimate the size of the reaction zone at the interface? If so, would he provide an example.

Dr. C. Tondre (*University of Nancy, France*) said: I would like to address a question to Prof. Albery concerning the different ways of monitoring the rate of interfacial reactions in the rotating diffusion cell. If the ring–disc method appears to be well suited to provide a continuous record of the rate of reaction without perturbing the kinetics, I do not understand how a titration of the acidic or basic products of the reaction using the pH-state method can be performed directly in one of the cell compartments without influencing the measured flux.

Prof. W. J. Albery (*Imperial College, London*) said: First I must take issue with Prof. Astarita on the difference between an interfacial reaction and a reaction that takes place in a thin reaction layer in the aqueous phase. Prof. Astarita suggested in an unrecorded remark that this distinction was only a semantic one. In our view this is not the case. The distinction is real and, as discussed in our paper, will lead to different rate laws for the variation of the observed rate with the concentration of aqueous reactant. For an interfacial reaction the transition state must be located within a few ångström of the interface. The reactants are partially solvated by the two different solvents. For the reaction layer, the transition states are entirely in the aqueous phase and are located all over the reaction layer. In answer to Dr. Noble, the thickness of this layer can be as large as 10^{-2} cm. Its thickness, δ, is determined by the balance between the diffusion of the oxime and the rate of its reaction with Cu^{2+}:

$$\delta = (D/k[Cu^{2+}])^{1/2}. \tag{1}$$

The thickness of the layer varies with $[Cu^{2+}]$, and this is why a different rate law will be observed. For very fast reactions the thickness δ can approach molecular dimensions of 2 Å or so; under these circumstances the distinction between the reaction layer and the interface becomes blurred. However, such a thin reaction layer requires $k[Cu^{2+}]$ to be as large as 10^{11} s^{-1}. Such a large value can only be found if the homogeneous reaction is diffusion controlled and the concentration of Cu^{2+} is larger than 1 mol dm^{-3}. These conditions are not found for our system. Hence in our system (and most similar systems) the reaction layer would have to be at least several μm thick and the distinction between the interfacial reaction and the reaction layer is a real one.

I agree with Dr. Hughes that in his paper[1] he and his coworkers locate the reaction in a thin reaction layer in the aqueous phase. They discuss how the reaction-layer thickness depends on 'the balance between the rate of reaction and the rate of transfer of the species concerned'. This balance is given by eqn (1) above. Hence there is no doubt that these authors[1] are suggesting that the reaction takes place in a reaction layer. As argued above, the thickness of this layer must be at least several μm. In that case, as stated in our paper, all the transition states must be 'entirely in the aqueous phase'. The diffuseness of the liquid/liquid interface can only extend over a matter of ångströms and therefore cannot really affect the distinction between the interfacial reaction (ångström) and the reaction layer ($> \mu$m). We do not claim that an interfacial reaction takes place in a simple unimolecular layer. The reorganisation of the solvents must extend through several layers, but the reaction zone is nevertheless much smaller than 1 μm.

As regards surface impurities and surface-active agents Dr. Fisk and Mr. Choudhery have found that Teepol completely blocks the reaction, while traces of sodium lauryl sulphate can increase the rate by up to 100%. Prof. Kreevoy has found the same effect with octyl sulphate.[2]

We have examined the data on the protonation of the oxime ligands.[3] In the pH range 0-5 there is very little alteration in the interfacial tension for our P50 ligand. The authors[3] themselves conclude that the protonation of this oxime is slight. We agree.

Answering Dr. Hughes' questions about our experimental plots, first the gradients in fig. 3 of our paper are calculated Levich gradients allowing for the diffusion of both Cu^{2+} and of the oxime ligand. The lines in fig. 5 cannot pass through zero. In fig. 5 we are plotting a reciprocal flux. So a value of zero corresponds to an impossible flux of infinity! As stated in our paper, the intercepts are in reasonable agreement with a term describing rate-limiting diffusion of the ML_2 complex through the membrane. The experimental points in fig. 6 are perfectly clear; there is hardly room to stop the line before the y-axis.

The results of Djugumović et al. are most interesting. It is particularly gratifying that the same conclusions have been reached by two different groups using quite different experimental techniques. Taken together these results should settle the controversy as to where the reaction takes place.

Answering Dr. Robinson's questions, first we are studying other metal ions besides copper. Preliminary results on Ni^{2+} show that the reaction rates are an order of magnitude slower. We would expect a correlation between the rate constant for a homogeneous reaction in solution and the corresponding process taking place on a liquid/liquid interface. Indeed we have found such behaviour in our studies of modified electrodes. However, there are experimental difficulties in measuring the homogeneous reaction when the ligand is very insoluble.

The idea of varying the alkyl chain length to distinguish between k_1 and k_3 is a good one. Thank you. We have described above some results with surfactants. More interesting work remains to be done with such systems.

In answer to Dr Tondre, the great advantage of the pH-stat method is that the pH of the outer compartment does not change, and hence has no effect on the flux. The two methods are complementary. For extraction the copper concentration is large but the effect of the release of H^+ on the pH can be followed. For stripping the H^+ concentration is large, but the flux of Cu^{2+} can be followed with the ring electrode.

Dr. Steinchen, Prof. Sanfeld and Prof. Meares raise the question of the surface area, its curvature and the effect of surface tension on the area. We agree that these are interesting questions, but at the moment the experimental data are not sufficiently precise to enable one to see any effects arising from such variations in the surface area. The maximum possible difference is a factor of 2 between a flat disc of πr^2 and a hemisphere of $2\pi r^2$. In practice the variation of the area with the composition of the solution will be much less than this. We therefore believe that these are second-order effects compared with the variation of the rate with the concentration of reactants. At the present state of the art we should concentrate on the first-order effects and establish the mechanisms of reaction.

Turning to the second point raised by Prof. Meares, as to whether the rotating-disc motion can introduce circulatory motion in the pores, we agree with Dr. Noble that this is impossible. The thickness of the hydrodynamic boundary layer for the rotating disc is of the order of several mm. This is much larger than the radius of the pores of the Millipore filter (ca. 10^{-5} cm). Hence the Millipore surface will appear to be uniform. Finally, we are intrigued by the problem of interfacial turbulence. At present we have no evidence for such effects, but it may be sensible to look for these effects with systems which are known to suffer (or enjoy?) large Marangoni instabilities.

---

[1] R. J. Whewell, M. A. Hughes and C. Hanson, in *ISEC'77* (Canadian Institute of Mining and Metallurgy, 1979), vol. 21, p. 185.

[2] M. M. Kreevoy, personal communication.

[3] T. A. B. Al-Diwan, M. A. Hughes and R. J. Whewell, *J. Inorg. Nucl. Chem.*, 1977, **38**, 1419.

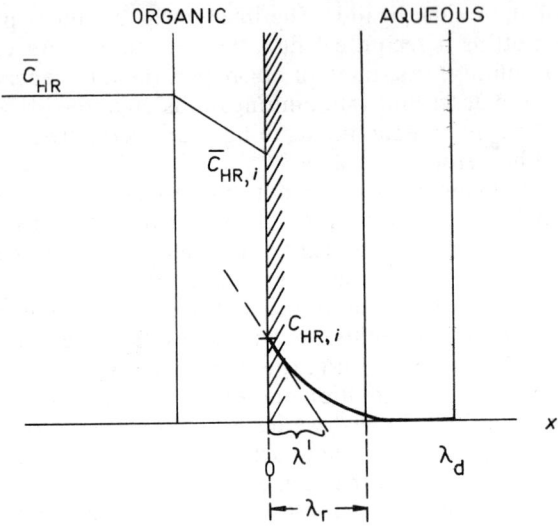

Fig. 3. Concentration profile about an interface.

Dr. R. D. Noble (*N.B.S., Boulder, Colorado*) asked whether Dr. Hughes could estimate the size of the reaction zone at the interface for his model? Could he provide an example?

Dr. M. A. Hughes (*University of Bradford*) replied: The definition of the thickness of the zone is somewhat arbitrary, since the model predicts an asymptotic concentration profile of the extractant HR in the aqueous film, as shown in fig. 3.

The interfacial flux N_{HR} can be expressed as follows.
(i) For the organic side

$$N_{HR,i} = \bar{k}_{HR}(\bar{C}_{HR} - \bar{C}_{HR,i}). \tag{1}$$

(ii) for the aqueous side

$$N_{HR,i} = -D_{HR}\left(\frac{\partial C_{HR}}{dx}\right)_{x=0} = D_{HR}\frac{C_{HR,i}}{\lambda'} = \frac{D_{HR}}{P_{HR}}\frac{\bar{C}_{HR,i}}{\lambda'}. \tag{2}$$

By eliminating $\bar{C}_{HR,i}$ from these equations we obtain the distance λ', e.g.

$$\lambda' = \frac{D_{HR}}{P_{HR}}\left(\frac{\bar{C}_{HR}}{N_{HR,i}} - \frac{1}{\bar{k}_{HR}}\right). \tag{3}$$

We define the thickness of the reaction zone as $\lambda_r = 3\lambda'$; hence

$$\lambda_r \approx \frac{3D_{HR}}{P_{HR}}\left(\frac{\bar{C}_{HR}}{H_{HR,i}} - \frac{1}{\bar{k}_{HR}}\right). \tag{4}$$

Now the thickness of the diffusional film is related to the mass transfer coefficient by

$$\lambda_d = D_{HR}/k_{HR} \tag{5}$$

so the relative thickness is given by

$$\xi = \frac{\lambda_r}{\lambda_d} \approx \frac{3}{P_{HR}} \left(\frac{k_{HR} \bar{C}_{HR}}{N_{HR,i}} - \frac{k_{HR}}{\bar{k}_{HR}} \right) \qquad (6)$$

which is our eqn (18) in the original paper.

We can now estimate the thickness, λ_r, using the following typical values:

$$D_{HR} = 10^{-9} \, m^2 \, s^{-1}, \qquad P_{HR} = 10^4, \qquad \bar{C}_{HR} = 0.1 \, kmol \, m^{-3}$$

$$N_{HR,i} = 10^{-8} \, kmol \, m^{-2} \, s^{-1}, \qquad k_{HR} = 10^{-5} \, m \, s^{-1}$$

$$\lambda_r = \frac{3 \times 10^{-9}}{10^4} \left(\frac{10^{-1}}{10^{-8}} - \frac{1}{10^{-5}} \right) \approx 3 \times 10^{-6} \, m.$$

If the 'solubility' of the extractant in the aqueous phase is very low and the extraction rate is high even at low extractant concentrations then a very small thickness is predicted. Thus in the case of a substituted 8-hydroxyquinoline, used for copper extraction, it was found that $P_{HR} \approx 10^5$. $N_{HR,i} = 10^{-8} \, kmol \, m^{-2} \, s^{-1}$ and $\bar{C}_{HR} = 0.02 \, kmol \, m^{-3}$.

Then

$$\lambda_r = \frac{3 \times 10^{-9}}{10^5} \left(\frac{10^{-2}}{10^{-8}} - \frac{1}{10^{-5}} \right) \approx 3 \times 10^{-8} \, m.$$

Prof. W. Nitsch (*Technische Universität München*) asked: Can the model be altered to incorporate a surface reaction which occurs simultaneously with a reaction in the zone?

Dr. M. A. Hughes (*University of Bradford*) replied: The model could be extended to incorporate a reaction occurring at an interface simultaneously with the reaction taking place in the reaction zone. However, this would require additional parameters in the model and these parameters are difficult to estimate from the experimental results. Thus a surface chemical reaction rate and surface concentrations are required. We are not sure if it is feasible to determine directly what reacts at the interface and what reacts in the film.

Prof. M. M. Kreevoy (*University of Minnesota*) said: Can the model accommodate the observation in the stripping of copper where the flux is independent of chelate concentration at the interface?

Dr M. A. Hughes (*University of Bradford*) answered: We understand this to mean the proposal by yourself and Prof. Albery that CuR_2 accumulates at the interface during the stripping process. It is well known that with certain extractants a product can accumulate at the interface during extraction. Thus we have seen 'films' of copper–LIX complex formed at interfaces in diffusion cells and it is clearly demonstrated (by direct observation) that solid product precipitates at the liquid/liquid interface when oxime is contacted with aqueous copper under high-concentration conditions. In other words the rate of complex formation can exceed the rate of its removal from the interface by diffusion into the organic phase.

It is difficult to see how CuR_2 accumulates at the interface during the stripping process since conditions at the interface are favourable for the reverse reaction, whereby this component is consumed.

The basic formulation of our model in the form of differential equations describing concentration profiles is the same for both extraction and stripping processes. In the case of stripping the aqueous proton concentration is particularly high and the reverse extraction reaction equation can be neglected. The resulting relationships for flux calculations are quite simple in this case.

Dr. A. Steinchen (*Université Libre de Bruxelles, Belgium*) said: I have two questions for Prof. Nitsch.

(1) Could he comment on the instabilities observed in the uranyl nitrate extraction from water to hexane with tributylphosphate at very low stirring rate. Has he observed surface motion with these reactants on non-stirred system (kicking of drops or emulsification) and under what conditions?

(2) When the surface is covered by a surfactant how does one explain the formation of a rigid region in the centre of the interface after a critical threshold of the stirring velocity?

Prof. M. M. Kreevoy (*University of Minnesota*) said: Komasawa et al.[1] have reported a solubility in water of 3×10^{-6} mol dm^{-3} for 2-oximinobenzoyl-4-nonylphenol (LIX-65N).[2] This implies a much larger distribution coefficient than the one reported in Prof. Nitsch's table: 10^7 might be a good guess, since LIX-65N is very soluble in most organic solvents. This observation strengthens the argument.

[1] I. Komasawa, T. Otake and A. Yamada, *J. Chem. Eng. Jpn*, 1980, **13**, 130.
[2] LIX is a registered trademark belonging to Henkel-America, Inc.

Prof. W. J. Albery (*Imperial College, London*) said: It is interesting that the shapes of the curves in the paper by Barker et al.[1] are similar to those in our work[2] and that the same type of double reciprocal plot is found in their fig. 7 for the variation of flux with substrate concentration. Mr Choudhery and I have also

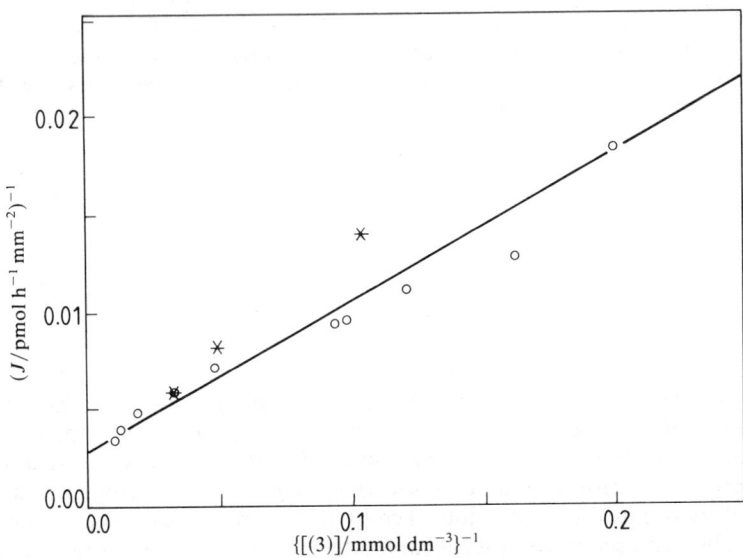

Fig. 4. Variation of flux with substrate concentration for compound (3), N,N-bis(2-hydroxybutyl)octadecylamine.

analysed the data in fig. 4 and 5 of the paper for the variation of flux with carrier concentration according to the same double reciprocal plot. Results for N,N-bis(2-hydroxybutyl)octadecylamine, compound (3), are presented in fig. 4. A reasonable straight line is found with a definite intercept that corresponds to a flux of 300 pmol h^{-1} mm^{-2}. Whereas the intercept in fig. 7 of ref. (1) could be caused by rate limiting transport through the membrane, the intercept in our figure, corresponding to infinite carrier concentration, cannot be caused by such rate-limiting transport. We suggest that it must be caused by rate limiting interfacial kinetics. Does Dr. Hadgraft agree?

[1] N. Barker, J. Hadgraft and P. K. Wotton, *Faraday Discuss. Chem. Soc.*, 1984, **77**, 97.
[2] W. J. Albery, R. A. Choudhery and P. R. Fisk, *Faraday Discuss. Chem. Soc.*, 1984, **77**, 53.

Dr. J. Hadgraft (*University of Nottingham*) replied: I was interested to see that the transport kinetics in Prof. Albery's work contained a significant interfacial barrier. At present we have no conclusive evidence to suggest that our facilitated transfer scheme is limited by a slow interfacial term. However, I agree with Prof. Albery and, in the light of previous work,[1] think that the intercept is caused by rate-limiting interfacial kinetics.

[1] W. J. Albery, J. F. Burke, E. B. Leffler and J. Hadgraft; *J. Chem. Soc., Faraday Trans. 1*, 1976, **72**, 1618.

Dr. D. Leahy (*ICI Pharmaceuticals, Macclesfield*) said: There are a few points that I would like Dr. Hadgraft to elaborate.
(1) Did he attempt any analysis of the intercept terms from $(k^{-1}, w^{-1/2})$ plots? This might give interesting information on the diffusion coefficients and the size of the specie moving within the IPM-saturated membrane support.
(2) Has he any other evidence to support the ion-pair transport mechanism? Could the RBR be partially neutralised as it moves within the membrane, the amine then acting as a proton source?
(3) Has he any thoughts on an explanation for the large differences in rate-enhancing ability between amines of such similar structure?
(4) He mentions that ionic strength changes modify transport rates. I would be grateful for an elaboration of this point.

Dr. J. Hadgraft (*University of Nottingham*) replied:
(1) The intercept in the $(k^{-1}, w^{-1/2})$ plots comprises two terms, diffusion through the membrane and interfacial transfer. In the absence of knowledge concerning the latter it is not possible to calculate the diffusion coefficients. An effective diffusion coefficient could be measured but this would be meaningless.
(2) We have conducted partitioning studies over a range of pH which suggest that the mechanism of transport is governed by ion-pair formation.
(3) We have observed that interfacial kinetics are modified in the presence of phospholipids such as distearoylphosphatidylcholine. Small changes in the alkyl chain length alter the magnitude of the interfacial kinetic term.[1] The exact mechanism for this is not understood.
(4) At an early stage we found that ionic strength altered the transport rates. However, we did not investigate this further. We chose in the experiments to maintain a constant strength.

[1] M. Ahmed, J. Hadgraft and I. W. Kellaway, *Int. J. Pharm.*, 1983, **13**, 227.

Dr. B. H. Robinson (*University of Kent*) asked: What is the thickness of the IPM lipid layer and has Dr. Hadgraft any idea of the concentrations in the membrane?

150

Is the permittivity of IPM such that dissociation of ion pairs or proton transfer will occur.

Dr J. Hadgraft (*University of Nottingham*) replied: The thickness of IPM layer is 150 μm. We have not attempted to calculate absolute concentrations of the ion pairs in the membrane. In view of the complex sets of equilibria possible which are not fully understood it would be difficult to do at this stage. The permittivity of IPM is such that we would not expect ion-pair dissociation to be significant.

Dr. M. Spiro (*Imperial College, London*) said: In discussing the transport-controlled rates obtained in high-concentration sulphuric acid media, Crooks and Chisholm concluded that diffusion is faster than predicted for stagnant conditions, with stirring taking place within $\frac{1}{2}$–1 μm of the liquid/liquid interface.

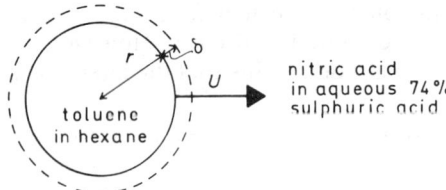

Fig. 5. Schematic representation of toluene droplets moving through the sulphuric acid medium.

I would like to show that this is what would be expected from a hydrodynamic analysis of the situation and that there is no need to invoke Marangoni effects. In a medium of *ca.* 74% sulphuric acid at 25 °C (see fig. 5), where the viscosity, η, is *ca.* 125×10^{-4} kg m^{-1} s^{-1},[1] the Stokes–Einstein equation gives the diffusion coefficient D of toluene as 0.70×10^{-10} m^2 s^{-1}. In this medium the authors dispersed droplets of hexane containing toluene and emulsifying agent, and I estimate their density ρ as *ca.* 0.76 g cm^{-3}. The mean radii r of the droplets is given as 1.21 μm in 73.6% H$_2$SO$_4$ (fig. 1 of the paper) and as 2.00 μm in 74.8% H$_2$SO$_4$ (fig. 2). The thermal velocity of these spheres is therefore

$$U = (3kT/m)^{1/2} = (9kT/4\pi\rho r^3)^{1/2} = \begin{cases} 1.48 \times 10^{-3} \text{ m s}^{-1} & (\text{for } r = 1.21 \ \mu\text{m}) \\ 0.70 \times 10^{-3} \text{ m s}^{-1} & (\text{for } r = 2.00 \ \mu\text{m}) \end{cases}$$

where m is their mass, k is Boltzmann's constant and T is the absolute temperature. That their motion is streamlined and not turbulent is shown by the small size of the Reynolds number (*ca.* 2×10^{-4}). To obtain the diffusion layer thickness we have first to evaluate two dimensionless hydrodynamic parameters for this system. The first is the Péclet number, given by

$$Pe = rU/D = \begin{cases} 25.6 & (\text{for } r = 1.21 \ \mu\text{m}) \\ 19.9 & (\text{for } r = 2.00 \ \mu\text{m}). \end{cases}$$

The second parameter, the Nusselt number Nu, follows from the equation of Brian and Hales[2]

$$Nu = \sqrt{(1 + 0.4802 \ Pe^{2/3})} = \begin{cases} 2.27 & (\text{for } r = 1.21 \ \mu\text{m}) \\ 2.13 & (\text{for } r = 2.00 \ \mu\text{m}). \end{cases}$$

This leads directly[3] to the effective thickness of the diffusion layer around the spherical droplets:

$$\delta_{\text{eff}} = r/Nu = \begin{cases} 0.53\ \mu m & (\text{for } r = 1.21\ \mu m) \\ 0.94\ \mu m & (\text{for } r = 2.00\ \mu m). \end{cases}$$

Although no allowance has been made here for disturbance by Brownian motion,[4] these values should be of the right magnitude. Both δ_{eff} values are $<1\ \mu m$, which nicely explains the findings of Crooks and Chisholm.

[1] J. H. Ridd, *Adv. Phys. Org. Chem.*, 1978, **16**, 1.
[2] P. L. T. Brian and H. B. Hales, *AIChE J.*, 1969, **15**, 419.
[3] M. Spiro and P. L. Freund, *J. Chem. Soc., Faraday Trans. 1*, 1983, **79**, 1649.
[4] W. B. Russel, *Annu. Rev. Fluid Mech.*, 1981, **13**, 425.

Dr. C. Tondre (*University of Nancy, France*) said: I wish to put two questions concerning the work on kinetics in emulsions presented by Dr. Crooks: (1) He said that the reaction between the aromatic molecule and the nitronium ion, NO_2^+, occurs in the aqueous phase because the last one is insoluble in the organic phase, whereas toluene can partition between the two phases. A comparable situation has been reported to occur for the reaction of oxygen with cobalt(II)-L-histidine complex in perfluorotributylamine emulsion,[1] where the complex is presumably in the water-continuous phase and the oxygen partition between the two phases with a marked preference for the perfluorinated phase. In this case the reaction kinetics are observed to be slower in emulsion in comparison with the homogeneous reaction in water and this result can be simply explained by considering a fast partition equilibrium of oxygen in addition to the reactions taking place in the homogeneous situation. Could Dr. Crooks say how his nitration reaction compares with the corresponding homogeneous reaction and would it agree with the preceding explanation? (2) Would Dr. Crooks think that changing the emulsifier in his experiments (the non-ionic Brij's) to an anionic surfactant would increase the rate of reaction owing to the expected increase of the local concentration of NO_2^+ around the electrically charged emulsion droplets?

[1] A. Berthod and J. Georges, *Anal. Chim. Acta*, 1983, **147**, 41.

Dr. J. E. Crooks (*King's College, London*) said: In answer to the first question I draw attention to curves (c) in fig. 2 of my paper. These show the computed rates of nitration for an emulsion in which diffusion is infinitely fast, which are the same as for the homogeneous system making due allowance for the partition of toluene between the two phases. If the system were homogeneous, the nitration would be faster still.

In answer to the second question, note that, as shown in fig. 1 of my paper, the reaction zone is quite large in terms of molecular dimensions. I would only expect the increase in NO_2^+ concentration due to adsorption at an anionic interface to extend over a small portion of the zone. The effect would be small unless the effect of the anionic surfactant was to drastically reduce the thickness of the zone. We have no experimental data.

Prof. W. J. Albery and Dr. P. R. Fisk (*Imperial College, London*) (*partly communicated*): We have used the rotating diffusion cell to study the nitration of toluene.

The reaction scheme is as follows:

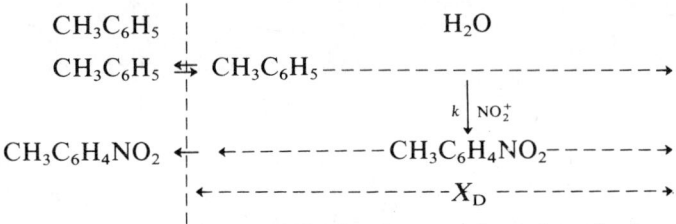

In this scheme X_D is the thickness of the diffusion layer and can be calculated from the Levich equation.[1] We obtain the following expression for the rate of reaction, $J/\text{mol s}^{-1}$:

$$J = ADc_0 \coth(X_D/X_k)/X_k \qquad (1)$$

where

$$X_k = (D/k)^{1/2}$$

and c_0 is the concentration of toluene at the aqueous side of the liquid/liquid interface.

The reaction was followed by measuring the increasing concentration of nitrotoluene in both the organic and the aqueous phases. Typical results for two different acid mixtures are shown in fig. 6 and 7.

For the aqueous phase we find that the number of moles of nitrotoluene, n_{aq}, is given by:

$$n_{aq} = Vc_0[1 - \exp(-ADt/VX_D)]. \qquad (2)$$

The 'rate constant' in the exponential term describes the supply of nitrotoluene across the diffusion layer into the volume, V, of the aqueous compartment. The steady-state concentration of nitrotoluene in the bulk aqueous phase matches that of the toluene at the liquid/liquid interface, c_0. This is because in the steady state the sum of the concentrations of toluene and nitrotoluene must be constant throughout the aqueous phase. In the bulk the concentration of toluene is negligible because it is destroyed by the reaction, and at the interface the concentration of nitrotoluene is negligible because it is being back-extracted into the toluene phase.

The shapes of the aqueous curves in fig. 6 and 7 show the exponential approach to the steady state described in eqn (2). Analysis of these results therefore allows us to determine the diffusion coefficient for nitrotoluene in the aqueous phase and the vital parameter c_0. Unlike other workers we do not have to assume that there is local equilibrium at the interface and that c_0 is simply determined by the partition coefficient. In fact we find that c_0 is close to its equilibrium value.

The number of moles of nitrotoluene in the organic phase n_{org} is given by

$$n_{org} = Jt = n_{aq} \qquad (3)$$

where J is given by eqn (1).

The results in fig. 6 and 7 show that the total number of moles of nitrotoluene increases linearly with time. From the gradient J can be found, and then using eqn (1) and the values of c_0 and D determined from eqn (2) one can find k. Results obtained by this technique are compared with those of other authors[2-5] in fig. 8. Reasonable agreement is found.

The behaviour of n_{org} in fig. 6 and 7 is different. In fig. 3 the nitrating mixture is strong (75% H_2SO_4 by weight) and the rate constant is large enough for $X_k < X_D$.

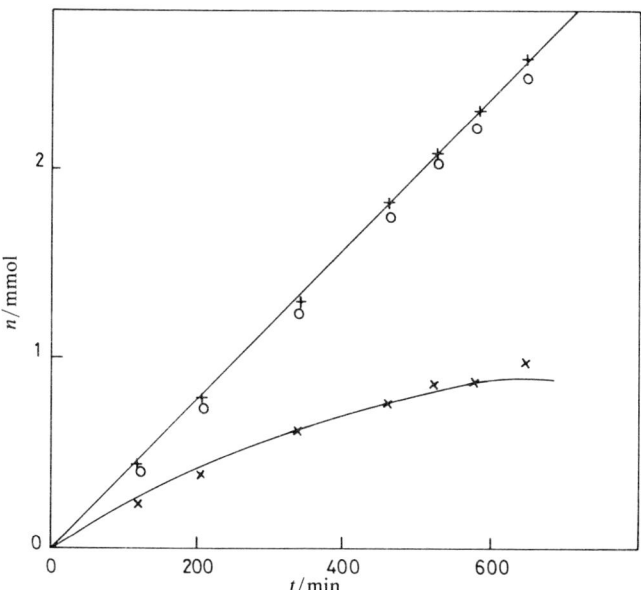

Fig. 6. Variation of n_{org} (○), n_{aq} (×) and n_{total} (+) with time for a nitrating mixture (by weight) of 74.5% H_2SO_4, 1.7% HNO_3 and 23.8% H_2O. Note that $n_{aq} \ll n_{org}$.

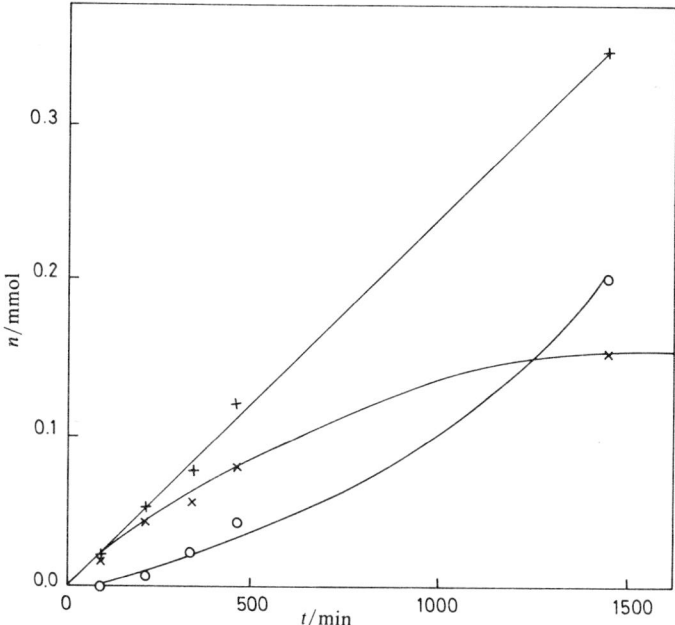

Fig. 7. Variation of n_{org} (○), n_{aq} (×) and n_{total} (+) with time for a nitrating mixture (by weight) of 56.8% H_2SO_4, 2.3% HNO_3 and 40.9% H_2O.

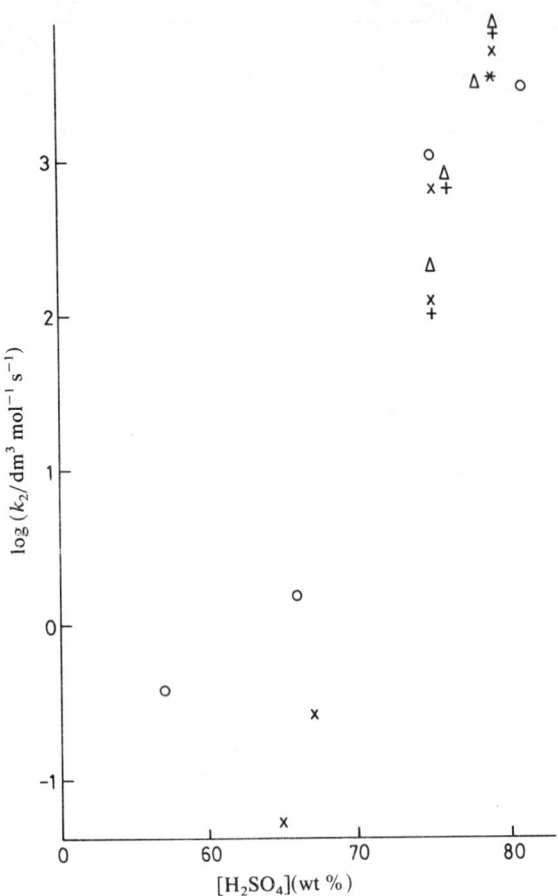

Fig. 8. Comparison of results for the second-order rate constant, k_2, where $k_2 = k/[HNO_3]$ for different nitrating mixtures: $\triangle$, ref. (2); $+$, ref. (3); $\times$, ref. (5) and $\bigcirc$, this work.

Hence the nitrotoluene is made in a thin reaction layer close to the interface. Most of it is therefore back-extracted into the toluene phase so that $n_{org} \gg n_{aq}$. In fig. 7, on the other hand, the nitrating mixture is weaker (57% H_2SO_4 by weight). Here $X_k > X_D$. Much of the reaction takes place in the bulk of the aqueous phase. Hence to start with there is no back-extraction. When the nitrotoluene has built up to its steady-state value in the bulk it diffuses back across the diffusion layer and then appreciable quantities appear in the organic phase. Results such as those in fig. 6 and 7 therefore confirm the accepted view that this reaction takes place in the aqueous phase. The rotating-diffusion-cell technique is particularly powerful in allowing one to control the balance between X_k and X_D so as to obtain the different partition behaviour of the product seen in fig. 6 and 7.

[1] V. G. Levich, *Physicochemical Hydrodynamics* (Prentice-Hall, Englewood Cliffs, N.J., 1962), p. 69.
[2] J. W. Chapman and A. N. Strachan, *J. Chem. Soc., Chem. Commun.*, 1974, 293.
[3] R. B. Moodie, K. Schofield and P. G. Taylor, *J. Chem. Soc., Perkin Trans. 2*, 1979, 134.
[4] G. F. Sheats and A. N. Strachan, *Can. J. Chem.*, 1978, **56**, 1280.
[5] A. N. Strachan, in *Industrial and Laboratory Nitrations*, A.C.S. Symp. Ser. No. 22, ed. L. F. Albright and C. Hanson (A.C.S., Washington D.C., 1976), p. 210.

Dr. B. H. Robinson (*University of Kent*) said: Has Dr. Tondre any idea of the droplet sizes and concentrations in the microemulsion phase and how these vary with composition? At Kent we have made some size measurements using the technique of photon correlation spectroscopy on the $C_{12}E_4 + H_2O$ + alkane system in the single-phase region of the phase diagram and it was found that the droplet sizes varied with water content in a complex way because of surfactant partitioning between the interface and the continuous oil phase.

Dr. Tondre's transport mechanism involves an initial transfer of the metal-ion picrate as an ion-pair into decane. Has he considered the possibility of an interface reaction with the droplet and, in this connection, could I invite him to speculate on what happens to a droplet when it diffuses to the interface? For example, does it open up or bounce back elastically? Does Dr. Tondre think it would be possible to transfer water (or T_2O) between his aqueous compartments?

Is the position of the maximum in fig. 5 of the paper at 5% water independent of the picrate concentration?

Dr. C. Tondre (*University of Nancy, France*): Concerning the first question of Dr. Robinson relative to droplet sizes and concentrations in the microemulsion phase, we unfortunately have little information available for the moment. We have tried to avoid the use of the word 'droplet', because it may be that spherical reversed micelles (as was found for instance with AOT systems) do not exist in the non-ionic systems investigated here, at least for low water contents. Some structural information is available for the system TEGDE + decane + water in the absence of hexanol. The addition of hexanol was found to be necessary in order to obtain Winsor II systems separating into two perfectly clear phases. When there is no hexanol present in the system, neutron scattering data[1,2] have been shown to be consistent with 'hank-like' or 'lamellar' structures incorporating from 20 to 1000 surfactant molecules at low water content, and with oblate ellipsoids incorporating *ca.* 1500 surfactant molecules at high water content (the longer axis of the ellipsoid would be of the order of 200 Å). The free surfactant concentration in the continuous phase was found to decrease rapidly on increasing the water content.[2,3] We do not know what is the effect of hexanol, but our results, on varying the water content of the system, do not seem in contradiction to such structural changes and we have not attempted to interpret these results other than in a very qualitative manner. The theoretical interpretation has been restricted to the case where the water content, and thus the structure of the dispersed phase is fixed.

Assuming then 'hank-like' or 'lamellar' aggregates (for system V, which was used for these experiments, the ratio of the number of water molecules per ethylene oxide group is 2.3, which should be just sufficient to hydrate the surfactant heads) one can speculate on an aggregation number ranging between 100 and 1000, giving, respectively, aggregate concentrations of *ca.* 3×10^{-3} and 3×10^{-4} mol dm^{-3}. If the latter were to be true the assumption made in our model that there is one alkali-metal picrate molecule by aggregate (or droplet?) would become questionable, but how could we understand then the linear dependence shown in fig. 6 of our paper?

I come now to the second question concerning the transport mechanism that we have postulated. Adopting an interfacial reaction between the metal-ion picrate and the droplet, instead of a reaction involving the transfer of the picrate as an ion-pair into decane, would not change the form of the flux equation as long as both mechanisms are considered to be fast compared with diffusion [see ref. (4) by Wong *et al.* in the paper under discussion]. The problem in the case of direct transfer is to define the correct boundary conditions in order to derive the dependence of the flux with the initial picrate concentration. In fact, considering that the interface is probably constituted of a layer of surfactant molecules, the situation is very similar to what happens when transfering pyrene or pyrene derivatives between two neutral vesicles. The solubility of pyrene in water is very poor, as is the solubility of picrates in decane. It seems nevertheless well established[4] that the transfer always occurs through the aqueous phase.

It is not easy to speculate on what happens to a droplet when it diffuses to the interface, but because there probably exists a layer of surfactant molecules at the interface I cannot imagine how the droplet could easily come in contact with the water phase in order to deliver some quantity of water or conversely to pick up a certain amount of water which could include picrate ions. On the other hand, the dynamic nature of these systems may be favoured by the presence of hexanol and perhaps should we not neglect the possibility of formation of 'holes' in the surfactant layer at the interface if Marangoni effects as described in Dr. Nakache's paper can take place?

I would think that the transfer of water between the two aqueous compartments is certainly possible, but checking it with tritiated water does not appear to be easy to do for different experimental reasons: (i) the simple fact of introducing T_2O in one compartment only will create a concentration gradient which initiates the transport by itself; (ii) in the present state of our experimental set-up it is not easy to devise a way of measuring the radioactivity, unless extremely small volume samples can be used.

Unfortunately we do not have enough results to give a definite answer to the third question concerning the position of the maximum when changing the picrate concentration. The transport experiments are not easy to perform and the effect of the water content was investigated at only one picrate concentration. Nevertheless we think that the position of the flux maximum is probably characteristic of the microemulsion structure and not of the picrate concentration: ^{13}C-n.m.r. relaxation times of the surfactant head groups also show a characteristic change at 5% water.[5]

[1] J. C. Ravey and M. Buzier, in *Surfactants in Solution*, ed. K. L. Mittal (Plenum Press, New York, 1983).
[2] J. C. Ravey, M. Buzier and C. Picot, *J. Colloid Interface Sci.*, 1984, **97**, 9.
[3] M. Buzier, *Thesis* (University of Nancy I, 1984).
[4] H. J. Pownall, D. L. Hickson and L. C. Smith, *J. Am. Chem. Soc.*, 1983, **105**, 2440.
[5] A. Xenakis, *Thesis* (University of Nancy I, 1983); C. Tondre, A. Xenakis, A. Robert and G. Serratrice, to be published.

Faraday Discuss. Chem. Soc., 1984, 77, 157-168

The Variational Principles of Onsager and Prigogine in Membrane Transport

BY GERHARD DICKEL

Institut für physikalische Chemie der Universität München, Sophienstraße 11,
8000 München 2, Federal Republic of Germany

Received 28th November, 1983

Whilst Onsager's principle of least dissipation of energy and Prigogine's principle of minimal entropy production both refer to the extremum value of a space integral, Hamilton's principle refers to a definite time integral. With the help of cyclic variables, Onsager's theorem can be transformed into a space-time variational problem. Using a method of Hilbert, a field of extremals can be obtained, composed of potential gradients μ_i and the coordinated geodesic slopes j_i. Only these field quantities furnish the validity of linear thermodynamics; the choice of arbitrary fluxes and forces, however, requires the introduction of excess terms resulting from a theorem of Weierstrass.

1. INTRODUCTION

In his first paper concerning reciprocal relations[1] Onsager pointed out that they may be understood as a consequence of Rayleigh's principle of least dissipation of energy.[2] Later, following Prigogine's recognition of the principle of minimal production of entropy,[3] Ono made attempts to clarify the relationship between these different principles.[4] Ono's conclusions provoked a widespread reaction, and finally Gyarmati[5] showed that Prigogine's principle is not independent, but rather a special case of Onsager's principle, valid for stationary states only. The energy of dissipation E can be obtained from an expression of the form

$$d^2 E / dV \, dt = \sum_{i=1}^{f} X_i J_i \qquad (1)$$

where X_i are the forces and J_i the conjugated fluxes.[1] In order to obtain further statements concerning the behaviour of such a system, further presumptions are necessary. Assuming a linear relation between fluxes J_i and forces X_i given by

$$J_i = \sum_{i=1}^{f} L_{ik} X_k \qquad (i = 1, 2, \ldots, f) \qquad (2)$$

where L_{ik} are indetermined coefficients, and introducing this into eqn (1), we obtain

$$d^2 E / dV \, dt \equiv \sigma = \sum_{i,k=1}^{f} L_{ik} X_i X_k. \qquad (3)$$

σ, called the local entropy production, is a quadratic expression of the forces X_i. Analogously, σ can be expressed by a quadratic expression of the fluxes.

A simple example concerning the diffusion of a binary gas mixture in the presence of a temperature gradient should demonstrate the principle of minimal entropy production as used by Prigogine. In this case the boundary conditions of the force

X_i and X_2 are given by

$$X_1 \text{ (fixed force)} = \Delta T = \text{constant}; \quad X_2 \text{ (free force)} = \Delta c \text{ (unknown)}. \quad (4)$$

Therefore eqn (1) and (2) can be differentiated with respect to X_2 but not X_1. The minimum condition with respect to the unique variable force X_2, given by Δc in eqn (3), i.e.

$$\frac{d\sigma}{dX_2} = 0 \quad (5)$$

yields, considering eqn (3),

$$(L_{12} + L_{21})X_1 + 2L_{22}X_2 = 0 \quad (6)$$

and further, regarding $L_{12} = L_{21}$ and eqn (2),

$$J_2 = 0. \quad (7)$$

This means that in the minimum of entropy production the diffusion fluxes vanish.

In this case we speak of free boundary conditions, because in contrast to ΔT the demixing effect Δc resulting from an unknown, 'kinetic' potential is not fixed by boundary conditions. Therefore Δc can be eliminated by differentiation. Any statement, however, concerning the demixing effect Δc itself cannot be obtained in this way.

Free boundary conditions, playing an important role in open systems, can be taken into account using Hamilton's theory of variation of the endpoint.[6] Hamilton's 'eiconal' is the base and the beginning of the potential theory. The term 'potential' was coined by Gauss later. Independent of this idea, Weierstrass founded the theory of fields of extremals by introducing the concept of field into the calculus of variations. Finally, this idea leads to Hilbert's independence theorem, which is the core of our considerations in the next section.

With the help of the boundary conditions resulting from the rigorous causal connection between forces and fluxes (potentials and extremals) in the general field theory, we turn in section 3 to an open system whose interfacial kinetics is ruled by free boundary conditions. The application of Routh's method will be successful in solving this problem.

In section 4 this theory is applied to a membrane problem (isotonic osmosis) investigated earlier. Finally, in section 5 the difference between an arrangement involving arbitrary forces and fluxes, as used in the thermodynamics of irreversible processes, and an arrangement using the forces and fluxes resulting from the theory of fields of extremals is discussed.

2. THE THEORY OF FIELDS OF EXTREMALS

The calculus of variations starts out from a definite integral. According to a statement of Gyarmati[7] the 'integral principle of thermodynamics refers to the stationary (extremum) value of a space integral, whereas the Hamilton principle refers to a finite time integral'. Fig. 1 explains the latter principle. Assuming Δt is the interval, if a ray runs from A to B along the straight line in fig. 1, we ask how the air–lens interface must be formed in order that all other rays, represented by broken lines, need the same Δt. This time law is not immediately evident using geometrical optics. Similar difficulties arise in the case of other time integrals, e.g. the principle of least action. The success, however, also suggests the introduction

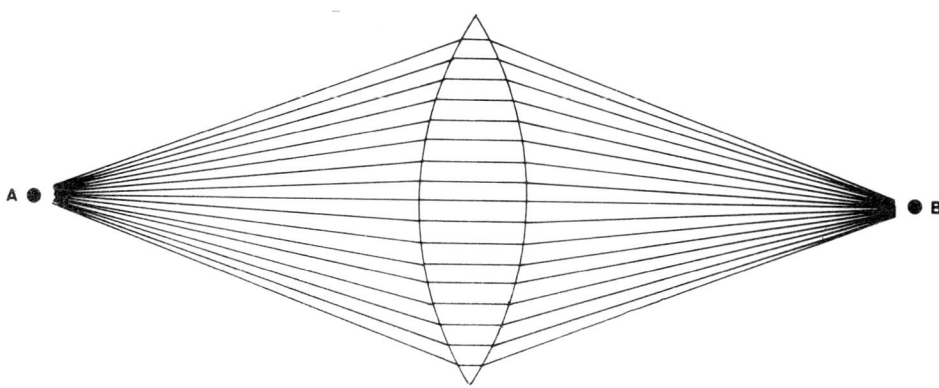

Fig. 1. Hamilton's principle of a system of rays. The condition that all rays pass from A to B in the same interval Δt can be fulfilled by varying the air/glass (lens) interface.

of a Hamiltonian form into thermodynamic transport problems. This can be performed with the help of a cyclic variable, a term coined by Helmholtz.[8]

A cyclic variable can be obtained by replacing fluxes by the number of objects or particles passing a fixed line or a cross-section. For example, a stream of cars will always be expressed by the number of cars passing a fixed line in unit time. Analogously a flux J_i of particles of type i will be measured by the number of moles dn_i/dt passing a cross-section $q = 1$ in unit time. The quantity $\dot{n}_i = dn_i/dt$, rather than J_i, is the correct variable in a variational problem. $\dot{n}_i$ is called a cyclic variable if the integrand of the variational problem depends on $\dot{n}_i$ but not explicitly on n_i.

To replace J_i in eqn (1) by a cyclic variable, let us start from a frictional model where N types of particle i are moving with a velocity v_i relative to the flux of the solvent, j_L. We obtain

$$d^2 E = \sum_{i=1}^{N} K_i \, dS_i = \sum_{i=1}^{N} f_i(v_i - v_L) \, dn_i(v_i - v_L) \, dt. \tag{8}$$

f_i is the frictional coefficient between the particle i and the solvent L. Considering

$$dn_i = c_i q \, dx \quad \text{and} \quad \dot{n}_i = c_i q v_i \quad (i = 1, 2, \ldots, N, L) \tag{9}$$

we obtain after integration of eqn (8) the integral form of eqn (1):

$$E = \int_0^{\Delta t} \int_0^{\Delta x} dL \, dt \tag{10}$$

where

$$dL(f_i, c_i, \dot{n}_i) = \sum_{i=1}^{N} f_i / c_i (\dot{n}_i/q - \dot{n}_L c_i / q c_L)^2 q \, dx \tag{11}$$

is the Lagrangian in the volume $dV = q \, dx$ and $\dot{n}_i$ represent cyclic variables. The remark that dL is a homogeneous function of second degree with respect to $\dot{n}_i$ and $\dot{n}_L$ and of degree -1 with respect to c_i is useful, because it results in essential statements to follow. E represents the energy dissipated in order to maintain a stationary state of motion of N types of dissolved particles i and solvent L in a volume $\Delta V = q \, \Delta x$ over the interval Δt.

According to the principle of least dissipation of energy we postulate a minimal value of the definite integral (10). This means the variation

$$\delta_t \delta_x E = \delta_t \delta_x \int_0^{\Delta t} \int_0^{\Delta x} \mathrm{d}L \, \mathrm{d}t = 0. \tag{12}$$

According to Hilbert[9] a solution of the double integral (12) can be obtained in two steps. In the first step we restrict ourselves to a volume element $q \, \mathrm{d}x$ and obtain the variational problem

$$\delta_t \int_0^{\Delta t} \mathrm{d}E = \delta_t \int_0^{\Delta t} \mathrm{d}L(\dot{n}_1, \dot{n}_2, \ldots, \dot{n}_{N_1} \dot{n}_\mathrm{L}) = 0. \tag{13}$$

The right-hand-side integral represents a line integral between the points $t_\mathrm{a} = 0$ and $t_\mathrm{e} = \Delta t$, whose value depends on the functions $n_i(t)$. The solution of a variational problem requires one to choose the function $n_i(t)$ in such a way that integral (13) assumes its minimal value in comparison with those values which would be obtained by using instead of $n_i(t)$ other functions $n_i^*(t)$ with the same values $t_\mathrm{a} = 0$ and $t_\mathrm{e} = \Delta t$. The functions $n_i(t)$, called extremals, can be conceived as integrals of the differential equation

$$\mathrm{d}n_i / q \, \mathrm{d}t = j_i \qquad (i = 1, 2, \ldots, N, \mathrm{L}) \tag{14}$$

where any j_i represent the slope at the point (t, n_i) of the unique extremal of the field of extremals. Generally j_i is called the geodesic slope, in contrast to any arbitrary slope, j_i^*, which does not furnish the extremum eqn (13).

The geodesic slope j_i can be obtained from the differential equation of the slope function. The latter is merely the Euler–Lagrange differential equation. However, we will adopt a different method. Let us assume we have found the geodesic slope. Taking the line integral (13) along the integral curves $n_i(t)$ of the differential equation (14), according to Hilbert's independence theorem we can transform the line integral (13) in an integral independent of the path of integration. Hilbert's theorem is given by[10-12]

$$\int_0^{\Delta s} \mathrm{d}L(\bar{n}_1, \bar{n}_2, \ldots, \bar{n}_\mathrm{L}) \, \mathrm{d}t = \int_0^{\Delta t} \left(\mathrm{d}L(j_1, j_2, \ldots, j_\mathrm{L}) + \sum_{i=1}^L (\dot{n}_i - q j_i)[\mathrm{d}L_{\dot{n}_i}]_{\dot{n}_i = q j_i} \right) \mathrm{d}t. \tag{15}$$

L in the upper limit of the sum means that the sum must be taken over all particles N and the solvent. Whilst the integral on the left-hand side must be taken along the integral curves $\bar{n}_i$ of eqn (14), the right-hand-side integral is independent of the path $n(t)$. An intuitively geometric interpretation of the independent integral will be given at the end of this section.

The right-hand side can be conceived as a Taylor series of $\mathrm{d}L$ in the neighbourhood of the geodesic slopes j_i, where all terms of higher order are zero. The remainder of this series is called the Weierstrass excess function,[11,12] the meaning of which is as follows. Taking the left-hand-side integral in eqn (15) not along the integral curves $\bar{n}_i$ of eqn (14) but along arbitrary paths, the difference between the left-hand-side integral and the independent integral is given by Weierstrass's excess function. This theorem will play a fundamental role in the last section of this paper.

To obtain the connection between Hilbert's theorem and Hamilton's theory let us introduce the canonical variables[12,13] defined by

$$[\mathrm{d}L_{\dot{n}_i}]_{\dot{n}_i = q j_i} = \pi_i q \, \mathrm{d}x \qquad (i = 1, 2, \ldots, N, \mathrm{L}). \tag{16}$$

Bearing in mind that we have restricted, in the first step, the variation with respect to time [eqn (13)] to the volume element $dV = q\,dx$, the canonical variable π depends on the position x of the 'strip dx'. The question now arises as to how the functions $\pi_i(x)$ must be chosen in order to furnish the minimal value of the double integral (12). This will be realized if

$$\tfrac{1}{2}q\pi_i(x) = -d\mu_i(x)/dx \equiv -\mu_i' \quad (i = 1, 2, \ldots, N, L) \qquad (17)$$

where $\mu_i(x)$ is any position function. μ_i' is the geodesic slope with respect to position. Substituting $d\mu_i$ for π_i in eqn (16) we obtain the extremal condition

$$\tfrac{1}{2}[dL_{\dot n_i}]_{\dot n_i = qj_i} = -d\mu_i \quad (i = 1, 2, \ldots, N, L). \qquad (18)$$

The $N+1$ position functions μ_i must be determined later with the help of physical conditions.

Applying Euler's theorem of homogeneous functions to dL on the right-hand side of eqn (15), and taking into account eqn (16) and (17), we obtain the function of state dG in the strip dx

$$dG = \int_0^{\Delta t} \left(\sum_{i=1}^{L} qj_i\, d\mu_i + \sum_{i=1}^{L} \dot n_i \pi_i q\, dx \right) dt. \qquad (19)$$

The second term in the integrand represents a total differential. As the addition of any total differential does not change the variation, we can omit this term.[14] Going over to the double integral and integrating eqn (19) with respect to position and time we obtain, taking into account eqn (14), the familiar equation of state:

$$\Delta G = \int_0^{\Delta x} \int_0^{\Delta t} \sum_{i=1}^{L} qj_i\, \Delta\mu_i = \sum_{i=1}^{L} \Delta n_i\, \Delta\mu_i. \qquad (20)$$

To understand Hilbert's theorem, eqn (15), we note that in the independent integral the three variables t, n_i and π_i must be taken as independent variables for a particle i. The slope j_i, however, must be expressed as a function of π_i, by solving eqn (16) for j_i.[13] This connection between potentials and fluxes is fundamental to the theory of fields of extremals. In contrast, in the left-hand-side line integral a particle i is represented by two variables, t and n_i, and the slope dn_i/dt.

The following interpretation should provide a physical statement of the independent integral in eqn (15). In fig. 2 a field of extremals, consisting of potential lines and lines of slope, is represented. Let us consider the real path of a particle through this field, given by the curved line. According to Hilbert's theorem, this curve represents an arbitrary path $n_i(t)$ in the field of extremals. Decomposing this path into components parallel to the lines of slope and potentials, we obtain the graph represented in fig. 2. Assuming an infinite number of particles and decomposing their paths in the same manner, we obtain lines covering the field of extremals. As energy must only be expended in the direction of the slope and not along the potential lines, it can be obtained from the state of energy of the field of extremals [eqn (20)]. However, the picture of the field of extremals is imaginary, and therefore the geodesic slopes j_i and μ_i' are not physical realities which can be introduced *a priori*.

Any theory which starts from fluxes and potentials should anticipate a result which follows *a posteriori* from the variational principle, eqn (12). This principle is based on the introduction of the cyclic variables $\dot n_i$ and the canonical variables π_i. However, the geodesic slopes j_i and μ_i' are the resulting extremals. The above-mentioned statement of Gyaramati[7] concerning the integral principles of

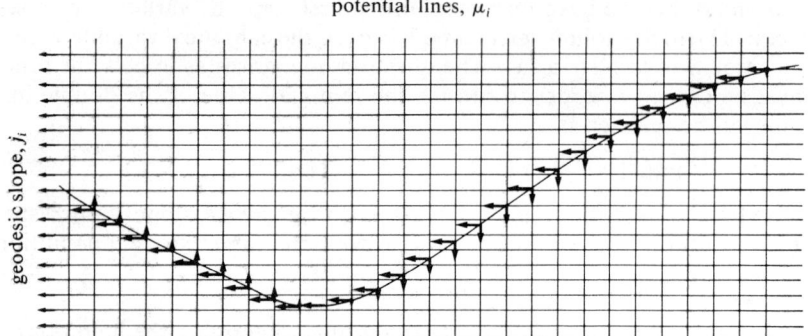

Fig. 2. The independent integral. The curve represents an arbitrary path of a single particle in the field of extremals, given by the geodesic slope j_i and the equipotential lines μ_i. Resolving the motion into components parallel to these lines, and regarding that energy must be expanded only in the direction of j_i, the independence of the path results.

thermodynamics represents a sharp-witted and correct analysis, stating that Hamilton's principle involving a definite time integral must be excluded if the geodesic slopes j_i are introduced *a priori*. In this case the rigorous causal connection between j_i and μ_i arising from the theory of extremals will be missed.

3. BOUNDARY CONDITIONS IN THE THEORY OF FIELDS OF EXTREMALS

Emphasizing that it is our task to take into account open systems, let us start from a strip dx bordering different media. An example is a permeable membrane separating two solutions. Assuming small concentration differences we apply the Erdmann–Weierstrass corner conditions[11,12,15] to the Lagrangian dL. We obtain

$$\left[dL - \sum_{i=1}^{L} \dot{n}_i \, dL_{\dot{n}_i} \right]^{I}_{\dot{n}_i = qj_i} = \left[dL - \sum_{i=1}^{L} \dot{n}_i \, dL_{\dot{n}_i} \right]^{II}_{\dot{n}_i = qj_i} \qquad (21)$$

$$[dL_{\dot{n}_i}]^{I}_{\dot{n}_i = qj_i} = [dL_{\dot{n}_i}]^{II}_{\dot{n}_i = qj_i} \qquad (i = 1, 2, \ldots, L). \qquad (22)$$

Eqn (22) yields, considering eqn (18),

$$d\mu_i^{I} = d\mu_i^{II} \quad \text{and} \quad \mu_i'^{I} = \mu_i'^{II} \qquad (i = 1, 2, \ldots, L). \qquad (23)$$

Regarding Euler's theorem for homogeneous functions, eqn (21) yields, considering eqn (18) and (23),

$$j_i^{I} = j_i^{II} \qquad (i = 1, 2, \ldots, L). \qquad (24)$$

By integrating eqn (23) from $x = 0$ to $x = \Delta x$ and choosing $\mu(x_0)^{I} = \mu(x_0)^{II}$, we obtain at any arbitrary position Δx

$$\mu_i^{I} = \mu_i^{II} \qquad (i = 1, 2, \ldots, L). \qquad (25)$$

These conditions, well known from thermodynamics, must be used in the case of a single phase boundary I/II.

Going over from an open system to a closed one, all fluxes vanish and the potential conditions given by eqn (25) remain. It therefore follows that in both closed and open systems the boundary conditions are determined by the same potentials. In addition, in an open system the conditions concerning the geodesic slope, j_i, given by eqn (24) apply. Because the values of these fluxes are unknown, we speak of free boundary conditions in open systems. The following example should illustrate the problem of free boundary conditions.

Fixing a chain at two points A and B we obtain a catenary. This function results from a variational problem by postulating the minimum value of the potential energy of the free hanging chain and will be given by a function $C(x, y, A, B)$. Fixing point A and varying $B(x, y)$ we obtain a family of catenaries with $B(x, y)$ as a parameter. Free boundary conditions can be obtained by providing the B-end of the chain with a roller, moving freely on a rail, whose slope is given by $y' = dy/dx$. In this case all points B under consideration lie on a curve of slope y'. In this case the catenary is given by $C = C'[x, y, A, B(y', x_e)]$, which represents a one-parameter family of curves (catenaries). This leads to the problem of determining x_e, lying on the rail, in such a way that the minimal value of the potential energy is furnished.

These considerations show that in our case the fluxes j_i in the boundary conditions (21) and (22) must be taken as parameters whose values must be determined in such a way that the minimum value of eqn (13) is furnished. It is convenient to differentiate dL with respect to $\dot{n}_i$, whilst taking all other fluxes as constant (partial differentiation). This, however, would be unrealistic, as in thermodynamics boundary conditions are given by potentials. These potentials can be taken as constant immediately and the fluxes only indirectly. Therefore the variation must be performed in such a way that all potentials are fixed, with the exception of the potential connected with the parameter being varied. This method was developed by Routh.[16]

To go over from the Lagrangian to the Routhian we apply Euler's theorem of homogeneous functions to the Lagrangian and obtain

$$dL = \frac{1}{2} \sum_{i=1}^{L} \dot{n}_i \, dL_{\dot{n}_i}. \tag{26}$$

Introducing the canonical variables of eqn (18) we obtain the Hamiltonian:

$$dH = \sum_{i=1}^{L} [\dot{n}_i \, d\mu_i]_{\dot{n}_i = qj_i}. \tag{27}$$

In order to perform the variation of the flux of the solvent we go over to the Routhian:

$$dR_L = \sum_{i=1}^{N} -[\dot{n}_i \, d\mu_i]_{\dot{n}_i = qj_i} + \frac{1}{2} \dot{n}_L \, dL_{\dot{n}_L} \tag{28}$$

by restricting the introduction of the canonical variables to $i = 1, 2, \ldots, N$. This Routhian represents a Lagrangian with respect to the solvent and a Hamiltonian with respect to the solute, or generally the sum of Hamiltonian potentials and a 'kinetic potential'. According to this transformation we consider instead of the Lagrangian problem (13) the Routhian problem

$$\delta \int_0^{\Delta t} dR_L(\mu_1, \mu_2, \ldots, \mu_N, \dot{n}_L) \, dt = 0. \tag{29}$$

We conceive immediately that the Routhian fulfils the above-mentioned condition that the variation of the flux $\dot{n}_i$ must be performed by fixing the values of the potentials of the dissolved particles.

In the same manner we get the Routhians

$$R_k(\mu_1, \mu_2, \ldots, \mu_{k-1}, \mu_{k+1}, \ldots, \mu_L, \dot{n}_k) = \sum_{i=1}^{L,k} -\dot{n}_i \, d\mu_i + \Delta \tfrac{1}{2}\dot{n}_k \dot{n}_k \, dL_{\dot{n}_k}$$

$$(k = 1, 2, \ldots, N). \quad (30)$$

The canonical variables of the Routhians (28) and (30) are given analogously to eqn (16) by

$$[(dR_i)_{\dot{n}_i \dot{n}_i}]_{\dot{n}_i = qj_i} = \pi_i q \, dx \quad (i = 1, 2, \ldots, N, L). \quad (31)$$

To obtain the connection between eqn (16) and (31) the following method, demonstrated with R_L, will be successful. Introducing eqn (28) into eqn (31), we obtain

$$\tfrac{1}{2}[dL_{\dot{n}_L} + \dot{n}_L \, dL_{\dot{n}_L \dot{n}_L}]_{\dot{n}_L = qj_L} = \pi_L q \, dx. \quad (32)$$

Differentiating eqn (26) with respect to $\dot{n}_L$ we get

$$dL_{\dot{n}_L} = \dot{n}_L \, dL_{\dot{n}_L \dot{n}_L} + \sum_{i=1}^{N} \dot{n}_i \, dL_{\dot{n}_i \dot{n}_L}. \quad (33)$$

Solving eqn (33) for $\dot{n}_L \, dL_{\dot{n}_L \dot{n}_L}$ and introducing this expression into eqn (32) we obtain

$$\left[dL_{\dot{n}_L} - \tfrac{1}{2} \sum_{i=1}^{N} \dot{n}_i \, dL_{\dot{n}_i \dot{n}_L} \right]_{\dot{n}_L = qj_L} = \pi_L q \, dx \quad (34)$$

or considering eqn (17)

$$\left[\tfrac{1}{2} dL_{\dot{n}_L} - \tfrac{1}{4} \sum_{i=1}^{N} \dot{n}_i \, dL_{\dot{n}_i \dot{n}_L} \right]_{\dot{n}_L = qj_L} = -d\mu_L. \quad (35)$$

The application of the same method to R_i yields

$$[\tfrac{1}{2} dL_{\dot{n}_i} - \tfrac{1}{4} \dot{n}_L \, dL_{\dot{n}_i \dot{n}_L}]_{\dot{n}_i = qj_i} = -d\mu_i. \quad (36)$$

PHYSICAL CONDITIONS

Whilst the Lagrangian dL involves a concrete physical statement, given by the right-hand side of eqn (11), the potentials resulting from a spatial variation of the formally introduced canonical variables represent undetermined physical quantities. Bearing in mind that the system of fluxes resulting from eqn (35) and (36) by the introduction of eqn (11), must satisfy general physical conditions, we have to find properties which must be fulfilled by these potentials. Two conditions should be taken into account.

(a) THE INERTIAL SYSTEM. Regarding a stationary state of fluxes through a strip dx, for example a membrane in any medium, the strip dx (membrane) must be at rest in the absence of a pressure difference. From this it follows that the sum of all forces K_i in the strip vanishes:

$$\sum K_i = 0. \quad (37)$$

(b) THE FRAME OF REFERENCE. If any particle k of a number of $N+1$ particles is taken as the frame of reference we postulate that j_k can assume any arbitrary constant value:

$$j_k = \text{constant}. \tag{38}$$

Regarding a closed system, e.g. a solution in a bottle, any variation in the fluxes at the boundaries must be excluded and we obtain

$$dL_{\dot{n}_i \dot{n}_L} = 0. \tag{39}$$

In this case eqn (35) and (36) go over into eqn (18). By introducing eqn (11) into eqn (18) and eliminating the constant flux j_L postulated by condition (b) we obtain the system of fluxes

$$\sum_{i=1}^{k} L_{ik} j_k = K_i \quad (i = 1, 2, \ldots, N) \tag{40}$$

where $L_{ik} = L_{ki} = f_i f_k$, if $i \neq k$, and K_i is a function of the potentials μ_i and μ_L.

In order to apply condition (a) we go over to specific forces by multiplying each side of eqn (18) by c_i. From eqn (37) we obtain

$$\tfrac{1}{2} \sum_{i=1}^{L} c_i \, dL_{\dot{n}_i} = \sum_{i=1}^{L} c_i \, d\mu_i = 0. \tag{41}$$

The zero on the right-hand side results from the fact that $\sum c_i \, dL_{\dot{n}_i}$ is a function of zero degree in c_i. As eqn (41) represents the Gibbs–Duhem equation we see, considering eqn (25), that the potentials resulting from the canonical variable are in agreement with the Gibbs potentials. We may therefore write $d\mu_i^\circ$ instead of $d\mu_i$.

A system of fluxes and forces of the form of eqn (40) cannot be applied to an open system, e.g. to transport through a membrane. In the latter case eqn (35) and (36) must be applied, and the force relation (37) yields

$$\tfrac{1}{2} \sum_{i=1}^{L} c_i \, dL_{\dot{n}_i} - \tfrac{1}{4} \dot{n}_L \sum_{i=1}^{N} c_i \, dL_{\dot{n}_i \dot{n}_L} - \tfrac{1}{4} c_L \sum_{i=1}^{N} \dot{n}_i \, dL_{\dot{n}_i \dot{n}_L} - \sum_{i=1}^{L} c_i \, d\mu_i = 0 \tag{42}$$

where $\dot{n}_i = q j_i$ must be considered. An application of this relation, however, requires a knowledge of the constraints resulting from the membrane itself. Therefore we will illustrate the application of this theory with a concrete example.

4. ISOTONIC OSMOSIS[14,17]

In a cation-exchange membrane, as used in the following investigations, the matrix of the membrane contains fixed ions consisting of phenylsulphonic acid molecules, whose hydrated ions can be exchanged by other cations, called counterions. Beneath these, in equilibrium with an electrolytic solution, anions are present as a consequence of the Donnan equilibrium. Whilst the fixed ions in the matrix itself represent the frame of reference for the counterions, the frame of reference for the Donnan ions, D, is the solvent. Therefore the variation $dL_{\dot{n}_D \dot{n}_L}$, vanishes and the second and third terms in eqn (42) involve counterions C only.

Forces resulting from the potential differences and the frictional forces are acting on all particles, including the fixed ions F. In order to restrain a flux j_F of the fixed ions, a counterforce arising from the matrix of the membrane must be taken into account in the sum of forces $\sum c_i \, d\mu_i$ in eqn (42). Assuming $c_L \, d\mu_L = c_L \, d\mu_L^\circ + dp$ and $c_F \, d\mu_F = c_F (d\mu_F^\circ + F \, d\psi)$, eqn (42) yields, setting $d\mu_F^\circ = 0$, and regarding eqn

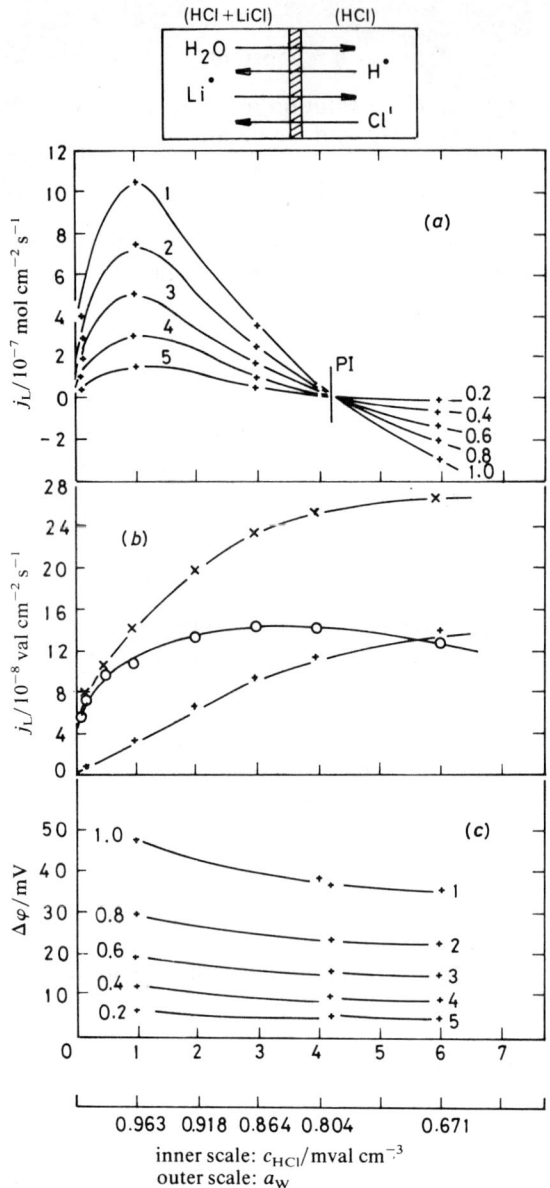

Fig. 3. Isotonic osmosis of the system HCl/HCl+LiCl. Top: The osmotic cell fluxes. Bottom: (a) Flux of water, j_L, as a function of the mole fraction Li/Li+HCl; (b) fluxes of the ions, j_L: ×, H^+; +, Cl^- and ○, Li^+; (c) electric potential. The abscissa shows the concentration and water activity of the solution. [Reproduced with permission from ref. (18).]

(11) and $\sum c_i \, dL_{\dot{n}_i} = 0$ the electromechanical equilibrium[3]

$$\tfrac{1}{2} \sum_{i=1}^{N} f_i j_i \, dx + \tfrac{1}{2} \sum_{i=1}^{N} f_i(c_i/c_L) j_L \, dx + c_F \mathscr{F} \, d\psi + dp = 0. \tag{43}$$

Here c_F is the concentration of fixed ions.

Eqn (43) should be applied in the following to an isotonic solution of $N-2$ univalent counterions, one coion D and a fixed ion. From eqn (11) and (36) it follows in the case of a compensated flux of solvent that

$$\tfrac{1}{2} \sum_{i=1}^{C} f_i j_i \, dx = -\tfrac{1}{2} \sum_{i=1}^{C} c_1 \mathscr{F} \, d\psi + \tfrac{1}{2} c_D \, d\mu_D. \tag{44}$$

The Gibbs–Duhem equation has been taken into account. Finally eqn (43) and (44) yield, assuming $j_L = 0$ and $d\mu_D = 0$,

$$dp = -\tfrac{1}{2}(c_F - c_D)\mathscr{F} \, d\psi. \tag{45}$$

The relation $\sum^{C} c_i = c_F + c_D$ has been taken into account.

Investigations using the isotonic system HCl/(HCl+LiCl) are represented in fig. 3.[18] Since in a solution of HCl+LiCl at any arbitrary concentration a hydrogen ion can be exchanged by a lithium ion without any change in the activity coefficient,[19] $d\mu_D = 0$ can be taken as true. By using different HCl+LiCl mixtures the driving force between the boundaries of the membrane can be varied.

Experiments involving measurements of the electric potentials have confirmed the validity of eqn (45). This represents a linear relation between the electro-osmotic pressure p and the electric potential ψ. The strong variation of the flux of water results from the variation of the coefficient $\tfrac{1}{2}(c_F - c_D)$ with the concentration of Donnan ions. If $c_D = c_F$, an inversion of the flux of water takes place, whilst the electric potential shows normal behaviour. Finally, the linearity between fluxes and forces follows immediately from

$$j_L = -\left[(c_F - c_D) \Big/ \sum_{i=1}^{C} c_i f_i \right] J \, d\psi/dx - \left[2c_L \Big/ \sum_{i=1}^{C} c_i f_i \right] dp/dx \tag{46}$$

which results from eqn (43), taking into account eqn (44) and (45).

Eqn (45) and (46), valid for isotonic solutions involving an arbitrary number N of solvated particles, must also be valid for a single electrolyte. In this case, instead of the diffusion potential occurring in binary isotonic osmosis, an electric field must be applied and so we go over to electro-osmosis. The only difference which was found between isotonic osmosis and electro-osmosis[20] was that in the latter case the flux of water vanishing in the point $c_F = c_D$ does not change sign and we have $j_L = 0$, if $c_D \geq c_F$. Taking into account that a potential assumes its minimum and maximum values at the boundaries, the following is true. Applying in the range of lower concentrations a fixed potential difference ΔE_b to the boundaries, a flux of water arises, generating a streaming potential ΔE_s opposite to ΔE_b. An inversion in the range of higher concentrations would mean an inversion of the streaming potential. In this case, in the membrane a potential difference $\Delta E_i = \Delta E_b + \Delta E_s \geq \Delta E_b$ would result, in contrast to the above-mentioned theorem. In isotonic osmosis, however, an electrochemical rather than an electric potential, involving the streaming potential, is applied to the boundaries.

5. DISCUSSION

A characteristic feature of the theory of fields of extremals is the rigorous connection between the potential gradient and the coordinated geodesic slope. This leads to two methods of obtaining 'the complete figures of the variational problem'.[13] We can start from the Euler–Lagrange equation as in our first paper,[14] or from

Hamilton's canonical variable, as presented here. Comparing the latter representation with the eiconal, the following difference is seen. Whilst in eiconal the velocities of the light are given, and the air–lens interface must be varied in order to satisfy Fermat's principle, in thermodynamics the interface is given and the fluxes must be varied in order to satisfy Onsager's principle. It must be emphasized that the boundary conditions $j_i^I = j_i^{II}$ [eqn (24)] are not trivial. The volume flux as used in membrane theories does not satisfy this condition! Moreover, gradients of electrical and chemical potentials are frequently taken as independent forces in place of the correct electrochemical potentials. Only the latter satisfy the boundary condition $\tilde{\mu}_i^I = \tilde{\mu}_i^{II}$.

Having chosen fluxes and forces arbitrarily, relations of the form $j_i \approx -(\mu' + R)$ must generally be taken into account, where R is the remainder of a Taylor series. This results from Weierstrass's theorem as follows. Assuming that we have found the unique slopes μ_i' and j_i of the field of extremals, R vanishes according to Hilbert's theorem. Taking $n_i^*(t)$, however, along an arbitrary path where $\dot{n}_i^* = qj_i^* \neq qj_i$, the line integral on the left-hand side of eqn (15) does not assume its minimum value and the remainder R must be taken into account. This means that fluxes and forces in linear thermodynamics cannot be chosen arbitrarily.

With regard to the results found with arbitrary arrangements of irreversible thermodynamics,[21] for many transport phenomena the range of approximate linearity extends far beyond equilibrium, whilst for most chemical reactions the linear approximation holds only very close to chemical equilibrium. This result underlines the importance of extending Hamilton's principle to Onsager's theorem of least dissipation of energy, in order to obtain fluxes and forces which furnish linear thermodynamics.

---

[1] L. Onsager, *Phys. Rev.*, 1931, **37**, 405.
[2] Lord Rayleigh, *Proc. Math. Soc. London*, 1873, **363**, 357.
[3] I. Prigogine, *Etude thermodynamique des phénomènes irreversibles* (*Thésis*) (Dunod, Paris and Desoer, Liège, 1947).
[4] S. Ono, *Adv. Chem. Phys.*, 1961, **3**, 267.
[5] I. Gyarmati, *Z. Phys. Chem.* (*Leipzig*), 1967, **234**, 371.
[6] W. R. Hamilton, *Theory of Systems of Rays* (*Irish Transactions*, 1828–1830, **15–17**).
[7] I. Gyarmati, *Non-equilibrium Thermodynamics* (Springer-Verlag, Berlin 1970).
[8] H. V. Helmholtz, *Crelles Journal für Mathematik*, 1884, **97**, 111.
[9] D. Hilbert, *Math. Ann.*, 1906, **62**, 351.
[10] D. Hilbert, *Arch. Math. Phys.*, 1901, **3**, Reihe, Bd. 1, 44.
[11] O. Bolza, *Lectures on the Calculus of Variations* (Chelsea Publishing Co., New York, 1973).
[12] J. C. Clegg, *Calculus of Variations* (Oliver and Boyd, Edinburgh, 1968).
[13] C. Carathedory, *Variationsrechnung*, in *Frank–Mises Differential-gleichungen der Physik I* (Friedrich Vieweg und Sohn, Braunschweig, 1930), chap. V, pp. 227–279.
[14] G. Dickel und G. Backhaus, *J. Chem. Soc., Faraday Trans. 2*, 1978, **74**, 115; 124.
[15] P. Funk, *Variationsrechnung und ihre Anwendung in Physik und Technik* (Springer-Verlag, Berlin 1970).
[16] E. J. Routh, *On the Stability of a Given State of Motion* (Macmillan, London, 5th edn, 1891–92).
[17] G. Dickel and H. Hönig, *Z. Phys. Chem.*, (*N.F.*), 1974, **90**, 198.
[18] R. Kretner, H. Hönig and G. Dickel, *Z. Phys. Chem.* (*Frankfurt am Main*), 1977, **106**, 30.
[19] H. S. Harned and B. B. Owen, *Physical Chemical of Electrolytic Solutions* (Reinhold, New York, 1950).
[20] G. Dickel and R. Kretner, *J. Chem. Soc., Faraday Trans. 2*, 1978, **74**, 2225.
[21] P. Chartier, M. Gross and K. S. Spiegler, *Applications de la thermodynamique du non-équilibre* (Herman, Paris, 1975).

ise. Chem. Soc., 1984, 77, 169-179

Motion Induced by Surface-chemical and Electrochemical Kinetics

By A. Sanfeld* and A. Steinchen

Chimie Physique II, Université Libre de Bruxelles, Campus Plaine, CP 231, Bd. du Triomphe, 1050 Bruxelles, Belgium

Received 30th November, 1983

It is well known that motion at a liquid–liquid interface may be generated by the transfer of matter. The constraint is the difference in concentration profiles between both bulk phases. In the same way, surface-chemical, sorption or electrochemical reactions may also induce convection. The constraint in this case is related to the non-equilibrium kinetic steps. A general theory based on a linear analysis of stability shows the role of mechanical (density, viscosity, interfacial tension, surface elasticity), chemical (kinetic constants, composition) or electrochemical (electrical field, dielectric constant) parameters in the onset of interfacial deformations and movements. The stresses acting on the system are mainly due to Laplace–Kelvin and Marangoni effects. Experimental evidence observed for solvent-extraction reagents are analysed within the framework of our theory.

Interfacial hydrodynamic instabilities leading to deformation and to surface motion have been observed in many fluid–fluid systems.[1,2] They also induce ordered behaviour with spatial and temporal patterns.[2-4]

These phenomena occur in non-equilibrium conditions owing to chemical and physical constraints: diffusional, thermal, mechanical, electrical and chemical effects. The non-linear character of the phenomena responsible for mechanical instability is directly related to the coupling between mechanical, electrical, chemical or thermal processes through the physico-chemical local properties of the interface (boundary conditions).[5,6]

From a fundamental approach, these processes may be reviewed as an example of the concept of dissipative structures developed by the Prigogine–Glansdorff–Nicolis group[7,8] and extended to electromagnetic fields and to interfaces by Steinchen and Sanfeld.[9]

Our purpose is to determine the constraints and the conditions responsible for the onset of surface-mechanical instabilities and their influence on the adjacent phases for single interfaces. Experimental evidence observed in solvent extraction and in spontaneous emulsification is analysed within the framework of our theory.

In order to obtain analytical predictions, we restrict our study to a linear stability analysis of a reference steady state at mechanical rest, at a constant and uniform temperature and under a pure electrostatic approximation.

BASIC RELATIONS

The non-autonomous character of the interface is responsible for propagation in the adjacent bulk phases of the dynamical instability generated in the surface. The interface between two immiscible fluids is a transition region in which the chemical composition and the related physical properties change abruptly. It is usually described by a geometrical surface model with singular surface properties: surface mass density, surface tension, surface charge, surface viscosities and surface diffusion. Out of mechanical equilibrium, the dynamical properties may also be described in terms of singular surface quantities[10] balancing the discontinuity of momentum fluxes from the neighbouring bulb phases.

Motion at a single fluid interface can be modelled by considering the transverse (T) and longitudinal (L) waves coupled by a coupling term C. The stability in then ruled by a general dispersion equation

$$L \times T - C = 0. \qquad (1)$$

The contribution of the transverse (longitudinal) displacement to C, together with the longitudinal mode L (transverse T) is the tangential (normal) stress. Tangential stress leads to Marangoni effects while normal stress is the Laplace–Kelvin generalized laws.[5,6,10–12] They are the main boundary condition acting on the interfacial layer. Other boundary conditions are the surface mass and charge balances and electrical continuity or discontinuity conditions. Finally, a surface-state equation is required for the closure of the system of equations.[5,11,12]

In the incompressible volume phases ($\nabla \cdot v = 0$) we write the following.

(1) Momentum balance:

$$\rho v + \nabla \cdot \mathbf{P} = \mathbf{F} \qquad (2)$$

with the pressure tensor

$$(\mathbf{P})^i_j = p\delta^i_j - \mu(v^i_{,j} + v^j_{,i}) \qquad (3)$$

and the body forces $\mathbf{F}$

$$\mathbf{F} = \rho g + \nabla \cdot \mathbf{T} \qquad (4)$$

with the Maxwell stress tensor $\mathbf{T}$ defined by

$$(\mathbf{T})^i_j = \frac{\varepsilon}{4\pi} E^i E_j - \frac{1}{8\pi} \sum_k E^2_k \delta^i_j. \qquad (5)$$

In this description p is the Kelvin pressure.

(2) Maxwell's equation:

$$\nabla \cdot (\varepsilon E) = 4\pi z \rho \qquad (6)$$

where ρz is the charge density.

(3) Mass balance:

$$\partial_t \rho_\gamma = -\nabla \cdot (\rho_\gamma v) - \nabla \cdot \mathbf{J}_\gamma + R_\gamma \qquad (7)$$

where $\mathbf{J}_\gamma$ are the diffusion-migration fluxes and R_γ the chemical sources.

In the absence of chemical reactions the Fick–Nernst laws read, for molar concentrations,

$$\partial_t C_\gamma = -\nabla C_\gamma \cdot v + D_\gamma \partial \cdot \left(\nabla C_\gamma + \frac{z_\gamma C_\gamma}{RT} \nabla \psi \right). \qquad (8)$$

To describe the dynamics of moving charged and polarized interfaces we have to take into account an electrical double layer composed of (i) a thin region of molecular dimensions (compact layer) containing adsorbed ions and (ii) an external continuous region (diffuse layer) in which adsorption forces are negligible.

The boundary conditions are as follows.
(1) Gauss' equation:

$$\Delta_s(\varepsilon E) \cdot n = 4\pi z \Gamma. \tag{9}$$

The surface charge density $z\Gamma$ is related to the surface concentrations of the ions adsorbed in the compact layer Γ_γ:

$$z\Gamma = \sum_\gamma z_\gamma \Gamma_\gamma. \tag{10}$$

(2) Jump in electrical potential, ψ:

$$\Delta_s \psi = P^s \cdot n \tag{11}$$

where the surface dipole density P^s includes both the contribution from oriented dipole moments of the adsorbed molecules and from the potential drop through the compact layer.

(3) Continuity of velocities in each phase β ($\beta = $ I, II):

$$v^\beta|_s = v^s. \tag{12}$$

(4) Surface momentum balance

$$\Gamma \dot{v}^s = \nabla_s \cdot \pi + F^s + \Delta_s(-P + T) \cdot n \tag{13}$$

where Γ is the total surface mass density, π is the intrinsic surface stress tensor and F^s is the total surface intrinsic body force.

In the horizontal plane eqn (13) is the Marangoni condition while along the normal coordinates z the same relation is the generalized Laplace–Kelvin condition. Along the horizontal coordinates x, y we assume for a two-dimensional Newtonian system:[13–15]

$$(\nabla \cdot \pi)_{\{y\}^x} = (\nabla \sigma)_{\{y\}^x} - \eta_{dil} \nabla_{\{y\}^x} v^s_{z,z} + \eta_{sh} \nabla^2_s v^s_{\{y\}^x} \tag{14}$$

where the phenomenological coefficients η_{dil} and η_{sh} are the intrinsic surface dilational and shear viscosities and σ is the interfacial tension. An analogous equation may be written in the curvilinear coordinates. Assuming the superposition of all contributions to the total force, we get

$$F^s = \Gamma g + F^s_E + F^s_m \tag{15}$$

where Γg is the surface weight, F^s_m is the excess chemical force due to very short-range interactions[16] and F^s_E is the electrical force.[17]

The surface tension, σ, is thermodynamically defined by a mechanical contribution σ_M due to the surface composition and an electrical contribution σ_E due to the influence of the double layer:[17]

$$\sigma_T = \sigma_M - \sigma_E \tag{16}$$

where

$$\sigma_E = \frac{1}{4\pi} \int \varepsilon E^2 (\approx \sqrt{g^*})_{,l} \delta x^l \tag{17}$$

with g^* the determinant of the matrix of the space fundamental tensor and l the coordinate curve of the field lines in the general curvilinear orthogonal coordinates.

(5) Surface mass balance:

$$\dot{\Gamma}_\gamma = -\Gamma_\gamma(\nabla \cdot v^s + a^*) - \nabla_s \cdot J^s_\gamma - \Delta_s\{J^\beta_\gamma\} \cdot n + R^s_\gamma \qquad (18)$$

where $\nabla_s \cdot v^s$ is the surface divergence of the surface velocity v^s, a^* is the change of the surface metric, R^s_γ is the source of surface chemical reactions, $\Delta_s J^\beta_\gamma \cdot n$ accounts for the interchange of mass between the adjacent bulk phases and the surface, J^s_γ is the singular diffusion–migration flux on the surface

$$\nabla_s \cdot J^s_\gamma = D^s_\gamma(\nabla_s \Gamma_\gamma + z_\gamma \Gamma_\gamma \nabla_s \psi) \qquad (19)$$

with D^s_γ the surface diffusion coefficient and Γ_γ the surface concentrations of the adsorbed ions. The sorption fluxes J^β_γ are related to the difference in electrochemical potentials between the surface and the sublayers.

(6) Change of interfacial tension:

$$\delta\sigma = -\varepsilon^s_d(\omega, k)\frac{\delta v^s_{z,z}}{\omega} + k\Psi(\omega, k)\frac{\delta v^s_z}{\omega} = -\sum_\gamma \alpha_\gamma \delta\Gamma_\gamma \qquad (20)$$

where ε^s_d is the dynamical surface elasticity,[11,12] related to the longitudinal displacement Dv^a_z, Ψ is a phenomenological quantity related to the normal displacement, and ω and k are the frequency and wavenumber of the perturbations. The longitudinal displacement is connected to the local variations of surface area A

$$\frac{\delta v^s_{z,z}}{\omega} = -\delta \ln A = \frac{1}{\omega} Dv^s_z. \qquad (21)$$

The phenomenological coefficients ε^s_d and Ψ are related to all the relaxation processes due to mass exchanges, chemical reactions and electrical effects. The coefficients α_γ are directly connected to the surface state equation.[18] For a surface perfect gas, $\alpha_\gamma = RT$.

Restricting our analysis to plane and spherical interfaces, we solved eqn (2)–(5) in terms of velocities along the normal or radial coordinates. Taking into account the boundary conditions [eqn (6)–(20)], we obtain the general characteristic equations eqn (1). Let us now analyse various situations related to the different constraints imposed on the reference state.

RESULTS

As we are mainly interested in surface chemical or electrochemical reactions for single interfaces, we now briefly summarize the results obtained for pure Fickian diffusion and sorption.[6,11,19,20]

PURE DIFFUSION

Let us first consider plane interfaces. A necessary and sufficient condition for the onset of surface monotonous motion is that diffusion of only one species between phase I and phase II occurs from the liquid with the smallest value of D to the fluid with the largest value of D.[21] Surface viscosity has a damping effect and the critical constraint of diffusion increases with viscosity, with surface elasticity and with the diffusion coefficient. Unstable periodic states are reached when

$$\frac{D^{II}}{D^I} < \frac{\nu^{II}}{\nu^I} \text{ with } \nu = \mu/\rho.$$

Moreover, the critical time for the onset of convective cells at the marginal state may be calculated easily from the diffusion coefficient, the composition and the surface viscosity. Our results are in good agreement with the experiment.[3,4,6,11]

For spherical interfaces the results are also analysed in terms of a critical value of the ratio of the diffusion coefficients. New possibilities of oscillatory instabilities appear for the motion *in toto* of a sphere and for a pendant drop. (Interesting predictions are obtained for two diffusion species.[22])

SORPTION KINETICS AND DIFFUSION

Pure diffusion kinetics in the bulk phase does not always account for the onset of motion in the interface, and relaxation mechanisms have to be considered. For example, non-equilibrium sorption processes may occur between sublayers and surface owing to orientation of polar head groups.[17,23-26] We only consider that no matter is accumulated in the sublayers in the reference state as well as in the perturbed state while no matter is accumulated in the surface in the reference state.

Transfer I → II by diffusion–sorption leads to unstable aperiodic regimes for $D^I < D^{II}$; II → I reveals new possibilities of instabilities.[26]

On the other hand, it is predicted that the potential barrier due to controlled sorption kinetics has a stabilizing effect in the oscillating regime.[24]

SURFACE-CHEMICAL REACTIONS AND SORPTION

From eqn (18) we may define the kinetic matrix element $C_{\gamma\beta}$ (surface-chemical reactions and sorption steps):

$$C_{\gamma\beta} = \frac{\partial R_\gamma}{\partial \Gamma_\beta} + \frac{\partial J_\gamma}{\partial \Gamma_\beta} \quad (\gamma =, \neq \beta). \tag{22}$$

The linear stability analysis leads to a general dispersion relation:[5,11,28-30]

$$H(\omega, \nu, k) + kR(\omega, \nu, k)\frac{\varepsilon}{\omega} = 0 \tag{23}$$

where H and R are functions of ω, ν and k. The quantity ε/ω is derived from the basic equations[28,29]

$$\frac{\varepsilon}{\omega} = \frac{\sum_{\gamma=1}^{N} \alpha_\gamma \det \mathbf{L}^{(\gamma)}}{\det \mathbf{L}} \tag{24}$$

where the matrices $\mathbf{L}$ and $\mathbf{L}^{(\gamma)}$ are, respectively, defined by

$$L_{\gamma\beta} = -C_{\gamma\beta} + \delta_{\gamma\beta}^{\text{Kr}}(\omega + k^2 D_\beta^s) \tag{25}$$

$$L_{\gamma\beta}^{(\gamma)} = L_{\gamma\beta} + \delta_{\gamma\beta}^{\text{Kr}}(\Gamma_\gamma^\circ - L_{\gamma\beta}). \tag{26}$$

The matrix $\mathbf{L}$ is only associated with non-convective surface kinetic processes while the matrix $\mathbf{L}^{(\gamma)}$ is closely related to the coupling between the convective and the chemical processes.

The general conditions for mechanochemical surface instabilities may summarized as follows. (1) Equilibrium surface-chemical reactions never induce mechanical instability. The only possibility of surface motion is then a drastic decrease in the interfacial tension, reaching transition zero or negative values due to very active surfactants. (2) For only one fluctuating species the chemical reaction in itself has to be unstable to obtain the onset of surface movements. This is the case for autocatalytic or cross-catalytic mechanisms.[5,11,31] (3) For two (or more) fluctuating species, the conditions are not so drastic. An intrinsically stable chemical reaction coupled with the hydrodynamic process may induce mechanical instability.[5,27,31] (4) A stable chemical reaction may be destabilized by mechanical constraints (for example a difference of densities).[31] (5) At spherical interfaces the onset of motion may lead to local deformations and to translation *in toto*.[27]

It is interesting to analyse the necessary conditions for the mechanochemical instability when: (i) the viscosities are negligible and (ii) the viscosities are taken into account.

(i) This situation means that we are now looking for the unstable condition $\varepsilon/\omega < 0$. Dalle-Vedove and Sanfeld[28–30] obtained simple analytical necessary conditions for two fluctuating species (1, 2).

(a) Non-auto- or non-cross-catalytic chemical steps ($C_{11} < 0$; $C_{22} < 0$; $C_{11}C_{22} - C_{12}C_{21} > 0$). The monotonous convective regime starts when

$$\Gamma_1^o(\alpha_2 C_{21} - \alpha_1 C_{22}^*) + \Gamma_2^o(\alpha_1 C_{12} - \alpha_2 C_{11}^*) < 0 \tag{27}$$

where $C_{jj}^* = C_{jj} - k^2 D_\gamma^s$. The time-oscillating regime starts when ·

$$\Gamma_1^o(\alpha_2 C_{21} - \alpha_1 C_{22}^*) + \Gamma_2^o(\alpha_1 C_{12} - \alpha_2 C_{11}^*) > 0 \tag{28}$$

$$\phi[\Gamma_1^o(\alpha_2 C_{21} + \alpha_1 C_{11}^*) + \Gamma_2^o(\alpha_1 C_{12} + \alpha_2 C_{22}^*)]$$
$$> \Gamma_1^o(\alpha_2 C_{21} - \alpha_1 C_{22}^*) + \Gamma_2^o(\alpha_1 C_{12} - \alpha_2 C_{11}^*) \tag{29}$$

with $\phi > 1$.

(b) Auto- or cross-catalytic chemical steps. An intrinsically unstable aperiodic (periodic) scheme may lead to monotonous (periodic) surface convection. However, oscillatory (aperiodic) solutions are also possible. For stable kinetics ($C_{11} + C_{22} < 0$; $C_{11}C_{22} - C_{12}C_{21} > 0$) the situation is comparable to the non-auto- or cross-catalytic kinetics. For example, oscillatory movements can be induced when conditions (28) and (29) are fullfilled.

(ii) The viscosities have a damping effect when the kinetics is intrinsically stable. They have a damping effect on the oscillatory (aperiodic) mechanochemical regime when the kinetics is monotonously (periodically) unstable. For large viscosities in the bulk phases, the influence of the chemical mechanism is dominant when the periodic (aperiodic) instability is due to chemical unstable periodic (aperiodic) schemes. For small viscosities in the bulk phase it seems possible to stabilize the aperiodic (periodic) mechanochemical regime when the kinetics is aperiodically (periodically) unstable by itself. Several examples have been discussed in previous papers.[28–34]

EXAMPLE OF INSTABILITY IN SOLVENT-EXTRACTION REAGENTS

We now focus attention on experimental observations in solvent-extraction reagents within the framework of our theory. During the experiments interfacial movements, kicking of drops and spontaneous emulsification are observed.[1]

In the liquid–liquid extraction of nickel from acid sulphate aqueous solutions using various mixtures of D2EHPAH and D2EHPANa in toluene, Durrani et al.[35] observed wakes. They suppose that this effect is due to emulsification by the sodium salt.

In the liquid–liquid extraction of copper from acid sulphate solutions using this time various mixtures of D2EHPAH and D2EHPANa in xylene, Dupeyrat and

Nakache[36] observed pendant drops, surface motion, organized cells and emulsification.

In order to analyse such complicated situations we need knowledge of the various steps in the extraction mechanism. This means not only the value of the kinetic constants but also those of the physico-chemical parameters: composition, densities, interfacial tension, viscosities and diffusion coefficients. An interesting speculative discussion on the extraction mechanism reveals the following possible rate-determining steps in the extraction of metals by organic extractants.[37] (1) Diffusion of solvated metal ions and organic solvated extractant (monomers and aggregates) from the bulk phase to the interface, (2) sorption kinetics between sublayers and surface, (3) orientations of the extractant with polar groups towards aqueous phase, (4) solvation–desolvation kinetics at the interface, (5) formation of acid–metal complexes (monomers and aggregate), (6) ionization processes in the interfacial region, (7) expulsion of cations in the aqueous phase, (8) hydration of cations, (9) reorientation of the metal complexes to carry the metal to the organic side of the interface, (10) desorption of the metal complexes or exchange of ligands and (11) diffusion of the metal complexes to the bulk organic phase.*

As an example we analyse a very simple model in the liquid extraction of copper from sulphate solutions of alkyl phosphoric acid (R are reservoirs).

$R_1 \xrightarrow{k_1} U$ (adsorption of copper from a reservoir R_1)

$R_3 \xrightarrow{K_2} 2R_4$ (dimerization equilibrium of the extractant)

$R_4 \underset{k_{-3}}{\overset{k_3}{\rightleftarrows}} Y$ (adsorption and Volmer desorption of the extractant)

$Y \overset{K_4}{\rightleftarrows} X + Z$ (dissociation of the extractant)

$U + Y \underset{k_{-5}}{\overset{k_5}{\rightleftarrows}} I + X$ (first complexation)

$I + Y \underset{k_{-6}}{\overset{k_6}{\rightleftarrows}} M + X$ (second complexation)

$I + Z \overset{K_{6'}}{\rightleftarrows} M$ (ionization equilibrium)

$M \xrightarrow{k_7} R_5$ (Volmer desorption of the complex)

$2R_5 \overset{K_8}{\rightleftarrows} R_6$ (dimerization of the complex)

$X \xrightarrow{k_9} X_a$ (desorption of the proton)

where $R_1 \equiv Cu^{2+}_{aq}$; $U \equiv Cu^{2+}_{interface}$; $R_3 \equiv [(RO)_2-POOH]_{2\,organic}$; $R_4 \equiv [(RO)_2-POOH]_{organic}$; $Y \equiv [(RO)_2-POOH]_{interface}$; $X \equiv H^+_{interface}$; $X_a \equiv H^+_{aq}$; $Z \equiv [(RO)_2-POO^-]_{interface}$; $I \equiv [Cu(RO)_2POO]^+_{interface}$; $M \equiv Cu[(RO)_2POO]_{2\,interface}$; $R_6 \equiv \{Cu[(RO)_2POO]_2\}_{2\,organic}$; $R_5 \equiv Cu[(RO)_2POO]_{2\,organic}$.

---

* Within this domain many important contributions have recently been published by various groups, particularly in the Proceedings of the International Solvent Extractions Conferences 'ISEC' (London 1974, Toronto 1977, Montreal 1979, Liège 1980, Denver 1983); in particular see the papers of Hughes, Hanson, Bauer, Danesi, Yagodin, Cox, Flett *etc.* and their coworkers.

As observed by several authors[38,39] the diffusion step seems not to be always the determining step. We will then assume that only competitions between sorption and chemical reactions may induce convective motion. The steady state is characterized by conservation of fluxes:

$$k_1 R_1 = k_7 M \frac{M_s}{M_s - M} \exp\left(\frac{M}{M_s - M}\right)$$

$$k_3 R_4 = k_{-3} Y \frac{Y_s}{Y_s - Y} \exp\left(\frac{Y}{Y_s - Y}\right)$$

where M_s and Y_s are the saturation concentrations of the complex and of the extractant, respectively. The right-hand side of the two steps is a Volmer desorption step. We further suppose that the time evolution of the fluctuations of X and Y may be considered as quasi-stationary.

The calculation of conditions (27)–(29) shows that the second complexation is responsible for the onset of mechanical instability, although the kinetics is always stable *per se*. For $k_6 > k_{-6}$ an aperiodic convective regime may be predicted. For $k_6 < k_{-6}$ a periodic convective regime is possible. Explicit competitions between step (6) and the desorption (k_7, k_{-3}, k_9), as well as the fluxes $k_3 R_4$ and $k_1 R_1$, may be important. Moreover, the role the parameters α_γ [eqn (20)] is discussed in connection with the surfactant character of the extractant and of the complex.

A much more realistic model would be the competition of the monomers Y and M with the dimers $D \equiv [(RO)_2 POOH]_{2\,\text{interface}}$ and

$$A \equiv Cu[(RO)_2 POO]^{2-}_{4,\text{interface}}; \qquad R_6 \equiv \{Cu[(RO)_2 POO]_2\}_{2,\text{organic}}.$$

For example, we may analyse the previous model using the following additional steps:

$R_3 \rightleftharpoons D$ (sorption of the extractant)

$D + U \rightleftharpoons M + 2X$ (complexation of the dimer with copper)

$M + D \rightleftharpoons A + 2X$ (complexation of the dimer with the monomer)

$A + U \rightleftharpoons UA$ (formation of the aggregate complex)

$UA \rightarrow R_6$ (desorption of the complex dimer).

The appropriate calculations have not yet been carried out.

Finally, we shall study the influence of the interfacial activity and of the viscosities which play an important role on the mechanism of extraction.[36,40–42]

ELECTRICAL AND ELECTROCHEMICAL CONSTRAINTS

All the results have been published in a general survey.[43]

(i) CONTINUOUS MODEL (DILUTE SYSTEMS)

The system considered consists of a charged plane interface, including a compact layer between two immiscible ionic solutions with dielectric constants ε^I and ε^{II}. The surface charge density is compensated by the integral charges of two electrical diffuse layers of thickness $\kappa^{-1} = (\varepsilon RT / 8\pi z^2 C_\infty)^{1/2}$ (with C_∞ the concentration in the bulk phase, $E=0$) extending into the neighbouring phases. The interfacial tension of such systems involves a mechanical term due to the free energy of the electrical double layers[17]

$$\sigma = \sigma_M - \sigma_E \tag{30}$$

with

$$\sigma_E = \int_{-\infty}^{\infty} \frac{\varepsilon E^2}{4\pi} \, dz. \tag{31}$$

Let us consider two situations.

(1) Restored Boltzmann macroscopic distribution in the diffuse layers. The constraint is the jump in the electrochemical potential between both phases, which remains uniform in each phase. From an analysis of the characteristic equation it is shown that only the negative contribution, σ_E, to the total surface tension, σ, may be responsible for the onset of surface motion. The marginal stability condition is then

$$\sigma = 0 \quad \text{or} \quad |\sigma_M| = |\sigma_E|. \tag{32}$$

Moreover, the viscosity increases the wavelength of the fastest rate of growth and reduces the fastest rate of growth; it thus has a stabilizing contribution.

(2) Non-restored Boltzmann distribution in the diffuse layers. In this case the relaxation of diffuse layers ($\kappa D/\lambda$) is of the order of magnitude of the characteristic time of the perturbation (ω^{-1}). The constraint is also the discontinuity of the electrochemical potential, but in the perturbed state this quantity does not remain uniform in each phase. We have restricted our analysis to ideally polarized systems (no net fluxes through the interface) and to uni-univalent electrolytes. Instabilities are obtained even for a non-vanishing total surface tension σ.

For low surface charge the general rules for the stability are

$$\rho^I \mu^I > \rho^{II} \mu^{II} \begin{cases} (\varepsilon^{II} C^{II})^{1/2} D^{II} > (\varepsilon^I C^I)^{1/2} D^I \text{ (stable)} \\ (\varepsilon^{II} C^{II})^{1/2} D^{II} < (\varepsilon^I C^I)^{1/2} D^I \text{ (unstable).} \end{cases} \tag{33}$$

Criterion (33) is in agreement with the experiments of Watanabe et al.[44] on electrical emulsification of the system $H_2O + KCl$ in contact with a solution of sodium dodecylsulphate in methylisobutylketone.

For a small potential drop or large surface charge the general rules are

$$\rho^I \mu^I > \rho^{II} \mu^{II} \begin{cases} \psi^{Is} - \psi^I_\infty > \psi^{IIs} - \psi^{II}_{-\infty} \text{ (stable)} \\ \psi^{Is} - \psi^I_\infty < \psi^{IIs} - \psi^{II}_{-\infty} \text{ (unstable).} \end{cases} \tag{34}$$

The system thus becomes unstable if the phase where the potential drop is largest is also the phase with the smallest $\rho\mu$. These effects could also partially explain the mechanical instabilities observed by Nakache and Dupeyrat.[2]

(ii) DISCRETE ELECTRICAL AND CHEMICAL INTERACTIONS

For large surface charges [e.g. for ionized monolayers spread at an oil (o)/water (w) interface] discreteness-of-charge effects have to be considered. The counterions are then located in an outer Helmholtz plane near the plane of primary charges (the inner Helmholtz plane). These two planes are separated by a layer of strongly oriented water molecules. The constraint is due to the absence of exchange between the inner Helmholtz plane and the solution. When dipoles are spread at an interface, discrete interactions also exist between them. The surface tension for these two types of discrete systems (charged or dipolar layers) also consists of a mechanical part σ_M and an electrical part σ_E due to the electrical interactions between charges or dipoles. Mechanical instabilities are predicted in such systems even for $|\sigma_E| < |\sigma_M|$. The general rules are

$$\varepsilon^w > \varepsilon^0 \begin{cases} \rho^w \mu^w > \rho^0 \mu^0 & \text{(stable)} \\ \rho^w \mu^w = \rho^0 \mu^0 & \text{(marginally stable)} \\ \rho^w \mu^w < \rho^0 \mu^0 & \text{(unstable)}. \end{cases} \quad (35)$$

The theoretical background of the influence of density, viscosity and dielectric constant on the onset of surface motion is discussed.[43] Excellent agreement with criteria [eqn (35)] is obtained for experiments performed on medicinal paraffin + water systems with cholesterol and sodium dodecylsulphate. The inequalities in relation (35) are also the conditions for emulsification and de-emulsification.

We thank Drs Dupeyrat, Nakache, Saumagne, Gentric, Bauer, Cote, Dalle Vedove and Adler for stimulating discussions. We also thank the C.E.E. (Actions de stimulation), the Belgian Government (National Education Ministery) and the F.N.R.S. Belgium for financial support.

[1] J. T. Davies, *Turbulence Phenomena* (Academic Press, New York, 1972).
[2] E. Nakache, M. Dupeyrat and M. Vignes-Adler, *J. Colloid Interface Sci.*, in press.
[3] H. Linde, P. Schwartz and H. Wilke, in *Lecture Notes in Physics*, ed. T. S. Sørensen (Springer-Verlag, Berlin, 1979), vol. 105, p. 75.
[4] A. Orell and J. W. Westwater, *AIChE J.*, 1962, **8**, 350.
[5] A. Sanfeld, A. Steinchen, M. Heunenberg, P. M. Bisch, D. Gallez and W. Dalle Vedove, in *Lecture Notes in Physics*, ed. T. S. Sørensen (Springer-Verlag, Berlin, 1979), vol. 105, p. 168.
[6] T. S. Sørensen, in *Lecture Notes in Physics*, ed. T. S. Sørensen (Springer-Verlag, berlin, 1979), 105, p. 1.
[7] P. Glansdorff and I. Prigogine, *Thermodynamics of Structure, Stability and Fluctuations* (Wiley-Interscience, London, 1971).
[8] G. Nicolis and I. Prigogine, *Self-organization in Non-equilibrium Systems* (Wiley Interscience, New York, 1977).
[9] A. Steinchen and A. Sanfeld, in *Modern Capillarity*, ed. F. C. Goodrich and A. I. Rusanov (Akad-Verlag, Berlin, 1980).
[10] L. G. Napolitano, *Acta Astronautica*, 1982, **9**, 1999.
[11] M. Hennenberg, *Thesis* (Université Libre de Bruxelles, 1980).
[12] P. M. Bisch, *Thesis* (Université Libre de Bruxelles, 1980).
[13] R. Aris, *Vectors, Tensors and the Basic Equations of Fluid Mechanics* (Prentice Hall, N.J., 1962).
[14] V. Mohan and D. T. Wasan, in *Colloid and Interface Science*, ed. M. Kerker (Academic Press, New York, 1976), vol. 4, p. 430.
[15] D. T. Wasan, N. F. Djabbarah, M. K. Vora and S. T. Shah, in *Lecture Notes in Physics*, ed. T. S. Sørensen (Springer-Verlag, Berlin, 1979), vol. 105, p. 205.
[16] A. Steinchen, *Thesis* (Université Libre de Bruxelles), 1970.
[17] A. Sanfeld, *Introduction to the Thermodynamics of Charged and Polarized Layers* (Wiley Interscience, London, 1968).
[18] R. Defay, I. Prigogine, A. Bellemans and D. H. Everett, *Surface Tension an Adsorption* (Longmans–Green, London, 1966).
[19] M. Hennenberg, T. S. Sørensen, and A. Sanfeld, *J. Chem. Soc., Faraday Trans. 2*, 1977, **73**, 48.
[20] T. S. Sørensen and M. Hennenberg, in *Lecture Notes in Physics*, ed. T. S. Sørensen (Springer-Verlag, Berlin, 1979), vol. 105, p. 276.
[21] C. V. Sternling and E. K. Scriven, *AIChE J.*, 1959, **5**, 514.
[22] A. Marquez, A. Sanfeld and W. Dalle-Vedove, in *Stability of Emulsion and Microemulsion* (Special Report Belgian Government, Brussels Region, 1980).
[23] P. Joos, G. Bleys and G. Petré, *J. Chim. Phys.*, 1982, 387.

[24] I. Panaiotov, A. Sanfeld, A. Bois and J. F. Baret, *J. Colloid Interface Sci.*, in press.
[25] M. Hennenberg, P. M. Bisch, M. Adler and A. Sanfeld, *J. Colloid Interface Sci.*, 1979, **69**, 128; 1980, **74**, 495.
[26] M. Hennenberg, A. Sanfeld and P. M. Bisch, *AIChE J.*, 1981, **27**, 1002.
[27] T. S. Sørensen, M. Hennenberg, A. Steinchen, A. Sanfeld, *J. Colloid Interface Sci.*, 1976, **56**, 191.
[28] W. Dalle-Vedove and A. Sanfeld, *J. Colloid Interface Sci.*, 1981, **84**, 318; 328; 1983, **95**, 299.
[29] W. Dalle-Vedove, *Thesis* (Université Libre de Bruxelles), 1984.
[30] W. Dalle-Vedove and A. Sanfeld, personal communication, 1983.
[31] A. Steinchen and A. Sanfeld, *Chem. Phys.*, 1973, **1**, 156; *Biophys. Chem.*, 1975, **3**, 99.
[32] A. R. Marquez, W. Dalle-Vedove and A. Sanfeld, *J. Chem. Soc., Faraday Trans. 2*, 1981, **77**, 2303.
[33] J. L. Ibanez and M. G. Velarde, *J. Math. Phys.*, 1977, **38**, 1479.
[34] T. S. Sørensen and J. L. Castillo, *J. Colloid Interface Sci.*, 1980, **76**, 399.
[35] K. Durrani, C. Hanson and M. A. Hughes, *Metall. Trans. B*, 1977, **8B**, 169.
[36] M. Dupeyrat and E. Nakache, personal communication, 1983.
[37] N. M. Rice and M. Nedved, *Hydrometallurgy*, 1976/1977, **2**, 361.
[38] G. A. Yagodin, I. S. Yu and V. V. Tarasov, in *ISEC Conf.* (Liège, 1980), p. 1.
[39] P. R. Danesi, in *ISEC Conf.* (Denver, 1983), p. 1.
[40] C. Hanson, in *ISEC Conf.* (Liège, 1980), plenary lecture, p. 1.
[41] M. Cox and D. S. Flett, in *ISEC Conf.* (Toronto, 1977), p. 63.
[42] G. F. Vandegrift and E. P. Horwitz, *J. Inorg. Nucl. Chem.*, 1980, **42**, 119.
[43] A. Sanfeld, M. Lin, A. Bois, I. Panaiotov and J. F. Baret, in *Adv. Colloid Interface Sci.*, in press.
[44] A. Watanabe, K. Higashitsuji and K. Nishizawa, *J. Colloid Interface Sci.*, 1978, **64**, 378.

Time-dependent Behaviour and Regularity of Dissipative Structures of Interfacial Dynamic Instabilities

By Hartmut Linde

Akademie der Wissenschaften der DDR, Zentralinstitut für physikalische Chemie, Rudower Chaussee 5, 1199 Berlin, German Democratic Republic

Received 5th December, 1983

Interfacial dynamic instabilities with self-amplifying and self-organizing convections driven by interfacial tension in a non-equilibrium two-phase fluid system show a surprising variety of dissipative structures. A complete investigation and description of them has to take into consideration at least four aspects. (1) First are the kinetic features of the convection behaviour and the deformation of the interface concerning stationary basic units of flow systems (b.u.f.) with the related deformations of the interface and different time-dependent behaviour (travelling quasi-stationary b.u.f. driven by long-range driving forces, relaxing oscillations with related autowave behaviour and classical mechanical waves). (2) The topological features include different spatial patterns, *e.g.* parallel, concentrically circular or spiral stripes, polygonal networks and hierarchically ordered structures. (3) The order-disorder features concern the size, shape and packing of the b.u.f. with respect to spatial regularity or spatial chaos. The time-dependent behaviour can be distinguished for harmonic, anharmonic and chaotic oscillations. (4) The driving faces (d.f.) and conditions for Marangoni instability I are heat-and/or mass-transfer and/or chemical reaction at the fluid interface. The resulting b.u.f., an interface-renewing flow, can behave (*a*) as stationary or quasistationary in travelling substructures, (*b*) as relaxing oscillations, *e.g.* travelling or spiral-shaped autowaves and (*c*) as classical longitudinal capillary waves. Marangoni instability II, with the same d.f. as instability I, induces in thin-layer amplification of the differences in the thickness of the layer. Marangoni instability III, with shear stress as the d.f. at a tenside-covered fluid interface, shows stationary or oscillatory hair needle-like or elliptical eddies in the plane of the interface itself. Meniscus instability, resulting from the viscous pressure as the d.f. at a travelling meniscus, is an excellent example of amplification as well as of stationary spatial deformations of the meniscus-shaped interface in a determinate way and travelling spatial deformations of substructures, which are caused by a repeated stochastic process.

One of the most manyfold spectra of dissipative structures (d.s.)[1-3] is caused by interfacial dynamic instabilities for the driving forces of heat- and/or mass-transfer and/or chemical reaction and/or shear stress in two-phase systems with fluid interfaces, if the systems have an internal feedback mechanism and exceed critical conditions. Stability theories use the Navier–Stokes equations, the laws governing the transport of matter and heat and some additional boundary conditions with respect to the usual hydrodynamics. The Gibbs–Marangoni effect is expressed by the stress balance at the interface

$$\mu_a \left(\frac{\partial v}{\partial x}\right)_a - \mu_b \left(\frac{\partial v}{\partial x}\right)_b = \frac{\partial \sigma}{\partial y} \quad \text{for } x = 0 \tag{1}$$

where $(\partial v/\partial x)_{a,b}$ is the derivative of tangential velocity in a direction normal to the interface in phase a or b ($x = 0$), $\mu_{a,b}$ is the dynamic viscosity in phase a or b, σ is the interfacial tension and x, y are the cartesian coordinates perpendicular and tangential to the flat interface. Eqn (1) shows that a fluid interface responds to the difference in the interfacial tension, which is due to local adsorption of surface active agents (tensides) or to local heating, by interfacial convection from areas of lower tension to areas of higher tension, inducing the related shear stress $\mu(\partial v/\partial x)$ in both phases. The inversion of action and reaction can be observed in a system with shear stress, which leads to a difference in the interfacial tension at a tenside-covered interface.

Secondly, we have to take into consideration the reaction of the interface to a space-dependent viscous pressure difference in both phases at the interface. For an originally flat interface the normal stress (pressure) balance at the interface is

$$p_b - p_a + 2\left[\mu_a\left(\frac{\partial u}{\partial x}\right)_a - \mu_b\left(\frac{\partial u}{\partial x}\right)_b\right] = \sigma\left(\frac{1}{r_1} + \frac{1}{r_2}\right) \qquad (2)$$

for $x=0$, where $p_{a,b}$ is the hydrostatic pressure at the interface in phase a or b, $(\partial u/\partial x)_{a,b}$ is the derivative of the velocity in a direction normal to the interface in phase a or b ($x=0$) and r_1, r_2 are the principal curvatures of the slightly deformed interface. The viscous pressure causes a deformation (r_1 and r_2 have finite values) even if the interfacial tension is not space dependent [contrary to the Gibbs–Marangoni effect, eqn (1)]. At least four different features play an important role in the investigations and therefore also in the description of the d.s. of interfacial dynamics: (*a*) kinetics, (*b*) topology, (*c*) order–disorder transitions and (*d*) driving forces.

KINETIC FEATURES OF THE CONVECTION AND DEFORMATION BEHAVIOUR

The stationary state of the basic unit of flow (b.u.f.), shown in fig. 1, is due to interfacial convection enclosing circulating flows in both phases. Periodic reamplification and breakdown of the b.u.f. of interfacial convection (relaxation oscillation of the intensity) occurs. Travelling b.u.f. convection systems originate at a leading line or centre[4] and behave as autowaves (no reflection at a wall, no interference but annihilation of colliding wavefronts). The periodic tangential or normal deformations of the interface (with related convections of the adherent fluid layers) behave as classical travelling or standing waves with reflection and interference. Deformations occur in the shape of the interface, which can be stationary or autowave-like travelling along the interface.

To distinguish classical waves and autowaves, note that classical elastic waves of small amplitude can be described by Helmholz's wave equation (a hyperbolic differential equation, invariant with time inversion):

$$\frac{\partial^2 x}{\partial t^2} = c^2 \Delta x$$

where Δ is the Laplace operator and c is the velocity of the waves. For longitudinal waves, x is the characteristic parameter of the scalar field, e.g. density in sound waves or the concentration of surface-active agents, $c_{s.a.a.}$, at the interface in longitudinal (capillary) waves. Autowaves can be described by a system of parabolic differential equations (variant with time inversion)

$$\frac{\partial x_i}{\partial t} = F_i(x_i) + D_i \Delta x_i, \qquad i=1,\ldots,n$$

in which non-linear functions $F_i(x_i)$ are necessary, and where, in chemical systems, x_i are the volume concentrations of autocatalysing or inhibiting chemical species,

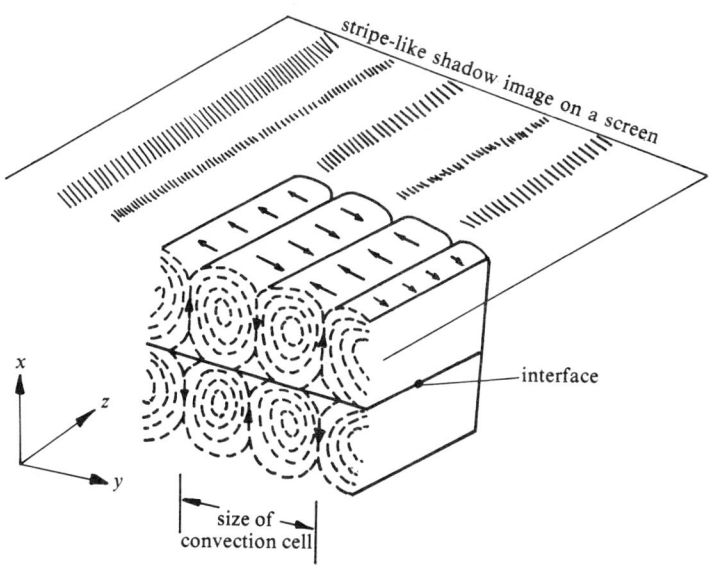

Fig. 1. Basic unit of flow.

D_i are the diffusion coefficients of the chemical species. x_i, in interfacial dynamic systems, are the concentrations of surface-active agents, $c_{s.a.a.}$, at the interface.

Note that in the same interfacial system both classical elastic (capillary) waves and autowaves are possible, with variation of the direction of mass- or heat-transfer, respectively, of

$$\frac{d\sigma}{dc_{s.a.a.}} \quad \text{or} \quad \frac{d\sigma}{dT}.$$

TOPOLOGICAL FEATURES OF TWO-DIMENSIONAL IMAGES OF BASIC UNITS OF FLOW OR OF INTERFACIAL DEFORMATION

Straightly parallel stripes are sometimes interrupted by splitting or by unification of stripes [fig. 2(a)] or by the free ends of newly formed stripes [fig. 2(b)]. Concentrically circular [fig. 2(c)] or irregular parallel curvaceous stripes can be stationary or travelling. Spirals may exist with one [fig. 2(d)] or more arms [fig. 2(e)] moving radially whilst the centre is rotating. Polygonal networks of stationary or travelling roll cells also occur [fig. 2(f)]. Structures of higher order (structure hierarchy by substructuration) occur if two or more of the basic structures expanding from different leading centres collide. The collision areas form new structural elements of higher order [fig. 2(g)]. Finally, there may be circular, elliptical [fig. 2(h)] or hair needle-like [fig. 2(i)] eddies in the plane of the interface itself, *i.e.* another convection system rather than the above-mentioned b.u.f.

ORDER–DISORDER FEATURES

Order–disorder features consist of (a) spatial structures characterized by one or more wavelengths, which can be highly ordered with single- or multi-peak

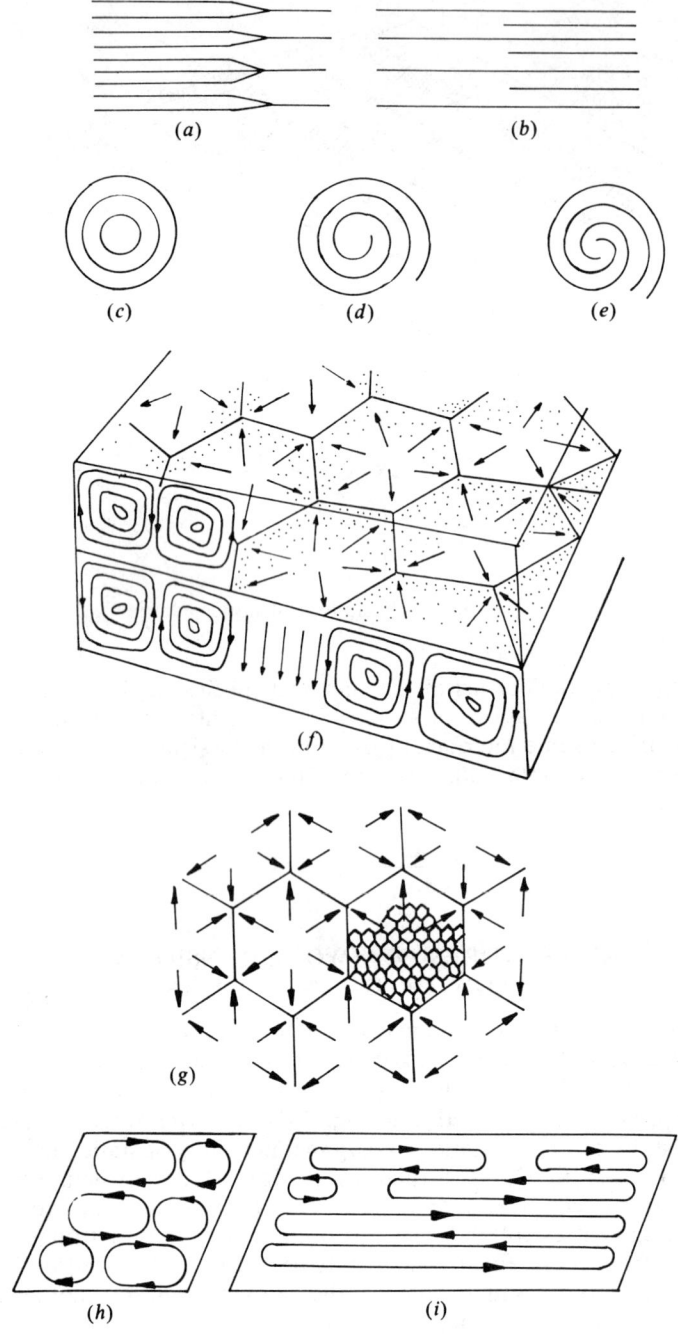

Fig. 2. (a) Splitting and unification of stripes; (b) free ends of newly formed stripes; (c) concentrically circular stripes; (d) spiral stripes with one arm; (e) spiral stripes with two arms; (f) polygonal networks (of roll cells); (g) structure hierarchy by substructuration; (h) elliptical eddies in the plane of the interface; (i) hair needle-like eddies in the plane of the interface.

wavelengths distributions or disordered with wide distributions of wavelengths or sizes, (b) spatial structures characterized by the shapes of the topological features, (c) high-order systems with polygonal networks of regularly shaped and regularly packed stripes or polygons or disordered structures with irregularly shaped polygons and therefore also with irregularities in their packing, (d) time-dependent behaviour of oscillations: harmonic oscillations, or anharmonic oscillations, e.g. sawtooth oscillations or chaotic oscillations.

DRIVING FORCES, CONDITIONS AND ESSENTIAL BEHAVIOUR

DISSIPATIVE STRUCTURES OF MARANGONI INSTABILITY I

Heat- and/or mass-transfer of surface-active agents leads to Marangoni instability I and spontaneous interfacial convection. Both the stationary regime and classical wave-like oscillation (longitudinal capillary waves) are predicted from linear stability theory[5] and observed in real physical systems.[1,2]

From an analysis including the thickness of the adjoining fluid layers[6] conditions for the critical Marangoni numbers of both basic regimes are predicted; the influence of additional exchange of latent heat on the Marangoni numbers and on the kind of regime has also been analysed recently.[7]

In reality the 'stationary' regime is more complicated than its theoretical description: i.e. the 'stationary' regime is expected to be time-independent like fig. 1 or fig. 2(f) and to exhibit order–disorder features with a sharp single peak. At the lowest level of instability, roll cylinders or roll cells of the theoretically predicted smallest size show relaxation oscillation, which is the reason for the recently recognized autowaves with travelling parallel stripes, concentrically circular waves and even spiral leading centres. These autowaves, like autowaves in autocatalytic reaction[4] systems, are expected to be propagated by a local driving force along the interface. Whether these travelling autowaves can form a stationary stable structure of higher order because of the structure of the collision area is, even in the case of travelling autowaves of the Belousov–Zhabotinski reaction,[4] unknown. The next highest level of instability allows roll cylinders or polygonal cells as a network in stationary convection systems, which remain in place, especially in thin liquid layers.

With a further increase in roll cell instability, characterized by an increase in the Marangoni number, convection systems again show travelling waves but with higher velocity than the above-mentioned pure autowaves. From the point of view of origination at a leading area[8] and annihilation at a wall or at a collision area, they behave like autowaves; but their greater velocity is caused by a long-range interfacial tension gradient from a second-order convection system. Roll cells of first, second and higher order are observed to form a hierarchical system. An approach using non-linear terms[9] was able to show the existence of second-order roll cells.

From this behaviour we find selection of different kinetic regimes with different topological features: e.g. hexagonal patterns are preferred for stationary structures and tetragons; ladder structures with different diameters between rungs and beams are preferred for travelling systems, if the movement is caused by hydrodynamic (or interfacial dynamic) shear stress. High velocity and high shear stress can cause chevron (herring-bone) patterns to be formed by deformation of the ladder structure.

With increasing driving force the irregularities increase and lead to chaotic distributions of sizes and roll cells of irregular shapes: i.e. the first transition to chaotic behaviour in Marangoni instability.

Roll cells of second and higher order can degenerate by relaxation oscillation, giving a wide range of diameters of chaoticaly spreading and oscillating cells (diameters decreasing and frequencies increasing with increasing driving force). Autowaves of roll cells of higher order have also been observed. The chaotic interaction of an ensemble of these stationary roll cells is characterized by a fast annihilation of every roll cell after or during the first period of oscillation by the consecutively very often nucleating adjoining cells. The chaotic features of spatial and time-dependent behaviour may be due to the stochastic behaviour of nucleation resulting in irregular anharmonic oscillations, which are not synchronized.

There is a strong contrast between the three kinetic features of the 'stationary' regime (real stationary roll cells, structure hierarchy by travelling substructures and relaxation oscillations, which lead on the one hand to autowave behaviour and on the other to chaotic oscillations) and the classical oscillatory regime of the Marangoni instability. The latter is identical to undamped longitudinal capillary waves and can occur for travelling or standing classical waves with reflection at a wall and with interference. For lower supercritical driving forces, the regularity is very high and the topological features correspond to fig. 2(a) or (b). We observe with increasing driving force a transition from travelling straight waves (stripes) to standing waves (with mutually penetrating stripes) and from these one-modal waves to two-modal and even three-modal waves [with angles between the corresponding parallel stripes of 90° (two modal) or 60° (three modal)]. For very high driving forces the irregularities increase again and lead to chaotic behaviour.

The experimental results are in good agreement with the predictions of stability theory[5-7] and with the more detailed description obtained by computer simulation.[10] The classical wave can be amplified as mentioned before by the driving forces caused by heat- and/or mass-transfer and/or chemical reaction at the interface: *i.e.* autowaves and classical waves are distinguished theoretically by the above-mentioned differential equations and the related kinetic behaviour concerning reflection and interference and not by the kind of energy supplied by local sources at the interface or by the entire interface itself.

MARANGONI INSTABILITY II[11]

Marangoni instability II results from the same driving forces as Marangoni instability I, but is connected with increases in the thickness differences in thin layers, *i.e.* increases in deformations normal to the interface. Thin layers, jets and droplets can be destabilized and broken into smaller compartments.

MARANGONI INSTABILITY III[12-14]

Marangoni instability III results from shear stress in the liquid and/or the gas phase at a tenside-covered liquid/gas interface and is characterized by stationary (with small driving force, *e.g.* in systems where flow in the gas phase is the only cause of this dissipative structure) or oscillatory (with higher driving force) hair needle-like [fig. 2(i)] or elliptical eddies[14] [fig. 2(h)] in the plane of the interface itself. This means that there is no surface renewal convection in this pattern of shear flow on the surface (which is, of course, accomplished by related convection in both boundary layers). As with the other dissipative structures, this instability occurs only if we exceed critical conditions, which are lower than the values of the critical conditions for transition from laminar to turbulent flow at a solid wall.

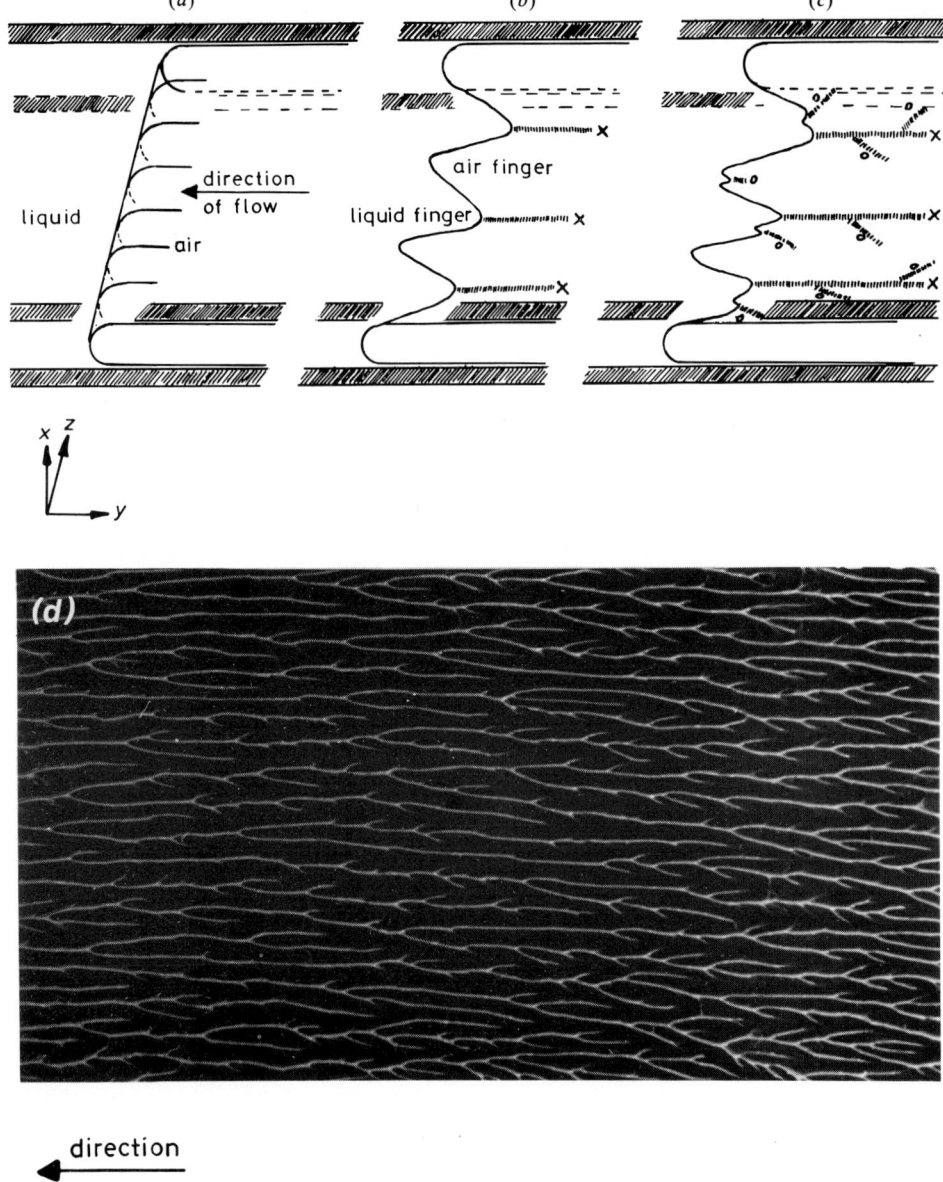

Plate 1. (a) Moving meniscus between two plates; (b) first-order deformation of a moving meniscus; (c) second-order deformation of a moving meniscus; (d) fixed consecutive structure of first-order, ×, and second-order, ○, meniscus deformations.

MENISCUS INSTABILITY[15]

This last example of surface instabilities results from the viscous pressure being the driving force, which operates at a travelling meniscus [plate 1(a)] and is separate from the above-mentioned Marangoni instabilities. This condition obtains if a Newtonian or non-Newtonian liquid in the gap between the two plates is displaced by the penetration of air, forming a moving half-cylinder-shaped meniscus.

The first regime of meniscus instability is characterized by a stationary deformation of the meniscus in the shape of a waveline in the z direction [plate 1(b)]. The fixed consecutive structure is a system of equidistant stripes of smaller and larger thickness in both layers, remaining at the surface of both plates [fig. 2(a) and (b)], and whilst the 'air fingers' produce thin stripe-like layers the 'liquid fingers' produce thick layers stripe-like layers [plate 1(b)]. The distance between these stripes (to avoid the expression wavelength, because we have to distinguish between classical waves and a variety of other wave-like phenomena) depends strictly on the viscosity, the velocity of the moving meniscus and the distance between the two plates (the radius of meniscus), so we observe regular parallel stripes (at small driving force) and a transition to a stripe system with smaller or larger distances if we change the plate distance during the experiment.[16] The transition to narrower stripes is caused by origination of new stripes of fig. 2(b), and the opposite transition is caused by the unification of stripes, fig. 2(a), i.e. in the first case there are new liquid fingers and in the second case there is unification of two liquid fingers [plate 1(c)].

This instability is a good example of a chance process, if with increasing driving force by a second bifurcation additional deformations (of second order) [plate 1(c) and (d)] with smaller characteristic length than the first-order deformation of the meniscus occur. The second-order deformations originate at the top of the air fingers as small liquid fingers, and they travel to the base of the air finger and then disappear. The nucleation of this second-order deformation is the cause of its chance mechanism: from both its timing and its direction of travel the process has an event distribution of probability. Thus we can recognize another basic mechanism causing chaotic behaviour, which follows the condition 'sensitive to initial conditions'.[17] The resulting fixed consecutive structure shows relatively regular first-order stripes and the irregular branches of the second-order stripes [plate 1(d)].

---

[1] H. Linde, *Marangoni Instabilities*, in *Dynamics and Instability of Fluid interfaces*, Lecture Notes in Physics no. 105, ed. T. S. Sørensen (Springer-Verlag, Berlin, 1979).

[2] H. Linde, in *Convective Transport and Instability Phenomena*, ed. J. Zierep and H. Oertel Jr (G. Braun-Verlag, Karlsruhe, 1982).

[3] H. Linde, *Dissipative Strukturen der Grenzflächendynamik*, in *Fortschritte der experimentellen und theoretischen Biophysik Bd. 21*, ed. E. Kahrig and H. Beßerdich (VEB Georg Thieme-Verlag, Leipzig, 1977).

[4] A. M. Zhabotinsky and A. N. Zaikin, *J. Theor. Biol.*, 1973, **40**, 45.

[5] L. E. Scriven and C. V. Sternling, *AIChE J.*, 1959, **5**, 514.

[6] H. Linde and J. Reichenbach, *J. Colloid Interface Sci.*, 1981, **84**, 433.

[7] G. Frenzel and H. Linde, *Teor. Osn. Khim. Tekhnol.*, 1983/84, in press.

[8] A. N. Zaikin and A. M. Zhabotinsky, *Nature* (London), 1970, **225**, 535; A. N. Zaikin and A. L. Kawczynski, *J. Non-equilib. Thermodyn.*, 1977, **2**, 39.

[9] H. Linde and P. Schwartz, *Chem. Tech.* (*Leipzig*), 1974, **26**, 455; 1977, deposition system.

[10] H. Wilke, *Chem. Tech.* (*Leipzig*), 1974, **26**, 456; 1977, deposition system; *Z. Angew. Math. Mech.* (*ZAMM*), 1980, **9**, 437.

[11] C. Arcuri and D. W. De Bruijne, *Proc. Vth Int. Congr. Surface Active Substances*, Barcelona, 1968.

[12] H. Linde and P. Friese, *Z. Phys. Chem.* (*Leipzig*), 1971, **247**, 225.
[13] H. Linde and N. Shulewa, *Mber. Dtsch. Akad. Wiss. Berlin*, 1970, **12**, 883.
[14] H. Linde and P. Schwartz, *Teor. Osn. Khim. Tekhnol.*, 1971, **401**.
[15] L. Weh, *Dissertation* (Humboldt-Universität, Berlin, 1972).
[16] H. Linde, *Nova Acta Leopold.*, 1984, in press.
[17] J. A. Yorke and E. D. Yorke, in *Hydrodynamic Instabilities and the Transition to Turbulence* (Springer-Verlag, Berlin, 1981).

…

The Contribution of Chemistry to New Marangoni Mass-transfer Instabilities at the Oil/Water Interface

BY EVELYNE NAKACHE* AND MONIQUE DUPEYRAT

Laboratoire de Chimie Physique de l'Université Pierre et Marie Curie,
11 Rue Pierre et Marie Curie, 75231 Paris Cedex 05, France

AND MICHÈLE VIGNES-ADLER

Laboratoire d'Aérothermique du C.N.R.S., 4 ter Route des Gardes,
92130 Meudon, France

Received 29th December, 1983

Spontaneous interfacial motions appear at an interface between two immiscible phases in a state far from equilibrium, *e.g.* an aqueous phase of an organic acid or a complex ion and an organic phase of a long-chain surfactant molecule. The instabilities observed are related to variations in the interfacial tension. It is shown that they resemble the well known Marangoni effect by considering interfacial convection and the coupling between diffusion and convection fluxes, but they differ by the presence of chemical reactions. The notion of assisted desorption is defined in order to interpret the experiments.

The analysis of this phenomenon could be useful in determining the optimal conditions for obtaining convective interfacial transfer to enhance liquid–liquid extraction processes.

Mass transfer accompanying chemical reactions is involved in many industrial and biological processes. Under certain conditions it is associated with interfacial turbulence. Many authors[1-4] have shown the important role of the interface and the influence of surfactant molecules on the rate of liquid extraction.[5] Some of them, *e.g.* Dobson and Van der Zeeuw,[6] studying the extraction of copper by hydroxyoximes, observed that the rate is greater when using Shell SME 529 extractant than when 2-octyl-5-hydroxyheptophenone oxime is used, although the interfacial density of the latter is higher.

We discuss here the role of the surfactant and show that interfacial convection, which may enhance the extraction rate, can be attributed to the Marangoni effect:[7-9] the interfacial flow can be governed by gradients in the interfacial tension driven by local concentration variations resulting from mass transfer.

As an example of interfacial convection, this paper reports observations of large-scale, chemically driven interfacial motion, first observed in our laboratory.[10] It appears at an interface between two immiscible phases, kept at constant temperature, and seems to be due to the simultaneous transfer in opposite directions of two solutes, one of them surface active. For example, if an aqueous solution of a long-chain hydrophobic molecule is poured over a solution of various organic acids or complex ions in nitrobenzene, two different spontaneous motions can be observed provided that the concentrations of the two compounds are suitable. If the wall of the container is made of polyethylene, local contractions and expansions of the interface are seen owing to the appearance of an emulsion some time after both phases are put in contact. We confirmed that this emulsion was not necessary for the onset of the movement and that no noticeable perpendicular deformation

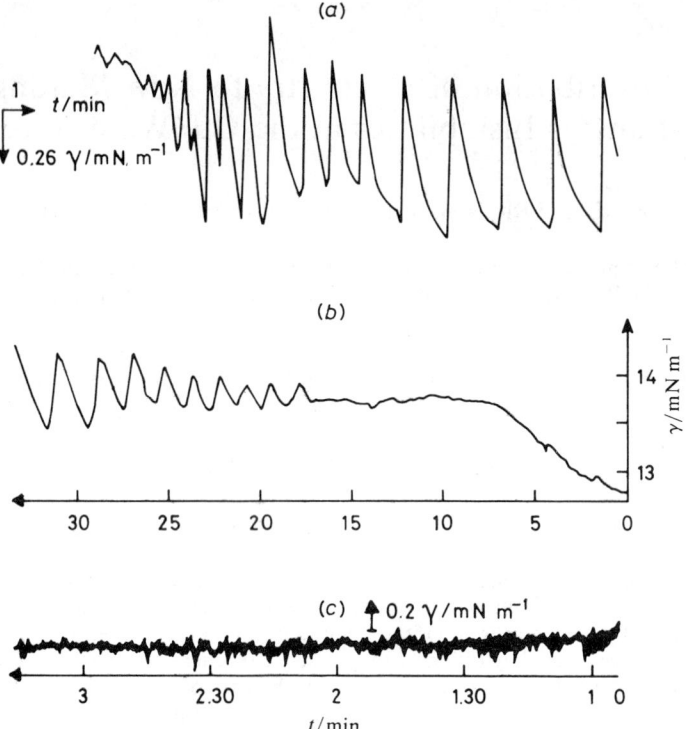

Fig. 1. Variation of γ with time: (a) interface (nitroethane) 3.5×10^{-4} mol dm^{-3} HPi/(water) 5×10^{-3} mol dm^{-3} C$_{12}$Br; (b and c) interface (nitrobenzene) 3.5×10^{-4} mol dm^{-3} HPi/(water) C$_{16}$Cl [(b) 4×10^{-4} and (c) 2×10^{-4} mol dm^{-3}].

occurred. In a glass container the instability turns into a strong deformation of the interface resulting in a wave of 1 cm in amplitude, propagating along the wall, followed by a total disturbance of the interface.

We have checked that the available models,[8,9,11] improved by adaptation to particular experimental conditions, could not account for the instabilities observed. The purpose of this work is to understand, from experimental data, the mechanism of the motion, in particular the influence of the interface on the hydrodynamical and physicochemical effects.

EXPERIMENTAL

We chose to study first movement in the plane of the interface, because the wall introduced wettability conditions which made the problem more complicated.

A solution of a classical surfactant, cetyltrimethylammonium chloride (C$_{16}$Cl, gift from a private laboratory), purified three times by crystallisation in methanol, was poured over a solution of picric acid (HPi, Merck), in nitrobenzene (chromatographic grade, UCB) or nitroethane (Carlo Erba). The solvents were previously saturated by each other.

We chose to follow the interfacial convection by measuring the variation in interfacial tension, which is correlated to the longitudinal deformation of the interface. The detachment method, using a stirrup instead of a ring, was used.

The curves of fig. 1 show several pseudoperiodic oscillations of γ as a function of time for various solute concentrations. In curve (a), where the pseudoperiod is very large, one

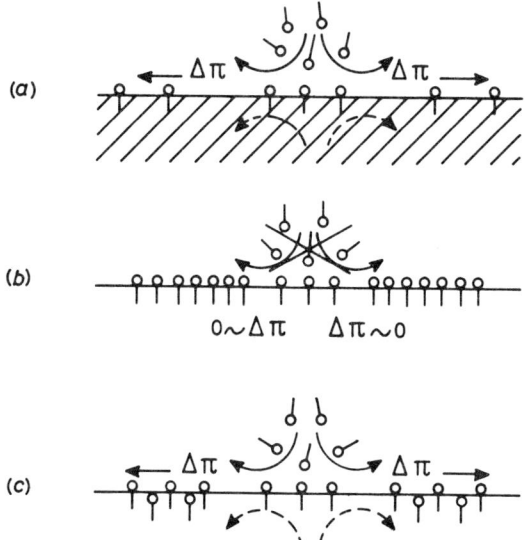

Fig. 2. Hydrodynamical effects related to the density of the adsorbed layer.

can see γ increasing progressively and then decreasing abruptly while an expansion of the interface, *i.e.* the arrival of surfactant molecules, is observed.

DISCUSSION

The transfer of these solutes from one phase to another may result in (i) hydrodynamical effects which depend on the transfer processes (diffusion, convection, adsorption–desorption) and on interfacial tension gradients and (ii) physicochemical effects, eventually involving chemical reactions.

HYDRODYNAMICAL EFFECTS

Our analysis will be made easier if we first try to interpret a well known instable extraction phenomenon studied by Davies and Haydon,[12] termed the 'kicking drop' by these authors: a stable pendant drop of water in toluene undergoes violent erratic movements or 'kicks', more or less rapidly damped by viscous drag when acetone is added to the toluene. Owing to the local convective currents a local increase of surfactant molecules near the interface produces at some point an increase in the surface pressure, $\Delta\pi$. The monolayer tends to spread further over the surface, dragging some adjacent liquid with it. If appropriate conditions of viscosity, diffusivity and concentration are fulfilled, according to Sternling and Scriven for example, an eddy of fresh solution would occur at this point, amplifying the movement which appears as an 'expansion', as shown in fig. 2(a). Then, owing to momentum transfer, eddies should also occur on the other side of the interface. If the density of surface-active material becomes such that the adsorbed layer is sufficiently condensed, $\Delta\pi$ tends to zero and the surface pressure of this monolayer resists the eddy [fig. 2(b)]. The interfacial convection is further inhibited by highly viscous dissipation. Transfer then becomes purely diffusive and of course slower. As acetone is not very surface active it can easily desorb, inducing a 'contraction' of the interface, and the latter step is not reached [fig. 2(c)]. If dodecyltrimethylammonium chloride ($C_{12}Cl$) is added, the movement is inhibited. Indeed, this very surface-active compound cannot desorb, so that the interfacial convection is inhibited by viscous dissipation.

We have observed that no motion occurs if our system consists of an aqueous solution of $C_{16}Cl$ in contact with pure nitrobenzene. The question then arises: how can the presence of HPi, a non-surface-active compound, bring about instability? The answer to this question requires further physicochemical information concerning transfer of the solutes.

PHYSICOCHEMICAL EFFECTS

In a recent paper[13] we have analysed the importance of the transfer of each species from the equilibrium concentrations computed from electroneutrality, partition coefficients and the dissociation equilibrium constant of HPi in nitrobenzene. At equilibrium this results in the complete dissociation of HPi and almost total exchange of H^+ with C_{16}^+ as the counter-ion of Pi^-. This means that, although the driving force comes from the fluxes of $C_{16}Cl$ and HPi in opposite directions, only C_{16}^+ and H^+ pass through the interface, Cl^- and Pi^- remaining in their respective phases. Indeed the structure of Pi^- is very similar to that of the nitrobenzene molecule, while H^+ is very polar, like water. Thus the very different affinities for the two phases of the species concerned permit a flux of C_{16}^+ by means of two opposite cation fluxes (H^+ and C_{16}^+).

This interfacial transfer can be symbolized by the following equation, possible in one direction only:

$$C_{16}^+Cl^-_{\text{water}} + H^+Pi^-_{\text{nitrobenzene}} \rightarrow C_{16}^+Pi^-_{\text{nitrobenzene}} + H^+Cl^-_{\text{water}}.$$

The transfer thus results in a decrease in $C_{16}Cl$ concentration and in the formation of $C_{16}Pi$.

Now, because of the very different structures of Cl^- and Pi^-, the partition coefficient between water and nitrobenzene ($P_{\text{water/nitrobenzene}}$) is much larger for $C_{16}Cl$ than for $C_{16}Pi$. Thus for the same initial concentration of $C_{16}X$ ($X = Cl^-$ or Pi^-) the concentration of C_{16}^+ in the aqueous phase is higher with the counter-ion Cl^-. The interfacial density of C_{16}^+ ions, which governs the interfacial tension, depends on the aqueous concentration of C_{16}^+. Therefore, if Cl^- ions are replaced by Pi^- ions in the bulk phases, the C_{16}^+ interfacial density decreases and the interfacial tension increases. The question is now to examine how the concentration of Cl^- and Pi^- can vary. Obviously this depends on the transfer processes. If we take into account that at this type of interface several authors[14,15] have shown that the adsorption–desorption phenomenon is fast compared with the diffusive–convective one, we can assume that the adsorbed monolayer is always in equilibrium with two small zones adjacent to the interface, so that the diffusion–convection process is the limiting step of the transfer. Therefore the concentration variation in these bulk zones rules the concentration variation at the interface. We can deduce that, if Cl^- ions are replaced by Pi^- ions in these adjacent layers, C_{16}^+ ions will immediately desorb towards the organic phase, clearing the interface and increasing the interfacial tension.

COUPLING BETWEEN THESE TWO EFFECTS

The mechanism of the observed phenomenon will be a coupling between the two effects.

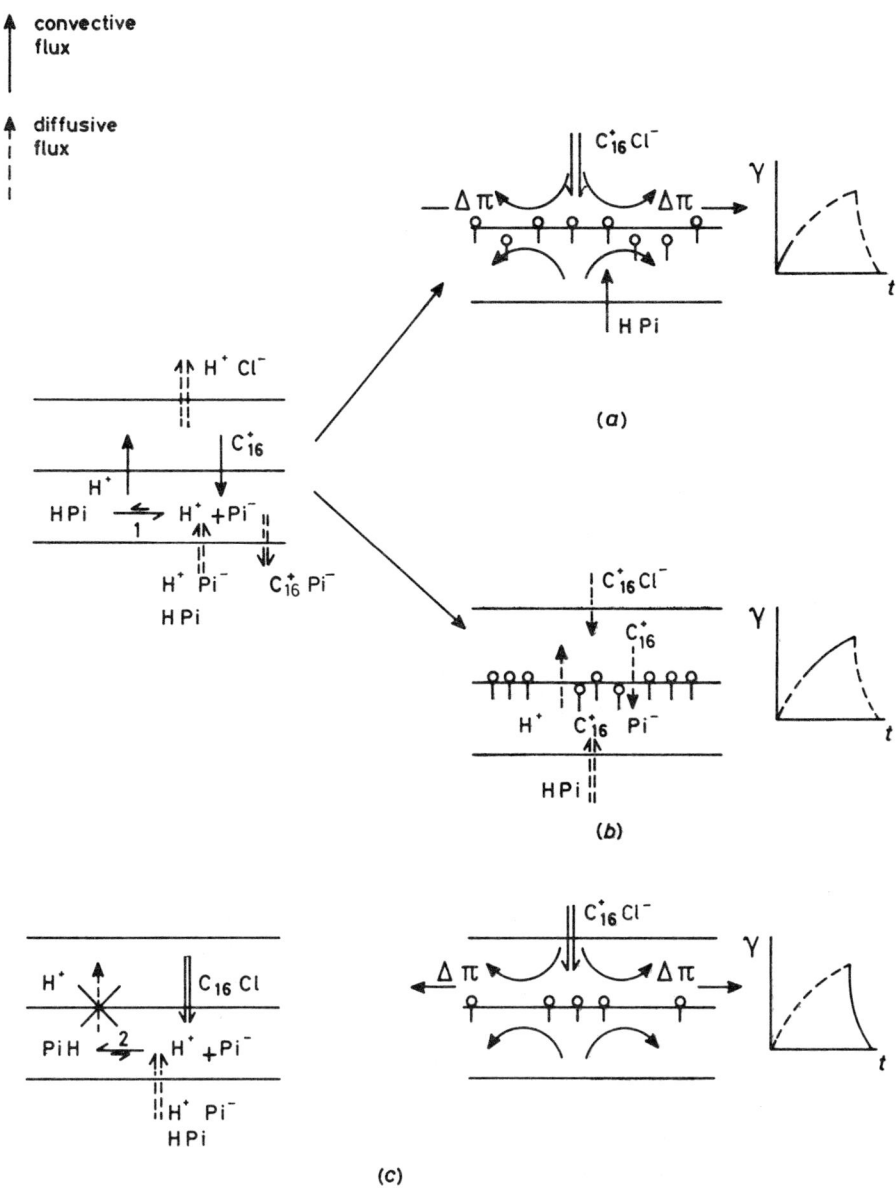

Fig. 3. Coupling between hydrodynamical and physicochemical effects.

The value of γ at the beginning of the phenomenon [fig. 1(a)] means that $C_{16}Cl$ is rapidly and massively adsorbed at the interface. The condensed layer thus formed tends to oppose the enhancement of convection, as shown before in fig. 2(b). At the same time $C_{16}Cl$ brought to the interface is progressively transformed by 'cation fluxes' in $C_{16}Pi$, increasing the interfacial tension. As convection is inhibited by the density of the adsorbed layer, the cation flux is diffusive. Owing to the very weak dissociation of HPi in nitrobenzene ($K_d = 10^{-6}$), the H^+ flux should be stopped

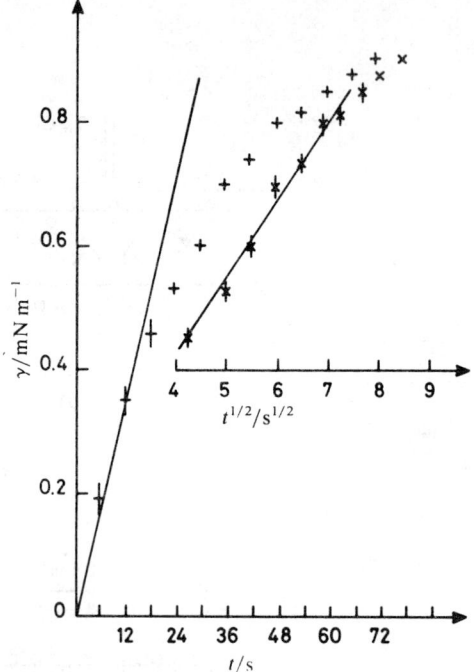

Fig. 4. Variation of γ with time and with square root of time. Interface (nitroethane) HPi 3.5×10^{-4} mol dm^{-3}/(water) C_{12}Br 5×10^{-3} mol dm^{-3}.

rapidly; however, it is sustained by the very fast dissociation of HPi according to the dissociation reaction in the forward direction:

$$\text{HPi} \rightleftharpoons \text{H}^+ + \text{Pi}^-.$$

During the time the Pi$^-$ concentration in the organic sublayer increases, more and more C_{16}Pi is formed, which in turn promotes the desorption of C_{16}^+ molecules as shown above. This increases the interfacial tension in a diffusive way, but also enhances the compressibility of the interface until a threshold is reached where natural convection can be amplified. Indeed we observed this convective behaviour at the beginning of an oscillation (fig. 4), where the increase of γ is proportional to time for *ca.* 15 s. {As this interfacial convection becomes predominant the cation exchange is very rapid and depletes the sublayers in H$^+$ and C_{16}^+, while C_{16}Cl and HPi are consumed [fig. 3(*a*)].}

At this stage the feeding of the sublayers in C_{16}Cl and HPi is ruled by diffusion from the bulk [fig. 3(*b*)], and the variation of the interfacial tension should be proportional to the square root of time. Fig. 4 shows, as expected, that in the second part of an oscillation γ increases with $t^{1/2}$.

However, while diffusion is continuing for a relatively long time, C_{16}Pi is accumulating in the organic phase. Therefore enhancement of Pi$^-$ concentration relative to that of H$^+$ favours the displacement of the dissociation reaction of HPi in the backward direction.

The recombination of HPi consumes H$^+$ ions. These are no longer available to pass through the interface and the 'cation fluxes' stop.

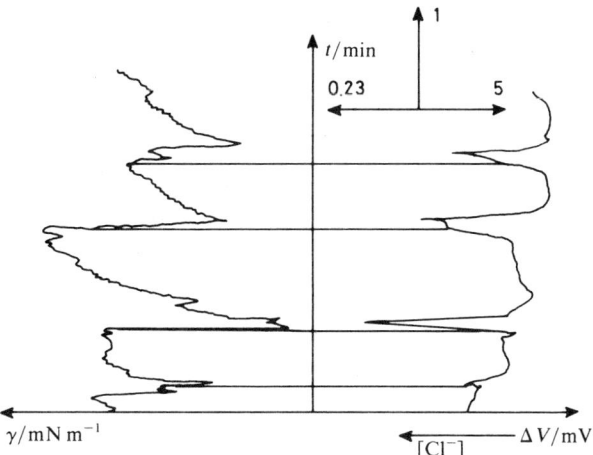

Fig. 5. Variation of γ and ΔV with time. Lines on the t axis refers to a synchronisation between a decrease in γ (increase in $C_{16}^+Cl^-$) and a decrease in ΔV (increase in Cl^-) near the interface.

At this stage the system increases its entropy by means of the flux of Cl^- and C_{16}^+ ions in the same direction. $C_{16}Cl$ then becomes predominant in the aqueous sublayer just as the accumulation of $C_{16}Pi$ in the organic sublayer induces a clearing of the interface, thus enhancing the compressibility of the interface until the critical threshold is reached where natural convection can be amplified. As the $C_{16}Cl$ flux arriving at the interface is convective, a rapid decrease in γ occurs, corresponding to the last part of the curve in fig. 3(c). This convection would restore the bulk concentration of the species on either side of the interface. Thus the cycle can start again.

EVIDENCE FOR SUCH A COUPLING

We report here only some evidence for the model.

(i) If the proposed mechanism is correct, the dissociation of HPi is crucial for the explanation of oscillation. We have observed that if HPi is replaced by KPi, a compound very similar but fully dissociated in nitrobenzene, no motion occurs.

(ii) During the descending part of the oscillation, an increase in the concentration of $C_{16}Cl$ should be correlated with a decrease in the interfacial tension. We have performed simultaneous measurements of Cl^- concentration (by means of a silver, silver chloride electrode) and of γ (fig. 5). It can be seen that the expected correlation is verified.

(iii) If the HPi concentration is kept constant while $C_{16}Cl$ decreases (fig. 1), the desorption is relatively more important for small values of $C_{16}Cl$ [curve (b)]. The convenient compressibility and the amplified interfacial convection appear more rapidly, and thus the period of oscillation is shorter and its frequency higher.

CONCLUSION

This system is only one particular case of the coupling which can occur at a liquid–liquid interface. We can now try to draw some more general conclusions about such heterogeneous systems and define the contribution of chemistry to these instabilities.

(1) A maximum in the interfacial tension as a function of time can be observed because there is possible competition between two paths for the transfer of material: fluxes of C_{16}^+ and Cl^- in the same direction or C_{16}^+ and H^+ in opposite directions, governed by a chemical reaction, the dissociation of HPi.

(2) Successive oscillations can be observed because under certain conditions the natural convection is amplified in the vicinity of the interface, resulting in the replacement of the usual diffusive interfacial transfer by a much faster convective one. This amplification is possible during a reasonable time only if a desorption process occurs allowing the adsorbed layer to reach critical compressibility. Desorption can be spontaneous with a short-chain surfactant which is easily desorbed. In this case the same molecule will be both adsorbed and desorbed. When the surfactant is not or only slightly desorbable (*i.e.* if the chains are long) desorption cannot be spontaneous, but it may be 'assisted' by an interfacial reaction which transforms a very adsorbable species into an easily desorbable one:

$$A_{adsorbable} + X \rightarrow D_{desorbable} + Y.$$

The kinetics of this 'assisted desorption', eventually taking into account solvatation or reorientation processes, has to be considered simultaneously with the kinetics of transfer in order to account for instability.[16]

This model seems to be relevant for liquid–liquid extraction and can explain the contradictory experiments reported by Dobson and Van der Zeeuw.[6] The specificity of interfacial transfer reported by many workers, but little studied up to now, could be approached fruitfully by this method.

[1] R. J. Whewell, M. A. Hugues and C. Hanson, *J. Inorg. Nucl. Chem.*, 1975, **37**, 2303.
[2] D. S. Flett, D. N. Okuhara and D. R. Spink, *J. Inorg. Nucl. Chem.*, 1973, **35**, 2471.
[3] R. Price and J. Tumelty, *Ind. Chem. Eng. Symp. Ser.*, 1975, **42**, 18.1.
[4] D. S. Flett, M. Cox and J. D. Heels, in *Proc. Int. Solvent Extraction Conf.* (Soc. Chem. Ind., London, 1974), vol. 3, session 24, p. 2560.
[5] W. Nitsh and L. Navazio, in *Proc. Int. Solvent Extraction Conf.*, 1980, vol. 80, p. 220.
[6] S. Dobson and A. J. Van der Zeeuw, *Chem. Ind.*, 1976, **5**, 176.
[7] L. E. Scriven and C. V. Sternling, *Nature (London)*, 1960, **187**, 186.
[8] C. V. Sternling and L. E. Scriven, *AICHE J.*, 1959, **5**, 514.
[9] H. Linde and M. Kunkel *Waerme Stoffubertrag.*, 1969, **2**, 60.
[10] M. Dupeyrat and J. Michel, *J. Exp. Suppl.*, 1971, **18**, 269.
M. Dupeyrat and E. Nakache, *C.R. Acad. Sci., Ser. C*, 1973, **277**, 599.
[11] M. Hennenberg, P. M. Bisch, M. Vignes-Adler and A. Sanfeld, *J. Colloid Interface Sci.*, 1980, **74**, 495.
[12] J. T. Davies and D. A. Haydon, *Proc. R. Soc. London, Ser. A*, 1958, **243**, 492.
[13] E. Nakache, M. Vignes-Adler and M. Dupeyrat, *J. Colloid Interface Sci.*, 1983, **94**, 187.
[14] J. T. Davies and E. K. Rideal, in *Interfacial Phenomena* (Academic Press, New York, 1963).
[15] L. Ter Minassian Saraga, *J. Chim. Phys. Phys. Chim. Biol.*, 1955, **52**, 181.
[16] M. Dupeyrat, E. Nakache and M. Vignes-Adler, in *Chemical Instabilities*, ed. G. Nicolis and F. Baras, NATO Advanced Science Institute, series C (D. Reidel, Dordrech, 1984), vol. 120.

Double Layers at Liquid/Liquid Interfaces

By Zdeněk Samec,* Vladimír Mareček and Daniel Homolka

J. Heyrovský Institute of Physical Chemistry and Electrochemistry,
Czechoslovak Academy of Sciences, Utováren 254, 10200 Prague 10, Czechoslovakia

Received 21st November, 1983

The electrical double layer at the interface between two immiscible electrolyte solutions (ITIES) has been studied by the fast-galvanostatic-pulse method for the system consisting of aqueous NaBr and a solution of tetrabutylammonium tetraphenylborate in nitrobenzene. The double-layer capacity has been evaluated as a function of the potential difference across the interface. The modified Verwey–Niessen model, in which a layer of oriented solvent molecules (the inner layer) separates two space-charge regions (the diffuse double layer), seems to provide a reasonable framework to interpret the experimental data, assuming (i) that the approximations to the Poisson–Boltzmann equation by Gouy and Chapman are removed and (ii) that the boundary between the space-charge region and the inner layer is considered to be diffuse rather than sharp. The use of the tetrabutylammonium cation as the reference ion in voltammetric studies of the water/nitrobenzene interface is discussed.

In many respects the development of models for the interface between two immiscible electrolyte solutions (ITIES) has paralleled that for metal/electrolyte or semiconductor/electrolyte interfaces. In the earliest model by Verwey and Niessen,[1] the ITIES was represented by a diffuse double layer (*i.e.* one phase contains an excess of the positive space charge and the other phase an equal excess of the negative space charge) which can be treated by the theory of Gouy[2] and Chapman[3] (GC). By analogy with Stern's modification[4] of the GC theory, Gavach et al.[5] introduced the concept of an ion-free layer of oriented solvent molecules which separates the two space-charge regions at the ITIES. On the other hand, Boguslavsky et al.[6,7] neglected the diffuse double layer and assumed an ionic bilayer at the ITIES formed by the specifically adsorbed ions on one side of the interface and by ions of the opposite charge on the other. A similar model based on ionic adsorption up to a monolayer thickness was adopted by Joos and Vanden Bogaert.[8]

The structure of the electrical double layer at the non-polarized[5,8-13] as well as at the ideally polarized[14-17] interfaces between two immiscible electrolyte solutions has been studied by measuring the surface tension[5,8-15] and impedance,[16,17] mainly for the systems water/nitrobenzene[5,8,10-16] and water/1,2-dichloroethane.[13,17] The experimental data are consistent with the modified Verwey–Niessen (MVN) model,[5] although the contribution of the inner layer to the double-layer properties is much less pronounced than in the case of metal/electrolyte interfaces. Thus the potential drop across the inner layer at the water/nitrobenzene interface has been found practically to be independent of the surface charge density[13,14] and equal to 1,[14] 0 ± 5[12] and 42 mV,[16] or its variation with the surface charge density was detected with the inner-layer potential drop being *ca.* 20 mV at the potential of zero charge.[15] In the latter case, however, the corresponding inner-layer capacity is rather large, as confirmed by the capacity data, which were inferred from the impedance measurements.[18] The evaluation of the surface excess of water at the water/nitrobenzene interface[13] implied that the inner layer provides only a fraction of the monolayer of the solvent molecules. This led the authors of ref. (13) to the conclusion that the concept of an ion-free layer at the ITIES is doubtful, while the continuous change in the composition from one phase to the other is a more realistic picture. Nevertheless, such an idea can be incorporated into the MVN model by considering the boundary between the space-charge region and the inner layer to be diffuse rather than sharp, by analogy with the treatment of the metal/electrolyte interface using the non-local electrostatic approach.[19]

In this paper we have tried to obtain insight into the structure of the interface between aqueous NaBr and a solution of tetrabutylammonium tetraphenylborate (TBATPB) in nitrobenzene on the basis of fast-galvanostatic-pulse measurements, which have recently been shown to provide reliable capacitance data for the ITIES.[20]

EXPERIMENTAL

CELL

A diagram of the four-electrode cell is shown in fig. 1. In contrast to the arrangement used in several laboratories,[21-24] the lower-density aqueous phase occupied the bottom of the cell, with the nitrobenzene phase above it. By means of a microsyringe connected to the cell, a flat and reproducible water/nitrobenzene interface with a geometric area of 22 mm^2 was adjusted in the round hole cut in the glass barrier B (cf. fig. 1). A remarkable improvement in the cell's response to voltage or current pulses was achieved by placing an isolated metallic wire inside the Luggin capillary for the nitrobenzene phase, so that only the metallic disc was exposed to the solution at the tip of the capillary, with the other end of the wire connected to silver of the reference electrode RE2. Prior to the experiment the cell was carefully washed with acetone, then with doubly distilled water and finally dried in a dry-box. During the experiment the cell was immersed in a water bath, the temperature of which was maintained at 25.0 ± 0.1 °C.

The potential difference E of the galvanic cell (I)

Ag/AgBr	$c°$(NaBr)	$c°$(TBATPB)	$c°$(TBABr)	AgBr/Ag'	
RE1	(w)	(o)	(w')	RE2	(I)

was controlled or measured in the potentiostatic or galvanostatic experiments, respectively, with $c° = 0.01, 0.05$ or 0.10 mol dm^{-3}. Since the potentials of the Ag/AgBr reference electrodes RE1 and RE2 practically cancel each other, the potential difference E can be written as

$$E = \Delta_{Ag'}^{Ag}\phi = \Delta_o^w\phi + \Delta_w^{Ag}\phi - \Delta_o^{Ag'}\phi$$
$$\approx \Delta_o^w\phi - \Delta_o^w\phi_{TBA^+}^\circ - (RT/F)\ln[\gamma_{TBA^+}(o)/\gamma_{TBA^+}(w')] \quad (1)$$

where $\gamma_{TBA^+}(o)$ and $\gamma_{TBA^+}(w')$ are the activity coefficients of the TBA$^+$ ion in nitrobenzene and in the reference aqueous phase, respectively. Consequently, E is the Galvani potential difference across the tested interface, $\Delta_o^w\phi = \phi(w) - \phi(o)$, and is related to the formal potential difference for TBA$^+$ ion:

$$\Delta_o^w\phi_{TBA^+}^{\ominus} = \Delta_o^w\phi_{TBA^+}^\circ + (RT/F)\ln[\gamma_{TBA^+}(o)/\gamma_{TBA^+}(w')]. \quad (2)$$

The standard potential difference $\Delta_o^w\phi_X^\circ$ for an ion X^z of charge z can be calculated from the standard Gibbs energy of transfer of the ion from water to the organic-solvent phase, $\Delta G_{tr,X}^{\circ,w\to o}$:[25]

$$\Delta_o^w\phi_X^\circ = \Delta G_{tr,X}^{\circ,w\to o}/zF \quad (3)$$

e.g. for the TBA$^+$ ion, $\Delta_o^w\phi_{TBA^+}^\circ = -0.248$ V.[25]

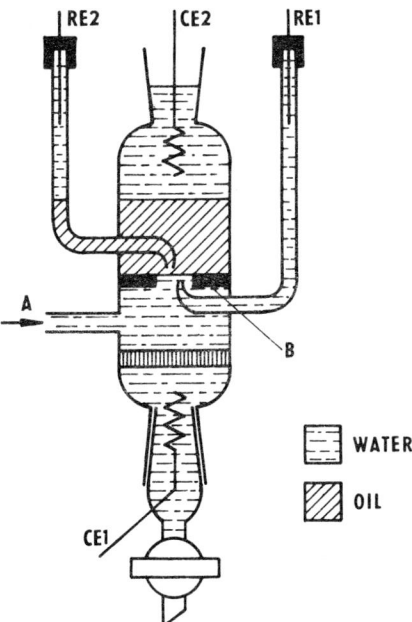

Fig. 1. Diagram of the electrolytic cell: RE1 and RE2 are the Ag/AgBr reference electrodes dipped in aqueous solutions of NaBr and TBABr, respectively; CE1 and CE2 are the platinum counter-electrodes dipped in an aqueous solution of NaBr; oil is the nitrobenzene solution of TBATPB; A is the connection to the microsyringe for the interface adjustment; B is the glass barrier in which a round hole has been cut (5.3 mm in diameter).

ELECTROLYTES

The aqueous electrolyte solutions were prepared from doubly distilled water and NaBr (p.a., Lachema) or tetrabutylammonium bromide (TBABr) (puriss, p.a., Fluka AG). NaBr was recrystallized from water prior to use. A fresh nitrobenzene solution of the electrolyte was prepared from nitrobenzene (p.a., Lachema) and tetrabutylammonium tetraphenylborate (TBATPB). Nitrobenzene was purified by fractional distillation under reduced pressure. TBATPB was prepared by the precipitation[26] of tetrabutylammonium iodide (Schuchardt) and sodium tetraphenylborate (puriss., p.a., Fluka AG). The product was purified by double recrystallization from acetone. The bromide contents of the aqueous solutions were checked by argentometric titration.

APPARATUS

Cyclic voltammetry was performed using a four-electrode potentiostat with ohmic potential-drop compensation based on positive feedback.[27] A block diagram of the potentiostat, which can be adapted from a conventional three-electrode potentiostat,[27] is shown in fig. 2. The potentiostat was controlled by a voltage-pulse generator and the current flowing through the cell was measured as the floating voltage drop across the measuring resistor R3 in fig. 2.

The electronic circuit used in the galvanostatic-pulse measurements is shown schematically in fig. 3. A single-current pulse of amplitude $I_0 = 25$, 50 or 100 μA was generated in the feedback loop of the operational amplifier OA1 by applying a square voltage pulse from a digital-to-analogue converter (12-bit DAC 12QZ, Analog Devices) across a 1 kΩ input resistor. The potential difference between the reference electrodes RE1 and RE2 was sampled by

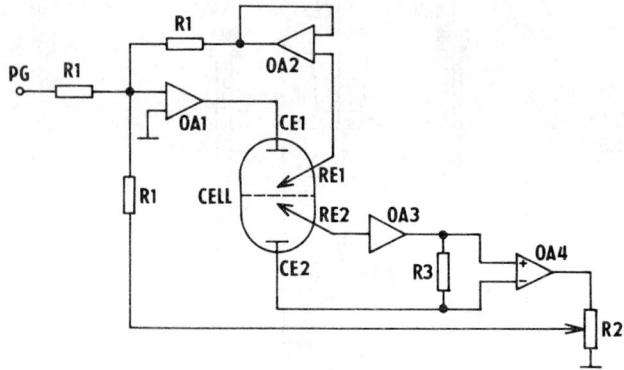

Fig. 2. Block diagram of the four-electrode potentiostat with positive feedback for the ohmic potential-drop compensation.

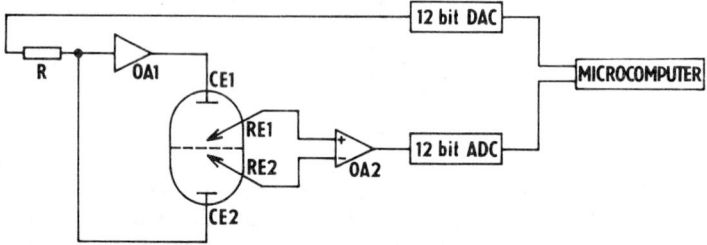

Fig. 3. Block diagram of the electronic circuit for the galvanostatic-pulse measurements.

means of a data-acquisition system (SDM 853, Burr Brown) every 100 μs for 5 ms before the current pulse was applied ($I = 0$) and for the next 5 ms pulse duration. All 100 points of the sampled potential difference were stored in the memory of a microcomputer (Intel 8080 microprocessor) which was also used to operate the digital-to-analogue converter and the data-acquisition system.

Throughout the paper the electric current connected with the transfer of the positive charge from the aqueous phase to the nitrobenzene phase will be considered as positive.

RESULTS AND DISCUSSION

CYCLIC VOLTAMMETRY

In the voltammetric experiment the principal aim was to find the potential range to use in the galvanostatic-pulse measurements, i.e. the range within which the boundary behaves as an ideally polarized interface. Fig. 4 shows a cyclic voltammogram of 0.01 mol dm^{-3} NaBr in water and 0.01 mol dm^{-3} TBATPB in nitrobenzene. Since Na$^+$ ion transfer from water to nitrobenzene is more favourable energetically ($\Delta G^{\circ,w \to o}_{tr,Na^+} = 34.2$ kJ mol^{-1})[28] than the transfer of the TPB$^-$ ion in the opposite direction ($\Delta G^{\circ,w \to o}_{tr,TPB^-} = -35.9$ kJ mol^{-1}),[28] the increasing positive current at ca. 0.45 V probably corresponds to the former ion transfer. Analogously, the increasing negative current at ca. 0.13 V is due to TBA$^+$ ion transfer from nitrobenzene to water ($\Delta G^{\circ,w \to o}_{tr,TBA^+} = -24.0$ kJ mol^{-1})[28] rather than to Br$^-$ ion transfer from water to nitrobenzene ($\Delta G^{\circ,w \to o}_{tr,Br^-} = 28.4$ kJ mol^{-1}).[28] In the potential range 0.13–0.45 V the current is controlled mainly by charging the interface, and hence in this range the system can be expected to possess the properties required.

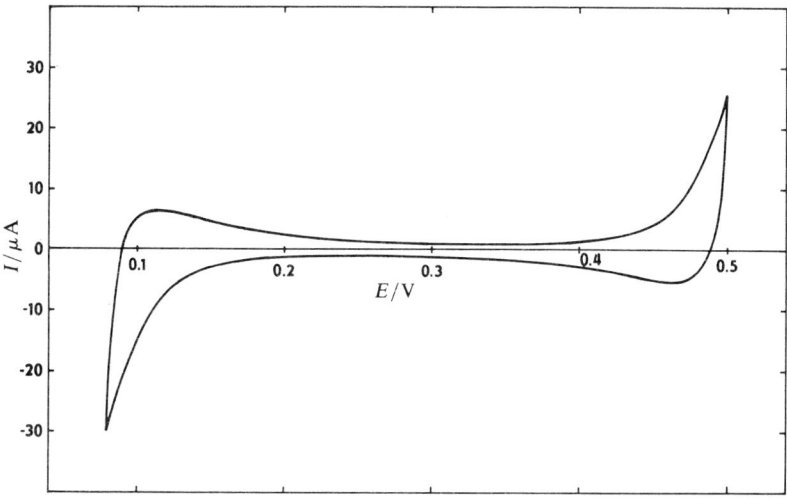

Fig. 4. Cyclic voltammogram of 0.01 mol dm^{-3} NaBr in water and 0.01 mol dm^{-3} TBATPB in nitrobenzene. Scan rate 0.1 V s^{-1}. Ohmic potential-drop compensation adjusted to 1.35 kΩ.

Table 1. Comparison of the standard potential differences $E_X^\circ = \Delta_o^w \phi_X^\circ - \Delta_o^w \phi_{TBA^+}^\circ$ from voltammetric (CV) and extraction (EXT) measurements for several ions

	$E_{1/2}^{rev}$/V	$D_X(w)/D_X(o)^a$	$E_X^\circ(CV)$/V	$E_X^\circ(EXT)$/V^b
tetramethylammonium	0.320[30]	2.64	0.308	0.283
tetraethylammonium	0.230[30]	2.03	0.221	0.189
tetrabutylammonium	—	1.69	(0)	(0)
Cs$^+$	0.413[27]	(4.4)	0.436	0.407
picrate	0.310[21]	1.87	0.318	0.295

^{a} Estimated as the ratio of the limiting ionic conductivities in water[31] and nitrobenzene.[32] The limiting ionic conductivity of the Cs$^+$ ion in nitrobenzene was assumed to be the same as for K$^+$. ^{b} Evaluated from the standard galvanic potential differences, which were calculated[25] from the extraction data.[28]

REFERENCE ION

Standard Gibbs transfer energies or standard potential differences for the individual ions are not accessible to direct measurement, since they are always related to the corresponding quantity for another ion, cf. eqn (1). Using an extrathermodynamic hypothesis, Rais[28] evaluated standard Gibbs energies of transfer of an ion from water to nitrobenzene from extraction data. Standard potential differences for ion transfer across the water/nitrobenzene interface were inferred from the voltammetric measurements using exclusively TBA$^+$ as the reference ion and have been compiled by Hung,[29] together with values obtained from the extraction data. However, a closer inspection of the voltammetric data reported in the literature indicates that the value $\Delta_o^w \phi_{TBA^+}^\circ = -0.248$ V[25] on which the evaluations of the standard potential differences have been based should be corrected. Table 1 summarizes the reversible half-wave potentials $E_{1/2}^{rev}$ which were determined for several ions

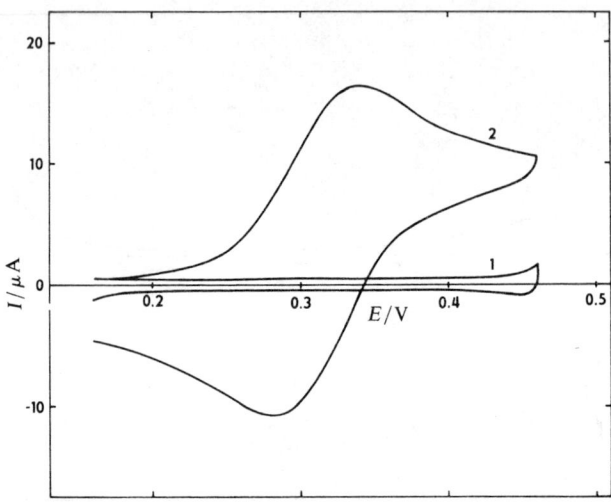

Fig. 5. Cyclic voltammogram of the transfer of tetramethylammonium ions across the water/nitrobenzene interface at 10 mV s^{-1}. Nitrobenzene phase: 0.01 mol dm^{-3} TBATPB; aqueous phase: 0.01 mol dm^{-3} NaBr (1) or 0.01 mol dm^{-3} NaBr and 0.8 mmol dm^{-3} TMABr (2).

by means of cyclic voltammetry.[21,27,30] The standard potential differences $E_X^\circ = \Delta_o^w \phi_X^\circ - \Delta_o^w \phi_{TBA^+}^\circ$ for the tetramethylammonium, tetraethylammonium and picrate ions have been calculated from the equation

$$E_{1/2}^{rev} = E_X^\circ + (RT/zF) \ln [D_X(w)/D_X(o)]^{1/2} + \Delta E' \qquad (4)$$

where z is the charge number of the transferred ion X^z. The ratio of the diffusion coefficients in the aqueous and the nitrobenzene phases, $D_X(w)/D_X(o)$, has been estimated as the ratio of the limiting ionic conductivities in water[31] and nitrobenzene[32] (cf. table 1). The activity-coefficient term $\Delta E'$ has been neglected, since its estimate from the extended Debye–Hückel equation amounts to $\Delta E' \leqslant 5$ mV. With respect to the appreciable association between the Cs$^+$ and TPB$^-$ ions in nitrobenzene, the standard potential difference for the Cs$^+$ ion must be evaluated from the equation[33]

$$E_{1/2}^{rev} = E_{Cs^+}^\circ + (RT/F) \ln [D_{Cs^+}(w)/D_{Cs^+}(o)]^{1/2}$$
$$- (RT/F) \ln \{1 + K_a c_{TPB^-}^\circ(o)[D_{CsTPB}(o)/D_{Cs^+}(o)]^{1/2}\} \qquad (5)$$

where $c_{TPB^-}^\circ(o)$ is the bulk concentration of TPB$^-$ ion in nitrobenzene, the association constant $K_a = 180$ dm^3 mol^{-1} at 20 °C and $D_{CsTPB}(o)/D_{Cs^+}(o) \approx 0.22$.[33]

The standard potential differences E_X° which have been obtained in this way are more positive by ca. 30 mV than those evaluated from the extraction measurements (cf. table 1), the mean being 27 ± 4 mV. Consequently, the reasonable agreement between the two sets of the data is reached by taking $\Delta_o^w \phi_{TBA^+}^\circ = -0.275$ V.

In this study we have used the tetramethylammonium cation as the reference ion *in situ*, i.e. after the galvanostatic experiment was complete, TMA$^+$ was dissolved in nitrobenzene or the aqueous phase and the cyclic voltammogram measured. A typical curve is shown in fig. 5. The conversion of the potential E into the Galvani potential difference $\Delta_o^w \phi$ has been made using the standard potential $E_{TMA^+}^\circ$, which has been inferred from the cyclic voltammogram, and $\Delta_o^w \phi_{TMA^+}^\circ = 0.035$ V.[25]

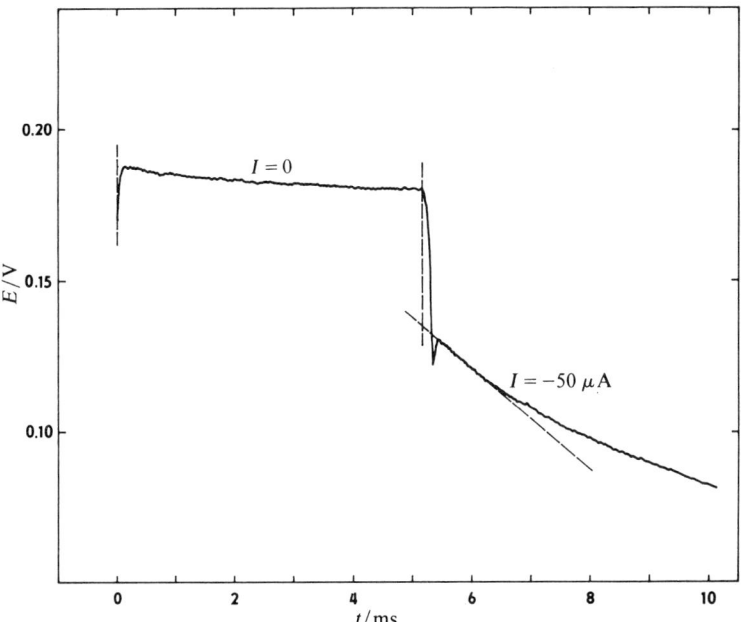

Fig. 6. Galvanostatic transient ($I_0 = -50$ μA) at the water/nitrobenzene interface. Aqueous phase: 0.01 mol dm^{-3} NaBr; nitrobenzene phase: 0.01 mol dm^{-3} TBATPB.

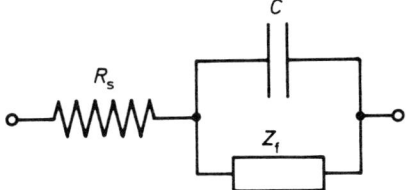

Fig. 7. Equivalent circuit for the water/nitrobenzene interface: C is the capacitance of the interface, Z_f is the faradaic impedance and R_s is the solution resistance between the tips of the Luggin capillaries.

GALVANOSTATIC-PULSE MEASUREMENTS

An example of a galvanostatic transient for the water/nitrobenzene interface is shown in fig. 6. In general, the current I_0 is the sum of the double-layer charging current I_c and the faradaic current I_f[35] (cf. the equivalent circuit in fig. 7)

$$I_0 = I_c + I_f = C(\mathrm{d}E/\mathrm{d}t) + I_f \tag{6}$$

where C is the capacity of the double layer. When a current step $\delta I = I_0$ is initialized at $t = t_0 = 5$ ms, the charging current I_c decreases with time while the faradaic current I_f increases. In the present case the faradaic current is due to the transfer of Na$^+$ or TBA$^+$ ions across the water/nitrobenzene interface. Since these processes are fast,[27] the quasi-equilibrium Nernst potential difference E_r should be established when the current flowing through the interface is zero, i.e.

$$E_r = E_{1/2}^{\mathrm{rev}} + (RT/zF) \ln \left[D_X^{1/2}(\mathrm{o}) c_X^*(\mathrm{o}) / D_X^{1/2}(\mathrm{w}) c_X^*(\mathrm{w}) \right] \tag{7}$$

where the reversible half-wave potential $E_{1/2}^{\text{rev}}$ is given by eqn (4) and $c_X^*(\text{o})$ or $c_X^*(\text{w})$ are the concentrations of the transferred ion at the interface on the nitrobenzene or aqueous side, respectively. For a small variation $\delta E \ll RT/zF$ of the potential difference E from E_r, I_c decreases with time according to[20]

$$\frac{I_c}{I_0} = \frac{RTC\{1 + \exp[(zF/RT)(E_r - E_{1/2}^{\text{rev}})]\}}{(zF)^2[\pi D_X(\text{o})\Delta t]^{1/2} c_X^*(\text{o})} \tag{8}$$

where Δt is time elapsed from $t = t_0$. Obviously, at very short times $(I_c/I_0) \to 1$ and

$$I_0 = C(\mathrm{d}E/\mathrm{d}t)_{\Delta t \to 0}. \tag{9}$$

Consequently, the interface behaves as a resistor R_s and capacitor C in series, cf. fig. 7. Under these conditions the ohmic potential drop $\delta E_0 = I_0 R_s$ appears as a step on the galvanostatic transient at $t = t_0$, while the slope of this transient at $t > t_0$ is controlled exclusively by the capacity C, eqn (9). The potential range over which eqn (9) holds with sufficient accuracy $(I_c/I_0 > 0.99)$ in the course of a pulse duration $\Delta t = 5$ ms can be estimated from eqn (7) and (8). This estimate shows[20] that eqn (9) is applicable when E_r (vs TBA$^+$) falls between ca. 150 and 450 mV or for the galvanic potential difference $\Delta_o^w \phi$ between ca. -150 and 150 mV.

The capacitance C of the water/nitrobenzene interface has been evaluated using eqn (9) from the slopes of the galvanostatic transient at $t = t_0 = 5$ ms and at $t = 10$ ms. In fig. 8 C is plotted as a function of the potential difference $\Delta_o^w \phi$ which was evaluated from the potential difference E_r as described in the previous section. E_r was estimated as the limit of $E(t)$ for $t \to t_0^-$.

We shall now examine the compatibility between the capacitance data obtained and the MVN model.

ZERO-CHARGE POTENTIAL DIFFERENCE

For the MVN model the galvanic potential difference $\Delta_o^w \phi$ can be written as the sum of three contributions:

$$\Delta_o^w \phi = \Delta_o^w \phi_i + \phi_2(\text{o}) - \phi_2(\text{w}) \tag{10}$$

where $\Delta_o^w \phi_i$ is the potential difference across the inner layer and $\phi_2(\text{o})$ or $\phi_2(\text{w})$ are the potential differences across the diffuse layers in the organic solvent and the aqueous phase, respectively. In the absence of specific adsorption in the inner layer, the double-layer capacitance, C, can be represented as a series combination of the inner-layer capacitance C_i and the diffuse-layer capacitances $C_{2,o}$ and $C_{2,w}$:[16]

$$C^{-1} = \mathrm{d}\Delta_o^w \phi/\mathrm{d}q(\text{w}) = C_i^{-1} + C_{2,o}^{-1} + C_{2,w}^{-1} \tag{11}$$

where $q(\text{w})$ is the surface charge density on the aqueous side of the interface, $C_i = \mathrm{d}q(\text{w})/\mathrm{d}\Delta_o^w \phi_i$, $C_{2,o} = \mathrm{d}q(\text{w})/\mathrm{d}\phi_2(\text{o})$ and $C_{2,w} = -\mathrm{d}q(\text{w})/\mathrm{d}\phi_2(\text{w})$. By solving the Poisson–Boltzmann equation of Gouy and Chapman[2,3] for 1:1 electrolytes[1,5] the capacitance of the diffuse double layer, C_d, can be expressed as[16]

$$\begin{aligned} C_d^{-1} &= C_{2,o}^{-1} + C_{2,w}^{-1} \\ C_{2,o} &= (FA^o/RT)\cosh[F\phi_2(\text{o})/2RT] \\ C_{2,w} &= (FA^w/RT)\cosh[F\phi_2(\text{w})/2RT] \\ A^{o(w)} &= [2RT\varepsilon^{o(w)} c^o]^{1/2} \end{aligned} \tag{12}$$

where $\varepsilon^{o(w)}$ is the dielectric constant and c^o the bulk electrolyte concentration.

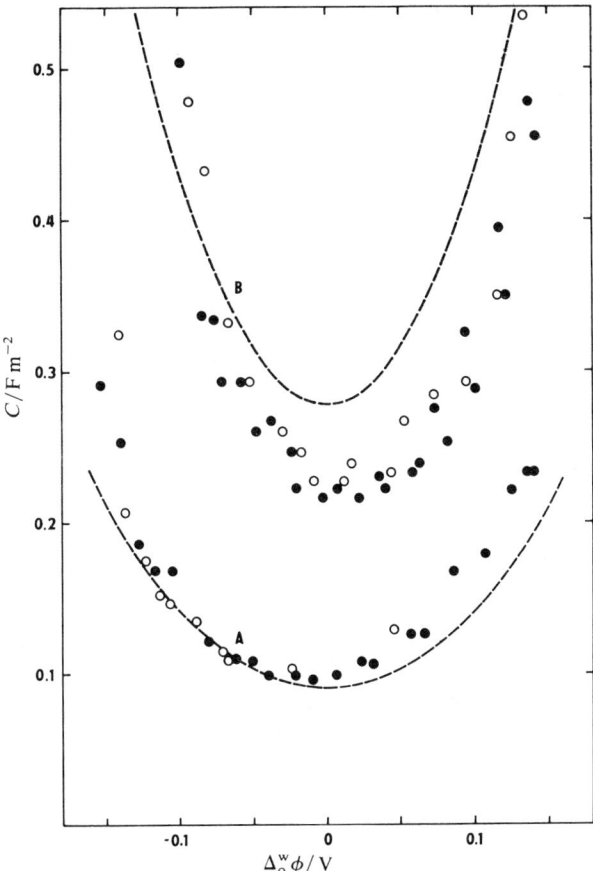

Fig. 8. Plot of the capacitance C of the water/nitrobenzene interface as a function of the potential difference $\Delta_o^w \phi$ for two concentrations of NaBr in water and of TBATPB in nitrobenzene: (A) 0.01 and (B) 0.1 mol dm^{-3}. Capacitance data evaluated from the slope of the galvanostatic transient at $t = 5$ ms (●) and $t = 10$ ms (○). Dashed lines show the capacitance of the diffuse double layer, C_d, calculated using Gouy–Chapman theory for $\Delta_o^w \phi_i =$ constant $= 0$ and relative permittivities $\varepsilon^w = 78.54$, $\varepsilon^o = 34.82$.

The zero-charge potential difference $\Delta_o^w \phi_{pzc}$ [$q(w) = q(o) = 0$] has been found from the potential difference corresponding to the minimum experimental capacitance at low electrolyte concentrations assuming that close to this potential difference $C_i \gg C_d$ and hence $C \approx C_d$. Since the minimum capacitance at $c^o = 0.01$ mol dm^{-3} is found at $\Delta_o^w \phi \approx 0$ (cf. fig. 8), we conclude that the zero-charge potential difference for the system under the investigation is $\Delta_o^w \phi_{pzc} \approx 0$. This result is in a good agreement with the literature data,[15,16] provided that the corrected value $\Delta_o^w \phi_{TBA^+}^o = -0.275$ V is used for the standard potential difference of TBA$^+$ ion transfer. In such a case the zero-charge potential difference reported for the system comprising LiCl in water and TBATPB in nitrobenzene by Kakiuchi and Senda[15] should read $\Delta_o^w \phi_{pzc} = -0.007$ V instead of 0.020,[15] while the value we have reported[16] should read $\Delta_o^w \phi_{pzc} = 0.015$ V instead of 0.042 V.[16]

The dashed lines in fig. 8 show the capacitance of the diffuse double layer C_d calculated from eqn (12) for $\Delta_o^w \phi_i = $ constant $= 0$. At low electrolyte concentrations the experimental capacitance practically coincides with the calculated value, which justifies the use of our assumption leading to the evaluation of the zero-charge potential difference. As the surface charge density increases, the experimental capacitance tends to rise above the theoretical value, particularly at higher electrolyte concentrations. Qualitatively this effect is in line with theories of the electrical double layer[34] in which the approximations involved in the Poisson–Boltzmann equation of Gouy and Chapman[2,3] have been removed.

CAPACITY OF THE INNER LAYER

Another trend which is apparent from fig. 8 is the drop in the experimental capacitance from the Gouy–Chapman value at low surface charge densities and high electrolyte concentrations. This effect can be ascribed to the formation of an inner layer at the water/nitrobenzene interface with a finite capacitance C_i, cf. eqn (11). In this case C_d must be evaluated from the surface charge density q obtained by integrating the plot of capacitance against potential difference. This has been done for the electrolyte concentrations $c° = 0.05$ and 0.10 mol dm^{-3}. Fig. 9 shows plots of the inner-layer capacitance C_i against the surface charge density $q(w)$; these are similar in shape and coincide to some extent. The capacitance of the inner layer is rather large and the corresponding variation of the potential difference across the inner layer $\Delta_o^w \phi_i$ rather small, although observable (cf. fig. 10). The evaluation of the capacitance of the diffuse double layer, C_d, using the modified Poisson–Boltzmann equation[34] would probably lead to a lower estimate of the inner-layer capacitance, C_i, and to a closer coincidence for different electrolyte concentrations, but it should hardly produce a significant change in the picture emerging from fig. 9 and 10. Note that the model in which the two space-charge regions are separated by a monolayer of solvent molecules is physically realistic, at least as the limiting case. The capacitance of such a monolayer can be approximated by the equation for a plate capacitor, i.e.

$$C_i = \varepsilon \varepsilon_0 d^{-1} \tag{13}$$

where ε is the relative permittivity, $\varepsilon_0 = 8.85 \times 10^{-12}$ F m^{-1} is the permitivity of a vacuum and d is the monolayer thickness. By taking $d = 0.31$ nm (the diameter of a water molecule) the minimum $C_i \approx 1$ F m^{-2} corresponds to $\varepsilon \approx 35$, which is roughly the value for monomeric water molecules ($\varepsilon \approx 25$) with random orientations in the absence of an electric field. However, on increasing the charge density, C_i should decrease owing to dielectric saturation. Since the contrary is observed (cf. fig. 9), the electrolyte ions probably gradually penetrate the solvent layer over some distance. Through the shielding of the electric field by ions, the high dielectric permitivity in the solvent layer can be preserved. Moreover, replacement of the solvent molecules by ions could make the solvent layer effectively thinner.

In summary, the MVN model represents a reasonable framework for interpreting double-layer data for liquid/liquid interfaces in the presence of spherical ions. However, its improvement along two lines is needed. First, a more elaborate treatment of the diffuse double layer seems to be desirable, such as that based on the modified Poisson–Boltzmann equation.[34] Secondly, a diffuse rather than sharp boundary between the space-charge region and the solvent layer should be considered. The picture resembles the spillover of metal electrons into the inner solvent layer which has been envisaged by Kornyshev et al.[19] for a metal/electrolyte interface.

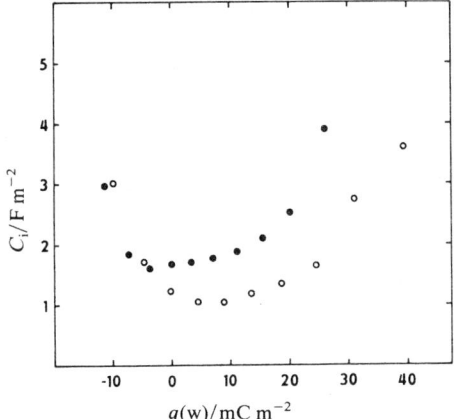

Fig. 9. Capacitance of the inner layer, C_i, at the water/nitrobenzene interface plotted against the surface charge density $q(w)$ on the aqueous side of the interface for two concentrations of NaBr in water and TBATPB in nitrobenzene: ●, 0.05 and ○ 0.1 mol dm^{-3}.

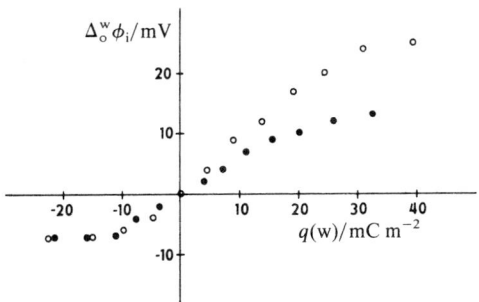

Fig. 10. Potential difference across the inner layer $\Delta_o^w \phi_i$ at the water/nitrobenzene interface plotted against the surface charge density $q(w)$ on the aqueous side of the interface for two concentrations of NaBr in water and TBATPB in nitrobenzene: ●, 0.05 and ○, 0.1 mol dm^{-3}.

---

[1] E. J. W. Verwey and K. F. Niessen, *Philos. Mag.*, 1939, **28**, 435.
[2] G. Gouy, *C. R. Acad. Sci.*, 1910, **149**, 654.
[3] D. L. Chapman, *Philos. Mag.*, 1913, **25**, 457.
[4] O. Stern, *Z. Elektrochem.*, 1924, **30**, 508.
[5] C. Gavach, P. Seta and B. d'Epenoux, *J. Electroanal. Chem.*, 1977, **83**, 225.
[6] V. S. Krylov, V. A. Myamlin, L. I. Boguslavsky and M. A. Manvelyan, *Elektrokhimiya*, 1977, **13**, 834.
[7] M. I. Gugeshashvili, M. A. Manvelyan and L. I. Boguslavsky, *Elektrokhimiya*, 1974, **10**, 819.
[8] P. Joos and R. Vanden Bogaert, *J. Colloid Interface Sci.*, 1976, **56**, 206.
[9] M. Kahleweit and H. Strehlow, *Z. Elektrochem.*, 1954, **58**, 658.
[10] L. I. Boguslavsky, A. N. Frumkin and M. I. Gugeshashvili, *Elektrokhimiya*, 1976, **12**, 858.
[11] P. Seta, B. d'Epenoux and C. Gavach, *J. Electroanal. Chem.*, 1979, **95**, 191.
[12] J. D. Reid, O. R. Melroy and R. P. Buck, *J. Electroanal. Chem.*, 1983, **147**, 71.
[13] H. H. Girault and D. J. Schiffrin, *J. Electroanal. Chem.*, 1983, **150**, 43.
[14] M. Gros, S. Gromb and C. Gavach, *J. Electroanal. Chem.*, 1978, **89**, 29.
[15] T. Kakiuchi and M. Senda, *Bull. Chem. Soc. Jpn*, 1983, **56**, 1753.
[16] Z. Samec, V. Mareček and D. Homolka, *J. Electroanal. Chem.*, 1981, **126**, 121.
[17] P. Hájková, D. Homolka, V. Mareček and Z. Samec, *J. Electroanal. Chem.*, 1983, **151**, 277.

[18] D. Homolka, P. Hájková, V. Mareček and Z. Samec, *J. Electroanal. Chem.*, 1983, **159**, 233.
[19] A. A. Kornyshev, W. Schmickler and M. A. Vorotynsev, *Phys. Rev. B*, 1982, **25**, 5244.
[20] V. Mareček and Z. Samec, *J. Electroanal. Chem.*, 1983, **149**, 185.
[21] D. Homolka and V. Mareček, *J. Electroanal. Chem.*, 1980, **112**, 91.
[22] T. Kakutani, T. Osaki and M. Senda, *Bull. Chem. Soc. Jpn*, 1983, **56**, 991.
[23] O. R. Melroy, W. E. Bronner and R. P. Buck, *J. Electrochem. Soc.*, 1983, **130**, 373.
[24] Z. Koczorowski and G. Geblewicz, *J. Electroanal. Chem.*, 1982, **139**, 177.
[25] J. Koryta, P. Vanýsek and M. Březina, *J. Electroanal. Chem.*, 1977, **75**, 211.
[26] C. Gavach and F. Henry, *J. Electroanal. Chem.*, 1974, **54**, 361.
[27] Z. Samec, V. Mareček and J. Weber, *J. Electroanal. Chem.*, 1979, **100**, 841.
[28] J. Rais, *Collect. Czech. Chem. Commun.*, 1971, **36**, 3253.
[29] Le Q. Hung, *J. Electroanal. Chem.*, 1980, **115**, 159.
[30] T. Osakai, T. Kakutani, Y. Nishiwaki and M. Senda, *Bunseki Kagaku*, 1983, **32**, E81.
[31] D. Dobos, *Electrochemical Data* (Akadémiai Kiadó, Budapest, 1975), p. 76.
[32] J. F. Coetzee and C. P. Cunningham, *J. Am. Chem. Soc.*, 1965, **87**, 2529.
[33] Le Q. Hung, *Ph.D. Thesis* (J. Heyrovský Institute of Physical Chemistry and Electrochemistry, Prague, 1980).
[34] C. W. Outhwaite, L. B. Bhuiyan and S. Levine, *J. Chem. Soc., Faraday Trans. 2*, 1980, **76**, 1388.
[35] R. Parsons, in *Advances in Electrochemistry and Electrochemical Engineering*, ed. P. Delahay (Wiley-Interscience, New York, 1970), vol. 7, p. 177.

Transfer of Alkali-metal and Hydrogen Ions across Liquid/Liquid Interfaces Mediated by Monensin

A Voltammetric Study at the Interface of Two Immiscible Electrolyte Solutions

By Jiří Koryta,* Guo Du,† Wolfgang Ruth§ and Petr Vanýsek

J. Heyrovský Institute of Physical Chemistry and Electrochemistry, Czechoslovak Academy of Sciences, Opletalova 25, 11000 Prague 1, Czechoslovakia

Received 21st November, 1983

Under certain conditions the interface between an aqueous phase and an organic phase (ITIES, interface of two immiscible electrolyte solutions) has certain properties analogous to a metal/electrolyte-solution interface. By polarization of the ITIES using potential-sweep voltammetry reproducible voltammograms corresponding to ion transfer across the ITIES are obtained. In the presence of cation-complexing ionophores in the organic phase cation transfer is facilitated. The acidic form HX of carboxylic ionophores, monensin A and B, acts in the nitrobenzene/water system as a sodium carrier while the complex of its anion with sodium or lithium cation (M^+) is a proton carrier. The equilibrium constants of the reactions $M^+ + HX \rightleftharpoons MHX^+$ and $MX + H^+ \rightleftharpoons MHX^+$ have been determined from the voltammograms. The stabilities of the complexes correspond to the series $Li^+ > Na^+ \gg K^+$. The substitution of methyl (monensin B) for ethyl (monensin A) on ring C increases the stability of the sodium complex by ca. 1 kJ mol^{-1} and has no effect on the acidity of the carboxy group.

Interest in electrochemical processes taking place at liquid/membrane surfaces has led to the development of a new method, voltammetry at the interface of two immiscible electrolyte solutions[1] (VITIES), which has various applications in mechanistic investigations as well as in analysis [for reviews see ref. (2)–(5)]. In this method, the interface between an aqueous phase and an organic phase is polarized by a triangular voltage pulse using a four-electrode potentiostat with positive feedback[1,6] in order to control the interfacial potential difference under electric current flow. Given a proper composition of both the phases, the interface (ITIES, interface between two immiscible electrolyte solutions) behaves similarly to a metal-electrode/electrolyte-solution interface.[4,7] When the organic phase contains a strongly hydrophobic indifferent ('base') electrolyte and the aqueous phase a strongly hydrophilic base electrolyte then, under polarization, a potential window is formed in the current–potential curve where the main contribution to the current is due to charging of the interface (a non-faradaic current), see fig. 1 (a). Consider that an ion whose concentration is considerably lower than that of the base electrolytes and which has a suitable value of the standard Gibbs energy of transfer between both the phases[8] is present in one of the phases. Under polarization in a triangular potential-pulse mode both positive and negative current peaks are observed on the polarization curves [fig. 1 (b)]. The rates of ion transfer are rapid so for low polarization rates the concentration of the ion concerned, close to the ITIES, follows the Nernst potential equation. In this 'reversible' case the current–potential dependence is exactly described by the Randles–Ševčík equation,[9,10] which was originally deduced for a mercury-electrode/electrolyte-solution interface.

† Present address: Institute of Applied Chemistry, Changchun, Jilin, China.
§ Present address: Section Chemie, Universität Rostock, Rostock, German Democratic Republic.

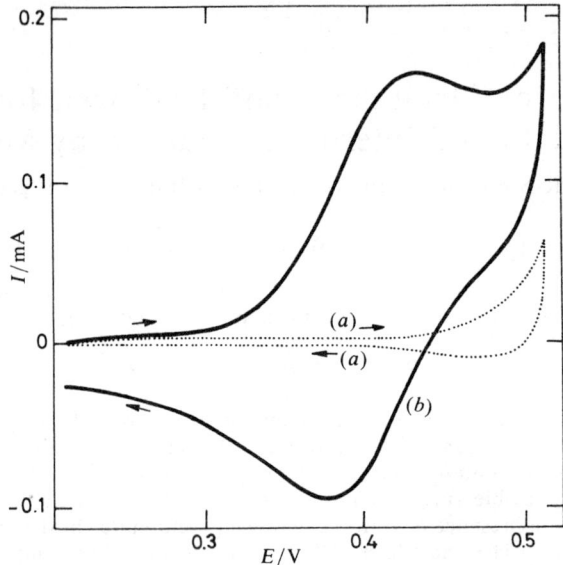

Fig. 1. Triangular-sweep voltammogram for the water/nitrobenzene interface polarized from $E = +0.21$ to 0.51 V and back; polarization rate $dE/dt = 0.005$ V s^{-1}; base electrolytes: 0.05 mol dm^{-3} LiCl in water, 0.05 mol dm^{-3} tetrabutylammonium tetraphenylborate in nitrobenzene. (*a*) Base electrolyte voltammogram, (*b*) 0.001 mol dm^{-3} Cs$^+$ in the aqueous phase. According to ref. (6).

The ITIES charge-transfer kinetics is closely connected with the structure of the electric double layer at the ITIES. According to the Verwey–Niessen model,[11] which correctly describes the situation at the ITIES in the absence of ion adsorption,[12] it consists of two diffuse double layers, the compact layer being absent. Under these conditions the influence of the electric potential difference across the ITIES is displayed as the change in the interfacial concentration of the transferred ion.[13-16]

A number of metabolites act as mobile sites (ionophores) for transport of alkali-metal or alkaline-earth-metal ions across biological membranes. These substances alone or, at least, when bound in a complex have a cyclic structure with polar groups stretching into the internal cavity and with a lipophilic outer envelope which helps to solubilize the complex in non-polar media [for a review see ref. (17)]. The action of these substances has also been modelled by VITIES.[2-5,18-20] In these experiments the aqueous phase usually contains a solution of the chloride of the cation concerned while the non-aqueous phase is a solution of the base electrolyte containing the ionophore at considerably lower concentration than the cation in the aqueous phase. Under these conditions the transfer of the cation can be observed at considerably lower potentials than in the absence of the ionophore, the potential shift being a function of the stability of the complex. The current-generating process is entirely controlled by diffusion of the ionophore to the ITIES and the diffusion of the complex formed from the ITIES into the bulk of the organic phase. The shape of the voltammogram usually corresponds to that of a fully 'reversible' process. This kind of ion transfer is termed 'facilitated' [*cf*. ref. (21)].

Monensin is a carboxylic ionophore,[22,23] three homologues of which are produced by *Streptomyces cinnamonensis*[24] (fig. 2). Although it is an acyclic acid it enters the

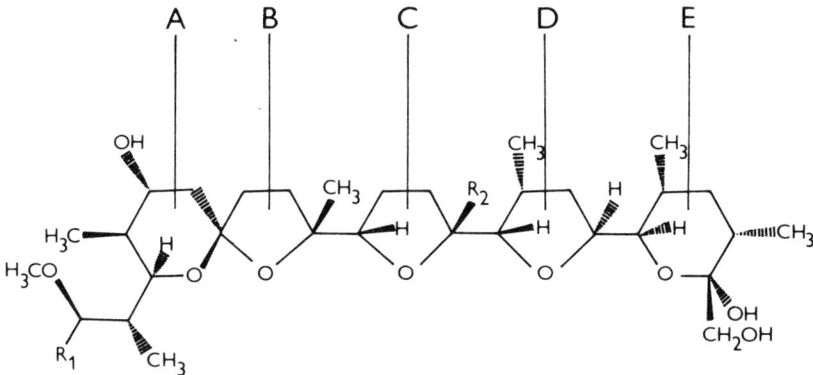

Fig. 2. Structures of monensins A, B and C [according to J. W. Westley in *Antibiotics*, ed. J. W. Corcoran (Springer-Verlag, Berlin, 1981), vol. 4, pp. 41-73]

	R_1	R_2
A	$CH_2(Me)CO_2H$	Et
B	$CH_2(Me)CO_2H$	Me
C	$(CH_2)_3CO_2H$	Me.

complexes in a cyclic form, the terminal groups closing the cycle with a hydrogen bond.[25] Among the alkali-metal cations it predominantly complexes sodium.[26-29] Because of its acidic properties monensin mediates the diffusion of sodium and hydrogen ions across biological membranes as well as bilayer lipid membranes, the transported species being the undissociated acid and the sodium complex of the monensin anion.[26,29] However, a sodium complex with the undissociated acid has also been found in the solid state.[30]

The mechanism of the transfer of an alkali-metal ion across the nitrobenzene/water interface and the determination of the equilibrium constant characterizing this process are the subjects of the present paper.

EXPERIMENTAL

The electrolytic cell is shown in fig. 3. The four-electrode potentiostat with positive feedback was the same as described in earlier communications.[1,6]

The organic phase contained 10 mmol dm^{-3} tetrabutylammonium tetraphenylborate as base electrolyte. The aqueous phase contained either alkali-metal nitrate together with dilute nitric acid or acetate, phosphate or borate buffer prepared from the salts of the alkali metals under investigation. The aqueous solution always contained 1 mmol dm^{-3} Cl$^-$. Before each measurement both the aqueous and the organic phase were brought into contact and equilibrated by shaking.

Monensins A and B (sodium salts) were the gift of Dr W. R. Fields, Lilly Research Laboratories, Indianapolis, Indiana, U.S.A. The mixture of monensins A and B (sodium salts) was the gift of Dr Z. Vaněk, Institute of Microbiology, Czechoslovak Academy of Sciences, Prague, Czechoslovakia. Nitrobenzene was a reagent grade product of Lachema, Brno, Czechoslovakia.

RESULTS

The voltammograms of the monensins recorded under the conditions described above (fig. 4) have the following characteristics: (i) the peak voltammograms appear

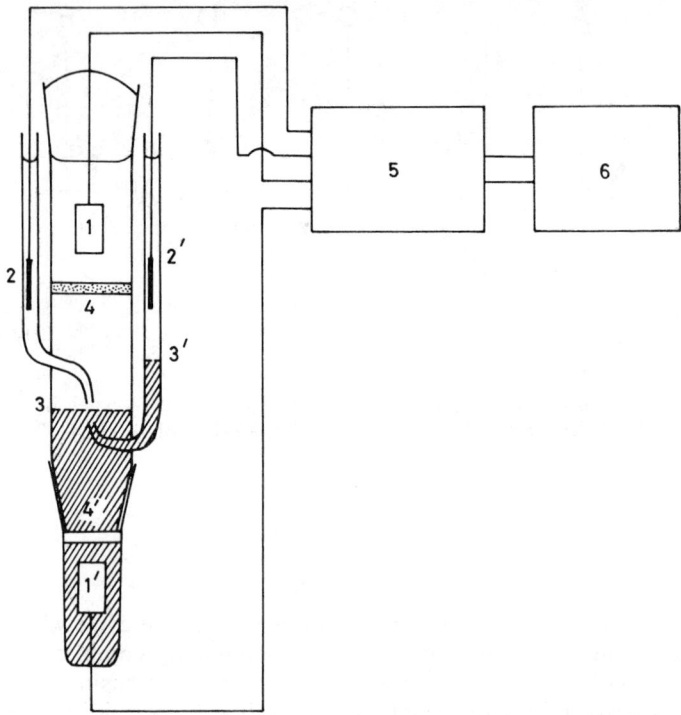

Fig. 3. Electrolytic cell. 1,1': Auxilliary platinum electrodes; 2,2': reference electrodes attached as close as possible to the ITIES by means of Luggin capillaries; 3: the ITIES; 3': the interface between the aqueous electrolyte of the reference electrode 2' and the organic electrolyte; 4,4': sintered glass diaphragms; 5: 4-electrode potentiostat and pulse generator; 6: recorder. The potential difference at 3' is kept constant as the organic electrolyte and the aqueous electrolyte of the reference electrode 2' have a common ion.

in aqueous solutions containing sodium or lithium as cations while potassium has no effect (the pH of the aqueous solutions must be <8); (ii) both the positive and the negative peak currents are proportional to the concentration of the monensin species; (iii) the difference in the potential of the positive and negative current peak, E_p^+ and E_p^-, is 60 ± 4 mV; (iv) the difference in the peak potential and half-peak potential, $E_p - E_{p/2}$, is 60 mV; (v) the peak potentials do not depend on monensin concentration; (vi) the peak current is directly proportional to the square root of the sweep rate, $v = dE/dt$; (vii) fig. 5 shows the dependence of the half-wave potential of the voltammogram, $E_{1/2} = 1/2\,(E_p^+ + E_p^-)$, on pH (obviously, at low pH values the half-wave potential is independent of pH while at higher pH it increases linearly with slope $\partial E_{1/2}/\partial \mathrm{pH} = 59$ mV); (viii) in acidic media the half-wave potential shows a linear dependence on the logarithm of the concentration of the alkali-metal ion c_{M^+} with slope $\partial E_{1/2}/\partial \log c_{M^+} = 59$ mV (at pH > 6 the half-wave potential is independent of the concentration of the alkali-metal ion).

From these findings we reach the following conclusions: (i) the voltammograms correspond to a reversible one-electron process of charge transfer[9,10] [for an instructive survey of the necessary criteria see ref. (31)]; (ii) while sodium and lithium show appreciable complex formation, the complexation of potassium is negligible;

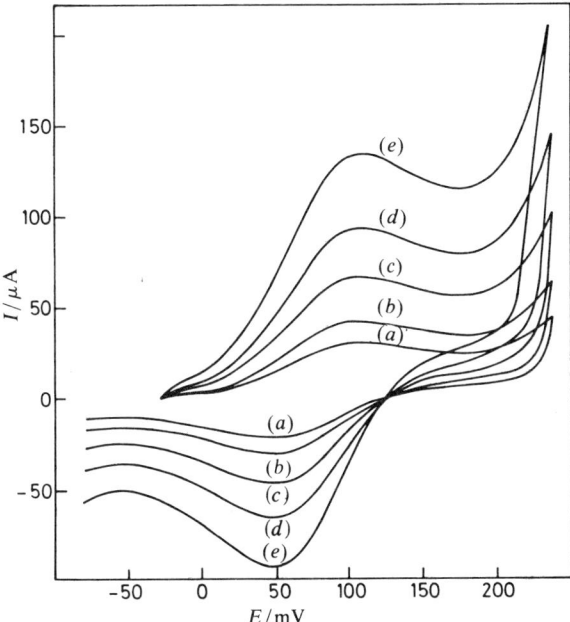

Fig. 4. Triangular sweep voltammograms of monensin. Aqueous electrolyte: 0.1 mol dm^{-3} sodium acetate, 3.2 mol dm^{-3} acetic acid, 0.01 mol dm^{-3} NaCl; nitrobenzene electrolyte: 0.01 mol dm^{-3} tetrabutylammonium tetraphenylborate, 0.001 mol dm^{-3} monensin. Polarization rates dE/dt: (a) 0.005, (b) 0.01, (c) 0.025, (d) 0.05 and (e) 0.1 V s^{-1}. The peak current is directly proportional to the square root of the polarizaion rate while the peak potential is constant.

(iii) the dependence of the half-wave potential on the concentrations of both the hydrogen and metal ions is described by the equation

$$E_{1/2} = (RT/F) \ln \left(\frac{k_1}{c_{M^+}(w)} + \frac{k_2}{c_{H^+}(w)} \right) \quad (1)$$

where $c_{H^+}(w)$ is the concentration of the hydronium ion in water.

DISCUSSION

The finding that the charge-transfer process is reversible, that the current is proportional to the concentration of the monensin species and that the half-wave potential depends in a characteristic way on the concentration of either the alkali-metal or the hydrogen ion is in agreement with the characteristics of ionophore-facilitated ion transfer across the ITIES.[2,4,5] At low pH values the monensin species facilitates alkali-metal-ion transfer across the ITIES, while at higher pH it mediates hydrogen-ion transfer. Thus, we assume that the monensin species at lower pH is the undissociated acid HX while at higher pH it is converted during equilibration into the complex MX, which then acts as an ionophore for the hydrogen ion. This transformation occurs in an electroneutral exchange reaction

$$M^+(w) + HX(o) \rightleftarrows MX(o) + H^+(w). \quad (2)$$

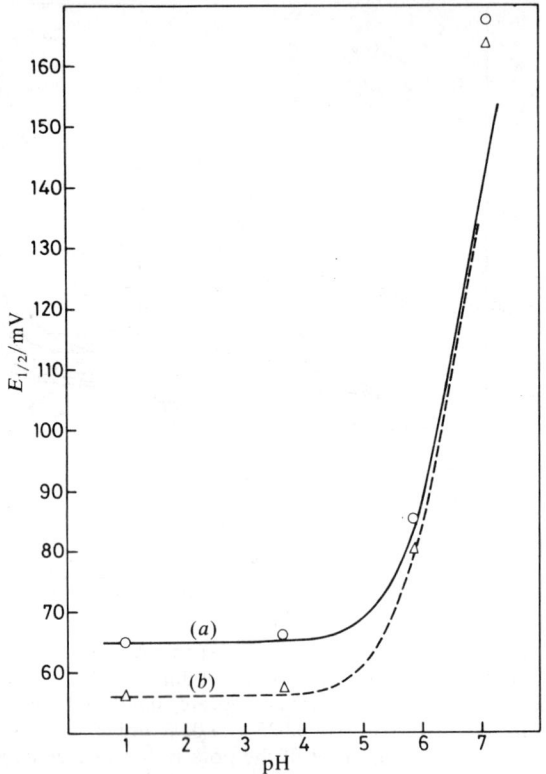

Fig. 5. Dependence of half-wave potential $E_{1/2}$ of (*a*) monensin A and (*b*) monensin B at $c_{Na^+} = 0.1$ mol dm^{-3} on pH. Because of overlapping of the facilitated transfer of Na$^+$ with the transfer of the base electrolyte ions the determination of $E_{1/2}$ at pH 7.1 is inaccurate.

Both the alkali-metal and hydrogen ions are transferred through the species MHX$^+$, whose existence in the solid state has been shown by X-ray crystallography.[30]

At low pH the potential difference across the ITIES is determined by the activities of the alkali-metal ion in the aqueous and organic phases, according to the Nernst equation

$$\Delta_o^w \phi = \Delta_o^w \phi^\circ + (RT/F) \ln c_{M^+}(o)/C_{M^+}(w) \tag{3}$$

where $c_{M^+}(o)$ and $c_{M^+}(w)$ are the concentrations of M$^+$ in the organic phase and in the aqueous phase, respectively, the activity coefficient being included in the standard potential term, and $\Delta_o^w \phi_{M^+}^\circ$ is the standard potential difference of transfer of M$^+$ from the organic phase to the aqueous phase related to the Gibbs standard energy of transfer of M$^+$ from the organic to the aqueous phase,[8] $\Delta G_{tr,M^+}^{\circ,o\rightarrow w}$, by the equation

$$\Delta_o^w \phi_{M^+}^\circ = -\Delta G_{tr,M^+}^{\circ,o\rightarrow w}/F. \tag{4}$$

The equilibrium in the organic phase is characterized by the equation

$$K_1 = c_{MHX^+}(o)/c_{M^+}(o)c_{HX}(o). \tag{5}$$

Table 1. Equilibrium constants K_1, K_2 and K_{exch} for monensin A and B (at 22 °C) based on the values of $\Delta_o^w \phi_{Na^+}^\circ = 0.354$ V, $\Delta_o^w \phi_{Li^+}^\circ = 0.395$ V and $\Delta_o^w \phi_{H^+}^\circ = 0.337$ V

ion	ionophore	log K_1	log K_2	log K_{exch}
Na$^+$	monensin A	5.9	10.4	−4.8
Na$^+$	monensin B	6.05	10.4	−4.65
Li$^+$	monensin A + B mixture	6.4	13.3	−5.16

After insertion of eqn (3) into eqn (5) we have

$$\Delta_o^w \phi = \Delta_o^w \phi_{M^+}^\circ + (RT/F) \ln c_{MHX^+}(o)/c_{HX}(o) K_1 c_{M^+}(w). \tag{6}$$

The condition for the half-wave potential ($\Delta_o^w \phi = E_{1/2}$) prescribes for the concentrations of the diffusing species, HX$^+$ and MHX$^+$, at the ITIES, that

$$D_{MHX^+}^{1/2} c_{MHX^+}(o) = D_{HX}^{1/2} c_{HX}(o) \tag{7}$$

where D_{MHX^+} and D_{HX} are the diffusion coefficients of MHX$^+$ and HX, respectively, in the organic phase. Because of the similar cyclic structure of both molecules,[30,32] $D_{MHX^+} \approx D_{HX}$. Thus, we obtain from eqn (6) and (7)

$$E^{1/2} = \Delta_o^w \phi_{M^+}^\circ - (RT/F) \ln K_1 c_{M^+}(w). \tag{8}$$

The excess concentration of M$^+$ in the aqueous phase, $c_{M^+}(w)$, is not affected by the charge-transfer process.

In a similar way we obtain for higher pH

$$E_{1/2} = \Delta_o^w \phi_{H^+}^\circ - (RT/F) \ln K_2 c_{H^+}(w) \tag{9}$$

where K_2 is given by

$$K_2 = c_{MHX^+}(o)/c_{H^+}(o) c_{MX}(o). \tag{10}$$

The significance of the constants k_1 and k_2 appearing in eqn (1) is given by

$$\begin{aligned} k_1 &= \exp(F\Delta_o^w \phi_{M^+}^\circ / RT) K_1^{-1} \\ k_2 &= \exp(F\Delta_o^w \phi_{H^+}^\circ / RT) K_2^{-1}. \end{aligned} \tag{11}$$

A more detailed analysis of the transport process has been published elsewhere.[33]

Finally the equilibrium constant of the exchange reaction (2) is given by

$$K_{exch} = c_{MX}(o) c_{H^+}(w)/c_{M^+}(w) c_{HX}(o) = K_2/K_1. \tag{12}$$

In table 1 the values of K_1, K_2 and K_{exch} are listed for sodium and hydrogen ion interactions with monensins A and B and for lithium and hydrogen ion interactions with the mixture of monensins A and B.

It is remarkable that the stability constants of complexes with HX, K_1, are in the order Li$^+$ > Na$^+$ ≫ K$^+$, which differs considerably from the order of membrane selectivities, Na$^+$ > K$^+$ > Li$^+$, found with decan-1-ol-based liquid membranes.[32]

The transfer of MHX$^+$ across the ITIES as a current-generating process obviously has no analogy with processes at bilayer lipid membranes or biological membranes. This may be ascribed to the fact that in the present process MHX$^+$ is stable in a medium of high permittivity (relative permittivity of nitrobenzene saturated with water is ca. 35 at 25 °C) while in a bilayer lipid membrane it can hardly exist in the non-polar interior of the membrane. This is the cause of the negligible change in the conductivity of a bilayer lipid membrane in the presence of monensin.[34]

The substitution of methyl (monensin B) for ethyl (monensin A) on ring C increases the stability of the sodium complex by *ca.* 1 kJ mol^{-1} while it has no effect on the acidity of the carboxy group.

[1] Z. Samec, V. Mareček, J. Koryta and W. Khalil, *J. Electroanal. Chem. Interfacial Electrochem.*, 1977, **83**, 393.
[2] J. Koryta, *Electrochim. Acta*, 1979, **24**, 293.
[3] J. Koryta, *Electrochim. Acta*, 1984, **29**, in press.
[4] J. Koryta and P. Vanýsek, in *Advances in Electrochemistry and Electrochemical Engineering*, ed. H. Gerischer and C. W. Tobias (Wiley-Interscience, New York, 1981), vol. 12, pp. 113-176.
[5] J. Koryta, *Ion-selective Electrode Rev.*, 1983, **5**, 131.
[6] Z. Samec, V. Mareček and J. Weber, *J. Electroanal. Chem. Interfacial Electrochem.*, 1979, **135**, 265.
[7] J. Koryta, P. Vanýsek and M. Březina, *J. Electroanal. Chem. Interfacial Electrochem.*, 1977, **75**, 211.
[8] A. J. Parker, *Electrochim. Acta*, 1976, **21**, 671.
[9] J. E. B. Randles, *Trans. Faraday Soc.*, 1948, **44**, 327.
[10] A. Ševčik, *Collect. Czech. Chem. Commun.*, 1948, **13**, 349.
[11] E. J. W. Verwey and K. F. Niessen, *Philos. Mag.*, 1939, **28**, 435.
[12] M. Gros, S. Gromb and C. Gavach, *J. Electroanal. Chem. Interfacial Electrochem.*, 1978, **89**, 29.
[13] E. D'Epenoux, P. Seta, G. Amblard and C. Gavach, *J. Electroanal. Chem. Interfacial Electrochem.*, 1979, **94**, 77.
[14] J. Koryta, *Anal. Chim. Acta*, 1982, **139**, 1.
[15] J. Koryta and K. Štulik, *Ion-selective Electrodes* (Cambridge University Press, Cambridge, 2nd edn, 1983), p. 18.
[16] Z. Samec, V. Mareček and D. Homolka, *Faraday Discuss. Chem. Soc.*, 1984, **77**, 197.
[17] Yu. A. Ovchinnikov, V. I. Ivanov and M. M. Shkrob, *Membrane Active Complexones* (Elsevier, Amsterdam, 1974).
[18] A. Hofmanová, Le Q. Hung and M. W. Khalil, *J. Electroanal. Chem. Interfacial Electrochem.*, 1982, **135**, 257.
[19] D. Homolka, V. Mareček, Z. Samec, O. Ryba and J. Petránek, *J. Electroanal. Chem. Interfacial Electrochem.*, 1981, **125**, 243.
[20] P. Vanýsek, W. Ruth and J. Koryta, *J. Electroanal. Chem. Interfacial Electrochem.*, 1983, **148**, 117.
[21] W. J. Ward, *AIChE J.*, 1970, **19**, 736.
[22] M. E. Haney and M. M. Hoehn, *Antimicrob. Agents Chemother.*, 1967, 349.
[23] A. Agtarap and J. W. Chamberlin, *Antimicrob. Agents Chemother.*, 1967, 359.
[24] M. Gorman, J. W. Chamberlin and R. L. Hamill, *Antimicrob. Agents Chemother.*, 1968, 363.
[25] M. Pinkerton and L. K. Steinrauf, *J. Mol. Biol.*, 1970, **49**, 533.
[26] B. C. Pressman, *Fed. Proc., Fed. Am. Soc. Exp. Biol.*, 1968, **27**, 1283.
[27] P. J. F. Henderson, J. D. McGivan and J. B. Chappel, *Biochem. J.*, 1969, **111**, 521.
[28] R. Ashton and L. K. Steinrauf, *J. Mol. Biol.*, 1970, **49**, 547.
[29] R. Sandeux, J. Sandeux, C. Gavach and B. Brun, *Biochim. Biophys. Acta*, 1982, **684**, 127.
[30] D. L. Ward, K-T. Wei, J. C. Hoogherheide and A. Popov, *Acta Crystallogr., Sect. B*, 1978, **34**, 110.
[31] A. J. Bard and L. R. Faulkner, *Electrochemical Methods, Fundamentals and Applications* (Wiley, New York, 1980).
[32] W. K. Lutz, H-K. Wipf and W. Simon, *Helv. Chim. Acta*, 1970, **53**, 1741.
[33] Guo Du, J. Koryta, W. Ruth and P. Vanýsek, *J. Electroanal. Chem. Interfacial Electrochem.*, 1983, **159**, 413.
[34] R. Sandeux, P. Seta, G. Jeminet, M. Alleaume and C. Gavach, *Biochim. Biophys. Acta*, 1978, **511**, 499.

GENERAL DISCUSSION

Prof. P. Meares (*University of Aberdeen*) said:
(1) In his paper Prof. 'Dickel has indicated that the contribution to the flux of Donnan co-ions causing a flux of solvent corresponds to complete coupling to the co-ions with the solvent, whereas the flux of counter-ions so produced is equal to the concentration of counter-ions multiplied by one half of the average velocity of solvent.

The solvent drag coefficient for the sodium counter-ions in the cation exchanger Zeo-Karb 315 was found to lie between 0.5 and 0.6 over the concentration range up to and just beyond that at which the concentrations of Donnan ions and fixed charges are equal.[1] The solvent drag coefficient of the Donnan ions lies between 1.3 and 1.4 over most of this concentration range. The value is believed to exceed unity because of the non-uniform distribution of the water stream in the polymer network of the membrane.

The coupling coefficients q_{ik} of non-equilibrium thermodynamics[2] have also been evaluated for the system. In the case of counter-ions and water the values lie between 0.5 and 0.6 but for Donnan ions the range is 0.2–0.4 over most of the concentration range and lower still in more dilute solutions. This anomalous behaviour of the co-ion coefficient is not readily interpreted.

Prof. Dickel's theory of isotonic osmosis leads to the expectation that the electro-osmotic flow of solvent will vanish when the concentrations of Donnan ions and counter-ions are equal. He has shown this to be so in a system he has studied. With Zeo-Karb 315 in NaBr solutions at this point of equal concentrations an electro-osmotic transfer of *ca.* 12 mol water/F takes place and in SrBr$_2$ solutions the transfer is *ca.* 6 mol water/F.[3]

(2) In his eqn (37) Prof. Dickel expresses the vanishing of forces on the membrane during steady flow. This, of course, is true when the membrane is at rest but it appears possible that the sum of forces should include a contribution from the flange of the transport cell in which the membrane is clamped since in unrestrained stationary flow one expects the centre of mass of the whole system of membrane + solutions to remain at rest.

[1] P. Meares, *J. Membr. Sci.*, 1981, **8**, 295.
[2] S. R. Caplan, *J. Phys. Chem.*, 1965, **69**, 3801.
[3] W. J. McHardy, P. Meares, A. H. Sutton and J. F. Thain, *J. Colloid Interface Sci.*, 1969, **29**, 116.

Prof. G. Dickel (*Universität München, West Germany*) said: From eqn (35) and (36) of my paper results for the Donnan ions and the counter-ions

$$J_D = J_D^\circ + (c_D/c_L)J_L \quad \text{and} \quad V_D = V_D^\circ + V_L$$

$$J_c = J_c^\circ + 1/2(c_c/c_L)J_L \quad \text{and} \quad V_c + V_c^\circ + \tfrac{1}{2}V_L$$

where J_D° and J_c° represent the fluxes in absence of a flux of solvent. From these equations we conclude that the action resulting from the flux of the solvent on the counter-ions equals half the action on the Donnan ions. Whilst this is the effect of reverse osmosis, the osmosis itself follows from eqn (45). For this reason we transform the latter into

$$p = (-\tfrac{1}{2}c_c + c_D)F\,d\varphi$$

and state that the action on the solvent (*i.e.* the electro-osmotic pressure) resulting from the flux of the counter-ions equals half the action resulting from the Donnan ions.

This statement concerning the reciprocal relation reveals further that the factor $\frac{1}{2}$ gives rise to a retardation resulting from an intrinsic interfacial equilibrium between the matrix of the membrane and the pore solution. Replacing the factor $\frac{1}{2}$ by 1 we get the equation discussed by Schlögl. This represents a Galilean transformation which does not yield the observed effect of the inversion of isotonic osmosis.

Concerning the condition of the sum of all forces vanishing, this is a necessary condition, first being valid only for the total system membrane + adjacent solutions. The postulate that the sum of all forces in the membrane itself must vanish is a good approximation. Indeed, we have observed in the course of our experiments alternating deflections of the membrane, indicating such forces. As the latter, however, are small with respect to the osmotic forces these can be disregarded.

I now direct attention to the following paper, by Prof. Sanfeld and Dr. Steinchen.

The effect of isotonic osmosis is a special case of the electrokinetic effects well founded on non-equilibrium thermodynamics. Thus eqn (45) in our paper is also valid in the case of a single electrolyte, if an outside electric field is applied to a membrane. Therefore the coefficient $\frac{1}{2}(c_F - c_D)$ is identical to the coefficient $L_{12} = L_{21}$ in the general structure of irreversible thermodynamics. In the numerous representations of this effect, however, we could find no indication of the variation and vanishing of this effect. My question is thus as follows: is it possible to give simple and useful criteria concerning the existence of such or similar effects, in order to avoid the application of the extensive method of least dissipation of energy.

Prof. H. Linde (*Academy of Sciences of the G.D.R., Berlin, East Germany*) (*communicated*): I would like Prof. Dupeyrat and Dr. Nakache to comment on the following remarks.

In your paper, the movements at the interface are characterized as pseudoperiodic oscillations. In Marangoni instability there occurs amplification of classical waves (longitudinal capillary waves) as well as of autowaves (the spatial and temporal synchronization of typical relaxation oscillations, for Marangoni instability has recently been recognized in our laboratory also). Classical waves show interference and reflexion at a wall, whereas autowaves show evidence of collision between two waves or between a wave and a wall. Can you connect your pseudoperiodic oscillations with the behaviour of classical waves or with autowaves?

Prof. M. Dupeyrat, Dr. E. Nakache and Dr. M. Vignes-Adler (*Université Pierre et Marie Curie, Paris, France*) (*communicated*): We have to note that the experimental procedure used up to now does not allow us to obtain homogeneous concentration gradients. Consequently it is not possible to obtain regular patterns such as those described by Linde and thus to develop a theoretical model which may be compared with the differential equations proposed for capillary and autowaves.

Experimentally we observed that the relaxation oscillations and cells depend on the size of the interface, and therefore on the reflexion at a wall, and on the deformation of the interface bound to the presence of a stirrup during the measurement of the interfacial tension. It is likely that these relaxation oscillations are enhanced by the curvature of the interface imposed by the measurement method.

We also remark that if HPi is not present, or if KPi replaces HPi (that is to say if there is no chemical reaction), these instabilities are not observed. Thus some chemical autowaves must also be important.

The phenomenon seems to be related to a coupling between capillary and autowaves.

Dr. C. Tondre (*University of Nancy, France*) said: May I ask Dr. Nakache to give some more details on a technical point of her work, the measurement of the time dependence of the interfacial tension γ (fig. 1 of the paper): could she explain how the measurement of γ using the detachment of a stirrup can be made fast enough to obtain the recordings shown in fig. 1? Is there no perturbation of the interface due to the measurement itself? Would not the optical methods using the intensity of the light scattered by liquid interfaces[1] be particularly appropriate for such measurements?

[1] J. Lachaise, A. Graciaa, A. Martinez and A. Rousset, *J. Phys. Lett.*, 1979, **40**, L-599.

Dr. E. Nakache (*Université Pierre et Marie Curie, Paris, France*) replied: The measurement of the interfacial tension, γ, with a stirrup can be made in two ways: (1) detaching the stirrup after its elevation and (2) elevating the stirrup to the maximum of the traction curve, as Guastalla[1] proposed. This author has shown that, under these conditions, there is a relation between the force measured and the interfacial tension. This method was used in this work because it permitted the measurement of γ as a function of time.

Preliminary experiments have shown that the influence of the curvature of the interface along the wall of the beaker was very important, probably because the convection process is enhanced.[2] Consequently the presence of the stirrup at the interface probably disturbs it by the two menisci created. It would be interesting to measure γ without disturbing the interface, for instance with the optical methods used by Lachaise *et al.* Unfortunately these methods are now too slow: according to Dr Langevin-Wallon[3] the time to obtain a measurement of γ is, at present, between one and several minutes.

[1] J. Guastalla, *J. Chim. Phys.*, 1971, **68**, 822.
[2] J. C. Berg and C. R. Morig, *Chem. Eng.*, 1969, **24**, 937.
[3] D. Langevin-Wallon, personal communication.

Dr. M. Spiro (*Imperial College, London*) said: There are certain aspects of the experimental arrangements used by Samec and coworkers that I do not understand. In the first place, how were they able to achieve a stable flat interface over a circular area >5 mm in diameter with the heavier nitrobenzene layer (density 1.20 g cm^{-3}) lying above the lighter aqueous layer (density 1.00 g cm^{-3})? My second set of questions concerns their method of placing an isolated metallic wire inside the Luggin capillary for the nitrobenzene phase. (*a*) What was the wire made of? (*b*) Since the wire is connected to the silver of the reference electrode RE2, does it not introduce an extra potential into cell I so that the potentials of the two reference electrodes no longer cancel each other out? (*c*) Does 'isolated' mean that the wire was insulated from both aqueous and the nitrobenzene solutions inside the Luggin capillary? (*d*) In what respect did the introduction of this wire achieve a 'remarkable improvement' in the cell's response? It would be helpful if Dr. Samec could give more details of the arrangement used.

Dr. Z. Samec (*J. Heyrovský Institute, Prague, Czechoslovakia*) said: The stable location of the water/nitrobenzene boundary in the round hole of the glass barrier B (see our fig. 1) was ensured both by the low surface tension between water and glass and by the liquid's incompressibility. In fact the inner space of the glass cell was made as hydrophilic as possible by washing with acetone, doubly distilled water and drying at a temperature of 120 °C in a dry-box. Eventually, the surface of the glass was treated by a concentrated solution of sodium hydroxide. Referring to fig. 1, the bottom space of the cell was filled completely with the aqueous phase. Concerning the reference electrode RE2 for the nitrobenzene phase, the copper wire of 0.4 mm in diameter was placed inside the Luggin capillary down to its tip. This wire was insulated from both the aqueous solution of TBABr and the nitrobenzene solution of TBATPB, and only the metallic disc of geometric area 0.13 mm^2 was exposed to the nitrobenzene solution just at the capillary tip. Since this wire was connected to the silver Ag' of the reference electrode RE2, a galvanic cell (II)

$$\text{Ag'} \mid \text{AgBr} \mid c°(\text{TBABr}) \mid c°(\text{TBATPB}) \mid \text{Cu} \mid \text{Ag'} \quad \text{(II)}$$
$$\text{(w')} \qquad \text{(o)}$$

must be considered in addition to cell (I) of the paper. It can be assumed that the copper electrode is ideally polarizable and hence the short circuit in cell (II) has no effect on the potential of the Ag|AgBr reference electrode. However, even if a faradaic process occurs at the copper/nitrobenzene interface, e.g. copper dissolution or nitrobenzene reduction, the potential of the Ag|AgBr reference electrode can hardly change. This is because the area of the Cu/o interface is at least two orders of magnitude smaller than the areas of the Ag'/w' and w'/o interfaces; the latter two interfaces are practically non-polarizable. On the other hand the introduction of the metallic wire had a remarkable effect on the response time of the cell. Referring to fig. 6 of our paper it is seen that the ohmic potential drop appears as a step in the galvanostatic transient, and in an ideal case the time necessary for its completion would be zero. However, it is apparent that the transient has a finite slope at $t = t_0$, which corresponds to a delay of ca. 0.1 ms for a potential change from 10 to 90% of the final value. Without the metallic wire inside the Luggin capillary, this delay was ca. 1 ms.

Dr. M. Spiro (*Imperial College, London*) said: Since the wire used was a copper one, and since pulses passed along the wire, there is surely the danger that a Cu^{2+}/Cu or Cu^+/Cu potential would be set up at the end of the wire. This would have affected the measured e.m.f. of cell (I). A platinum wire would have been better in this respect. However, a faradaic pulse passing through the tip of *any* wire would generate ions that could contaminate the nearby nitrobenzene/water interface. Moreover, even with a platinum wire the potential difference between the nitrobenzene solution and the platinum metal will not be identical to that between the nitrobenzene solution and the Ag|AgBr|Br$^-$(aq) electrode because only the latter includes a physical nitrobenzene/water interface with its ionic double layers.

Dr. Z. Samec (*J. Heyrovský Institute, Prague, Czechoslovakia*) said: Both reference electrodes were connected to the high-impedance inputs of the FET operational amplifiers (the voltage followers) and practically no current can flow through them during the change in the potential drop between the tips of the Luggin capillaries. However, Cu^{2+}/Cu or Cu^+/Cu potentials could be set up at the end of the wire owing to the short-circuit current in cell (II). As explained above, this would have only a small effect on the measured e.m.f. of the cell. In any case tetramethylammonium was always used as the inner standard for the evaluation of the interfacial potential difference. However, I agree with Dr Spiro that a platinum wire would be better in most respects, including the less probable contamination of the nitrobenzene phase by the metal ions.

Dr. E. Nakache (*Université Pierre et Marie Curie, Paris, France*) said: I would like to ask Prof. Koryta if he could provide more information about the adsorption of the different monensins at the nitrobenzene/water interface?

Prof. J. Koryta (*J. Heyrovsky Institute, Prague, Czechoslovalia*) replied: We have no information on monensin adsorption at the water/nitrobenzene interface. The only paper dealing with the influence of adsorption on ion transfer studied by voltammetry at ITIES is concerned with phospholipid adsorption.[1]

[1] J. Koryta, Le Q. Hung and A. Hofmanová, *Stud. Biophys.*, 1982, **90**, 25.

Ion-exchange Dynamics at the Zeolite/Solution Interface Studied by the Chemical-relaxation Method

By Tetsuya Ikeda, Minoru Sasaki and Tatsuya Yasunaga*

Department of Chemistry, Faculty of Science, Hiroshima University, Hiroshima 730, Japan

Received 17th November, 1983

The kinetics of the hydrolysis of surface hydroxyl groups on the aluminosilicate framework and ion exchange of NH_4^+, $CH_3NH_3^+$, $C_2H_5NH_3^+$, $n\text{-}C_3H_7NH_3^+$, $i\text{-}C_3H_7NH_3^+$, $(CH_3)_2NH_2^+$, $(CH_3)_3NH^+$ and $(CH_3)_4N^+$ for Na^+ at the zeolite/solution interface have been studied by using the pressure-jump relaxation method with electric-conductivity detection. Above pH 11.5 a single relaxation was found and was attributed to the hydrolysis of surface hydroxyl groups. Below this pH value two relaxations were found in aqueous suspensions of the systems comprising zeolite 4A with NH_4^+, $CH_3NH_3^+$, $C_2H_5NH_3^+$, $n\text{-}C_3H_7NH_3^+$ and $(CH_3)_2NH_2^+$, but no relaxation was observed in the systems involving zeolite 4A and i-$C_3H_7NH_3^+$, $(CH_3)_3NH^+$ and $(CH_3)_4N^+$. The fast and slow relaxations observed were attributed to diffusion of the alkylammonium ion on the surface of the particles, followed by adsorption of the alkylammonium ion on a site in the zeolite cage. From the kinetic and static experimental results it was found that the affinity of zeolite 4A for exchangeable cations is limited by the available intracrystalline space in the cage with larger cations unable to diffuse into the cage because of a lack of available intracrystalline space. The difference in the values of the rate constants associated with the alkylammonium ion entering the cage is interpreted in terms of a steric factor.

Crystalline zeolites having anionic aluminosilicate framework structures give rise to molecular-sieve action for adsorbing molecules as a shape-selective catalyst.[1] Such molecular-sieve materials have recently been synthesized in increasing quantities for use in adsorptive and catalytic applications because of their unique shape-selectivity and their framework structures.[2-4] The aluminosilicate framework of zeolites, which is terminated by surface hydroxyl groups,[3-7] is a major factor in the determination of their acidic properties.[8] In order to understand the dynamic properties of acid sites at the zeolite/solution interface, the chemical-relaxation method has been applied to the hydrolysis of surface hydroxyl groups on zeolite surfaces[9] and to acetate-ion adsorption on amorphous silica–alumina surfaces.[10] Another interesting property of zeolites is the existence of Na^+ in the cage structures, in which the exchange of organic and inorganic cations for Na^+ in the cage accompanies an ion-sieve effect or ion selectivity for exchangeable cations.[11-17] However, the mechanism of ion exchange, which includes the ion-sieve action, has not been well established on the basis of the kinetics because the ion-exchange reaction is too fast to be observed by ordinary methods, e.g. using radioactive tracers.[14] In the insertion process of the guest organic cation into the host cage through the channels of the zeolite, the diffusion rate is governed by a steric factor of the entering cation which may depend on both the aperture of the host cage and the size of the entering guest cation. In order to clarify size and shape correlations between the host cage and the guest molecule in the ion-exchange reaction, one must study the ion-exchange kinetics of various organic cations for Na^+ in the zeolite.

In this paper we present the results of pressure-jump relaxation experiments on the kinetics of the hydrolysis of surface hydroxyl groups on the aluminosilicate framework and of ion exchange of the various alkylammonium ions for Na^+ at the zeolite/aqueous-solution interface, and discuss the influence of ion size and shape on the ion-exchange properties.

EXPERIMENTAL

The details of the pressure-jump apparatus have been described previously.[18] The time constant of the pressure jump is 80 μs.

Zeolite 4A ($Na_2O \cdot Al_2O_3 \cdot 2SiO_2 \cdot nH_2O$) was supplied by the Toyo Soda Co. X-ray diffraction patterns were identical with those of zeolite 4A reported by Broussard et al.[19] Microscopic examination confirmed that the mean diameters of the dispersed zeolite 4A particles were <1 μm. Such small particles formed very stable suspensions with no sign of sedimentation during the kinetic measurements.

The amount of alkylammonium ion adsorbed was determined indirectly from the concentration change in the supernatant solution by means of colorimetric analysis with 2,4-dinitrofluorobenzene[20] for $CH_3NH_3^+$, $C_2H_5NH_3^+$, n-$C_3H_7NH_3^+$ and $(CH_3)_2NH_2^+$, and with ninhydrin[21] for NH_4^+. Prior to the measurements, samples of zeolite 4A suspensions containing each alkylamine hydrochloride were centrifuged for 30 min at 10 000g to settle the particles completely.

The particle concentration of all samples was 30 g dm^{-3}, and the samples were equilibrated for 72 h after preparation. All preparations and experimentation were done in a nitrogen atmosphere, and the temperature was controlled at 25.0 ± 0.1 °C.

RESULTS AND DISCUSSION

CHEMICAL RELAXATION IN AQUEOUS SUSPENSIONS OF ZEOLITE 4A

A single relaxation process of decreasing conductivity having a characteristic duration of the order of seconds was observed in basic aqueous suspensions of the system zeolite 4A–NaOH only above pH 11.5 by using the pressure-jump relaxation method. Values of the relaxation time were obtained from a semilogarithmic plot of the relaxation curve, and the dependence of the reciprocal relaxation time τ^{-1} on OH^- concentration in aqueous suspensions of the zeolite 4A is shown in table 1. The experimental results show that the value of τ^{-1} increases with increasing OH^- concentration. Since Na^+ can enter the cage of the zeolite 4A as mentioned above, a plausible mechanism for the relaxation phenomenon observed may be the base-catalysed adsorption–desorption of Na^+. However, when tetramethylammonium hydroxide was added as the base in the same manner, the same relaxation phenomenon was observed, even though the tetramethylammonium ion cannot enter the cage. Therefore it is suggested that the single relaxation observed may originate in the interaction between OH^- and the active site on the zeolite surface, i.e. the hydrolysis of surface hydroxyl groups on the aluminosilicate framework.

Below pH 11.1 another relaxation phenomenon shown in fig. 1(a) was observed in aqueous suspensions of systems comprising zeolite 4A with NH_4^+, $CH_3NH_3^+$, $C_2H_5NH_3^+$, n-$C_3H_7NH_3^+$ and $(CH_3)_2NH_2^+$ by using the pressure-jump method, where the relaxation signal again shows a decrease in the conductivity of the suspension during relaxation. No relaxation was observed in a zeolite 4A suspension of the same pH (pH < 11.1) in the supernatant solutions of the above systems and in aqueous suspensions of zeolite 4A with i-$C_3H_7NH_3^+$, $(CH_3)_3NH^+$ or $(CH_3)_4N^+$. Furthermore, measurements of the change in concentration of Na^+ in the bulk phase

Table 1. Kinetic and static data for aqueous suspensions of zeolite 4A at 25 °C

τ^{-1}/s^{-1}	[OH$^-$] /10^{-3} mol dm^{-3}	[SOH] /10^{-3} mol dm^{-3}	[SO$^-$] /10^{-2} mol dm^{-3}
0.88 ± 0.10	2.63	3.1	1.54
1.06 ± 0.13	4.57	2.2	1.63
1.47 ± 0.09	6.76	1.1	1.74
1.80 ± 0.13	9.12	1.0	1.75
1.91 ± 0.12	11.8	0.8	1.77

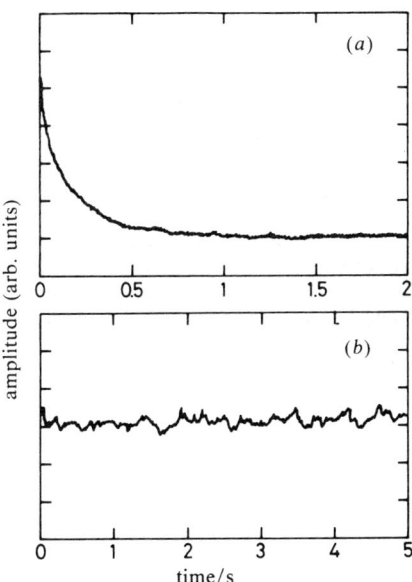

Fig. 1. Typical relaxation curves observed by using the pressure-jump relaxation method below pH 11.1 at a particle concentration of 30 g dm^{-3} and 25 °C for zeolite 4A with the following (a) NH$_4^+$, CH$_3$NH$_3^+$, C$_2$H$_5$NH$_3^+$, n-C$_3$H$_7$NH$_3^+$ and (CH$_3$)$_2$NH$_2^+$; (b) i-C$_3$H$_7$NH$_3^+$, (CH$_3$)$_3$NH$^+$ and (CH$_3$)$_4$N$^+$.

confirmed that release of Na$^+$ from the zeolite cage requires 48–72 h in the systems of zeolite 4A with NH$_4^+$, CH$_3$NH$_3^+$, C$_2$H$_5$NH$_3^+$, n-C$_3$H$_7$NH$_3^+$ and (CH$_3$)$_2$NH$_2^+$ as shown in fig. 1(b). These results suggest that for the alkylammonium ions relaxation reflects the difference in the size and shape correlations between the guest ion and the host cage. From the semilogarithmic plot of the relaxation curve it is seen that relaxation consists of two processes. The dependences of the fast and slow reciprocal relaxation times, τ_1^{-1} and τ_2^{-1}, on the concentrations of added ammonium chloride, methylamine hydrochloride, ethylamine hydrochloride, n-propylamine hydrochloride and dimethylamine hydrochloride in aqueous suspensions of zeolite 4A are shown in fig. 2 and 3, respectively. The value of τ_1^{-1} shows a steep increase, while the value of τ_2^{-1} shows an increase and then approaches a constant value. From a comparison of the values of the relaxation times for each alkylammonium ion in fig. 2 and 3 it can be seen that the values of both τ_1^{-1} and τ_2^{-1} decrease with

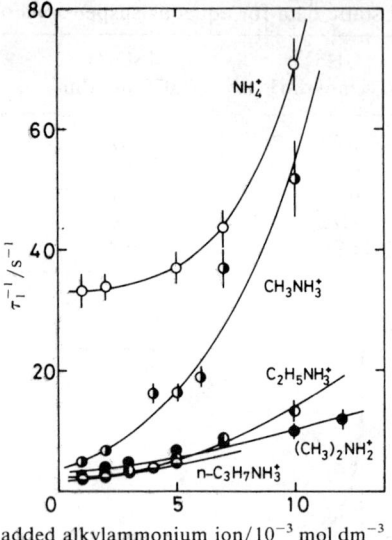

Fig. 2. Dependence of the fast reciprocal relaxation time on the added concentrations of ammonium chloride, methylamine hydrochloride, ethylamine hydrochloride, n-propylamine hydrochloride and dimethylamine hydrochloride in aqueous suspensions of zeolite 4A at a particle concentration of 30 g dm^{-3} and 25 °C.

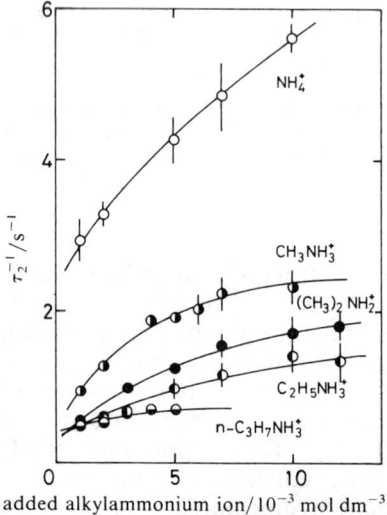

Fig. 3. Dependence of the slow reciprocal relaxation time on the added concentrations of ammonium chloride, methylamine hydrochloride, ethylamine hydrochloride, n-propylamine hydrochloride and dimethylamine hydrochloride in aqueous suspensions of zeolite 4A at a particle concentration of 30 g dm^{-3} and 25 °C.

increasing alkyl chain length. Taking into account the absence of relaxation in the supernatant solution, the concentration dependences of the relaxation times and the existence of an Na$^+$-release process, the results suggest that the relaxation observed may be due to ion exchange of the alkylammonium ion in the bulk phase for Na$^+$ in the cages on the zeolite surface. Furthermore, it appears from the above kinetic measurements that the ion exchange may consist of at least three elementary processes.

In order to clarify the mechanism of the two kinds of the reactions in aqueous suspensions of zeolite 4A, the kinetics of fast reactions were studied in detail, taking full account of both kinetic and static data.

HYDROLYSIS OF SURFACE HYDROXYL GROUPS ON THE ALUMINOSILICATE FRAMEWORK

For the single relaxation observed above pH 11.5, the mechanism of the hydrolysis of the hydroxyl groups on the aluminosilicate framework of zeolite 4A can be written as

$$\left[\underset{\text{SOH}}{\text{Al-O(H)-Si}} + \text{OH}^- \underset{k_b}{\overset{k_f}{\rightleftharpoons}} \underset{\text{SO}^-}{\text{Al-O-Si}} + \text{H}_2\text{O} \right] \quad (I)$$

where SOH and SO$^-$ denote the surface hydroxyl group and the dissociated hydroxyl group on the aluminosilicate framework, respectively, and k_f and k_b are the forward and backward rate constants, respectively. For the above mechanism τ^{-1} is given by

$$\tau^{-1} = k_f([\text{SOH}] + [\text{OH}^-]) + k_b' \quad (1)$$

with

$$k_b' = k_b[\text{H}_2\text{O}]. \quad (2)$$

In order to plot eqn (1) one must measure the equilibrium concentration of SOH. The concentrations of SOH together with SO$^-$ were determined from the adsorption isotherm of OH$^-$ and are listed in table 1.

The plot of τ^{-1} against the concentration term in eqn (1) yields a straight line, as shown in fig. 4. The linearity of this plot confirms the plausibility of mechanism (I). The values of the rate constants k_f and k_b' were determined from the slope and the intercept of the straight line and are listed in table 2. The value of the kinetic equilibrium constant K_H' was calculated from the ratio of the obtained rate constants and is given in table 2. The values of the static equilibrium constant calculated from $K_H' = [\text{SO}^-]/([\text{SOH}][\text{OH}^-])$ are also listed in table 2. As can be seen from a comparison of the kinetic and static equilibrium constants obtained, the kinetic equilibrium constant is in good agreement with the static equilibrium constant. Consequently, both the linearity in fig. 4 and the agreement of the equilibrium constants obtained from different sources lead to the conclusion that the single relaxation observed can be attributed to mechanism (I).

Kinetic parameters in the systems comprising zeolite X and zeolite Y with hydroxyl groups obtained previously are also given in table 2. Comparison of the three K_H' values in table 2 shows that the variation of K_H' with aluminosilicate

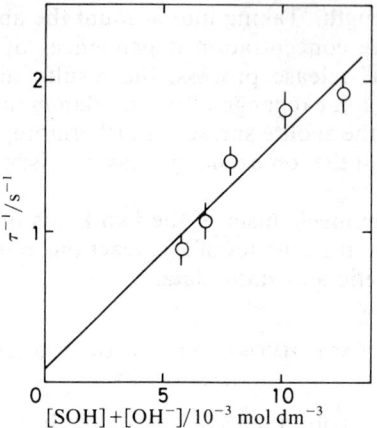

Fig. 4. Plot of τ^{-1} as a function of [SOH]+[OH$^-$] in eqn (1).

Table 2. Rate constants for the hydrolysis of hydroxyl groups on the surface of zeolites A, X and Y at 25 °C

zeolite	k_f /mol^{-1} dm^3 s^{-1}	k'_b /s^{-1}	K'_H/mol^{-1} dm^3	
			kinetic	static
A	1.6×10^2	8.7×10^{-2}	2.0×10^3	2.0×10^3
X^a	2.0×10^2	2.0×10^{-2}	1.0×10^4	1.3×10^4
Y^a	8.1×10	7.3×10^{-1}	1.0×10^2	1.3×10^2

^{a} Typical formulae of zeolites X and Y are Na$_2$O·Al$_2$O$_3$·2.5SiO$_2$ and Na$_2$O·Al$_2$O$_3$·4.8SiO$_2$, respectively. The values of k_f, k'_b and K'_H have been reported in ref. (9).

framework structure is representative of the composition difference of the zeolites and that the difference in acid strength is mainly reflected in the values of the backward rate constant, since the values of the forward rate constant are nearly equal in all cases studied.

ION EXCHANGE OF ALKYLAMMONIUM ION FOR Na$^+$

Consider the following mechanism of ion exchange of alkylammonium ion A$^+$ for Na$^+$ in zeolite 4A:

Interfacial reaction

$$S(Na) \underset{k_{-3}}{\overset{k_3}{\rightleftharpoons}} S^- \underset{k_{-1}}{\overset{k_1}{\rightleftharpoons}} S \cdot A \underset{k_{-2}}{\overset{k_2}{\rightleftharpoons}} S(A) \qquad (II)$$

$$\qquad\qquad Na^+ \quad A^+$$

step (a) step (b) step (c)
(very slow)

Bulk reaction

$$A + H_2O \xrightleftharpoons[\text{very fast}]{K_B} A^+ + OH^- \quad (III)$$

with

$$K_B = \frac{[A^+][OH^-]}{[A]} \quad (3)$$

where S(Na), S$^-$, S·A and S(A) denote the bound site of Na$^+$ in the zeolite cage, the vacant site, the bound state on the surface and the bound site of the alkylammonium ion which has entered the cage by intracrystalline diffusion, respectively. $k_{\pm 1,2,3}$ are the rate constants of each step.

Under the assumption that step (*a*) is much slower than steps (*b*) and (*c*), and that hydrolysis of the alkylamine in the bulk phase is a very fast reaction, the fast and slow reciprocal relaxation times are given by

$$\tau_{1,2}^{-1} = \tfrac{1}{2}\{a_{11} + a_{22} \pm [(a_{11} + a_{22})^2 - 4(a_{11}a_{22} - a_{12}a_{21})]^{1/2}\} \quad (4)$$

with

$$a_{11} = k_1\left([A^+] + [S^-]\frac{K_B + [A^+]}{K_B + [A^+] + [OH^-]}\right) + k_{-1} \quad (5)$$

$$a_{12} = k_{-1} \quad (6)$$

$$a_{21} = k_2 \quad (7)$$

$$a_{22} = k_2 + k_{-2}. \quad (8)$$

Eqn (4) cannot be solved explicitly for the general case. However, there is a procedure based on the following relationship which allows the evaluation of all four rate constants with good precision:[22]

$$\tau_1^{-1} + \tau_2^{-1} = k_1 C + k_{-1} + k_2 + k_{-2} \quad (9)$$

$$\tau_1^{-1}\tau_2^{-1} = k_1(k_2 + k_{-2})C + k_{-1}k_{-2} \quad (10)$$

with

$$C = [A^+] + [S^-]\frac{K_B + [A^+]}{K_B + [A^+] + [OH^-]}. \quad (11)$$

In order to plot eqn (9) and (10), one must measure the equilibrium concentrations of S$^-$ and A$^+$.

Fig. 5(*a*) shows the adsorption isotherms of NH_4^+, $CH_3NH_3^+$, $C_2H_5NH_3^+$, n-$C_3H_7NH_3^+$ and $(CH_3)_2NH_2^+$ in aqueous suspensions of zeolite 4A, where the concentration of each alkylammonium ion was estimated by using the pH value of each suspension and the dissociation constant K_a of each alkylammonium ion.[23] Furthermore, the amounts of i-$C_3H_7NH_3^+$ and $(CH_3)_3NH^+$ in aqueous suspensions of zeolite 4A were measured, but proved to be negligibly small. Since the lack of adsorption of i-$C_3H_7NH_3^+$ and $(CH_3)_2NH^+$ corresponds to a lack of relaxation as described above, it is clear that a size effect for the exchangeable cation is revealed in both the kinetic and static experimental results in the present study. The variation of pH accompanied by addition of each alkylamine hydrochloride in the bulk phase is in the range 10.5–11.1. This may be explained by the hydrolysis of the alkylamine

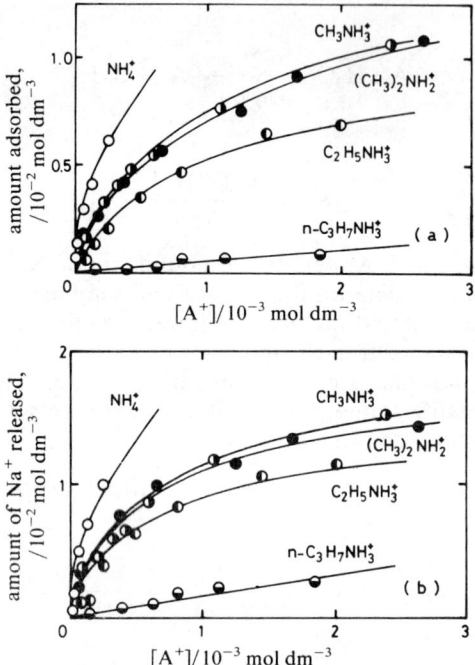

Fig. 5. (*a*) Adsorption isotherms of NH_4^+, $CH_3NH_3^+$, $C_2H_5NH_3^+$, $n\text{-}C_3H_7NH_3^+$ and $(CH_3)_2NH_2^+$. (*b*) The amounts of Na^+ released by the adsorption of each cation.

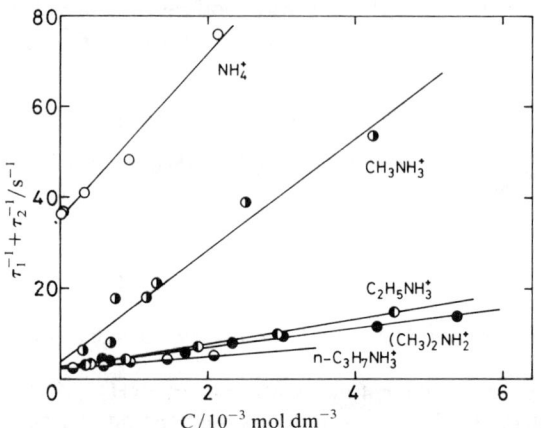

Fig. 6. Plots of $\tau_1^{-1}+\tau_2^{-1}$ against C in eqn (9).

and is not necessarily indicative of increased adsorption of OH^- caused by alkylammonium ion adsorption. The amounts of Na^+ released by the adsorption of each alkylammonium ion were measured by using a sodium electrode, and the results are also shown in fig. 5(*b*). For each system, a variation profile of the amount of Na^+ released is similar to that of the amount of alkylammonium ion adsorbed.

The plots of $\tau_1^{-1}+\tau_2^{-1}$ against C and $\tau_1^{-1}\tau_2^{-1}$ also against C for NH_4^+, $CH_3NH_3^+$, $C_2H_5NH_3^+$, $n\text{-}C_3H_7NH_3^+$ and $(CH_3)_2NH_2^+$ are shown in fig. 6 and 7. They yield

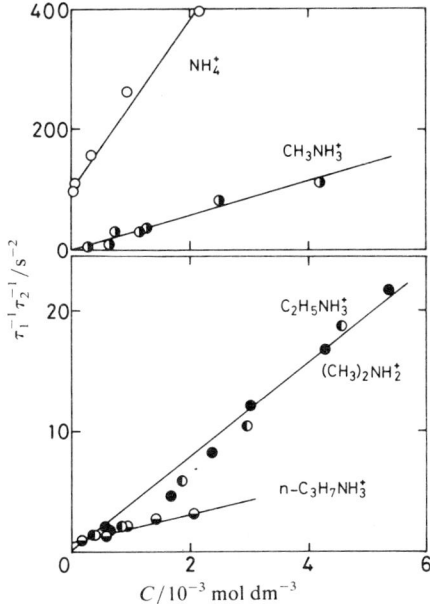

Fig. 7. Plots of $\tau_1^{-1}\tau_2^{-1}$ against C in eqn (10).

Table 3. Rate constants of steps (b) and (c) and the overall equilibrium constant of ion exchange of organic cations for Na^+ in zeolite 4A at 25 °C

cation	k_1 /mol^{-1} dm^3 s^{-1}	k_{-1} /s^{-1}	k_2 /s^{-1}	k_{-2} /s^{-1}	K'
NH_4^+	1.8×10^4	28	3.7	3.7	48
$CH_3NH_3^+$	1.2×10^4	1.8	2.3	—	25
$C_2H_5NH_3^+$	2.8×10^3	0.42	1.3	—	9
$(CH_3)_2NH_2^+$	2.1×10^3	0.73	1.8	—	17
$n\text{-}C_3H_7NH_3^+$	1.6×10^3	1.3	0.28	0.48	0.5

straight lines for each cation. These straight lines lead to the conclusion that the fast and slow relaxations can be attributed to steps (b) and (c) in mechanism (II), respectively. The four rate constants were evaluated from the slope and intercept of these straight lines. The values of the four rate constants obtained are listed in table 3. However, since the plots for $CH_3NH_3^+$, $C_2H_5NH_3^+$ and $(CH_3)_2NH_2^+$ in fig. 7 yield straight lines through the origin, the value of k_{-2} could not be evaluated.

With respect to step (a) in mechanism (II), equilibration for the release of Na^+ in the present study required 48–72 h. This result is in agreement with that reported by Sherry et al.[12] and Barrer et al.[16] Moreover, since the concentration dependences of the two relaxation times cannot be interpreted except for the case that step (a) is much slower than steps (b) and (c), the rate-determining step in the present ion-exchange reaction can be attributed to step (a). This result suggests that the affinity of the site in the cage for the small ion Na^+ may be stronger than that for the alkylammonium ion.

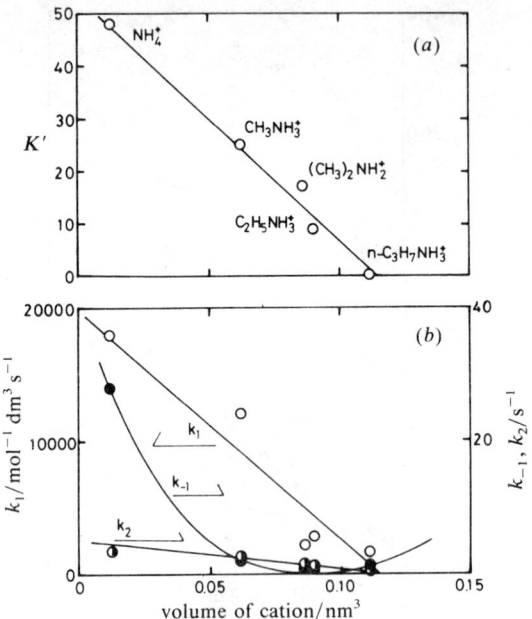

Fig. 8. (a) Plot of K' as a function of the volume of the cation. (b) Plots of $\bigcirc$, k_1; $\bullet$, k_{-1} and $\circledcirc$, k_2 as a function of the volume of the cation.

EQUILIBRIUM PROPERTIES OF STERIC AND ION-SIEVE FACTORS

The static equilibrium constant K' of the overall reaction in mechanism (II) is given by

$$K' = \frac{(\text{amount of A}^+ \text{ adsorbed}) [\text{Na}^+]}{[S(\text{Na})][\text{A}^+]}. \tag{12}$$

The values of K' calculated by using the static data for systems of zeolite 4A with NH_4^+, $CH_3NH_3^+$, $C_2H_5NH_3^+$, $n\text{-}C_3H_7NH_3^+$ and $(CH_3)_2NH_2^+$ are also listed in table 3. Fig. 8(a) shows a plot of K' as a function of cation volume in which the volumes of the alkylammonium ions were estimated from their van der Waals dimensions. The value of K' decreases with increasing volume of the entering alkylammonium ion. From fig. 8(a), since the upper limit of the volume of the exchangeable alkylammonium ion is ca. 0.12 nm^3, the affinity series suggests that the values of K' for $i\text{-}C_3H_7NH_3^+$, $(CH_3)_3NH^+$ and $(CH_3)_4N^+$ would be zero. In the light of the space requirement of the cations in the cage, such larger alkylammonium ions cannot be exchanged with Na^+ in the cage for the lack of available intracrystalline space. Consequently, the kinetic and static results for zeolite 4A with $i\text{-}C_3H_7NH_3^+$, $(CH_3)_3NH^+$ and $(CH_3)_4N^+$, i.e. no relaxation and no exchange of cations, can be reasonably interpreted in terms of space requirements. This may indicate that the volume steric effect described by Theng et al.[17,23,24] can limit the overall equilibrium constant of the ion exchange.

In ion exchange in zeolites water molecules together with Na^+ in the cage are displaced by adsorbed organic cations as described by Barrer et al.[25] If the alkylammonium ion exchange for Na^+ requires the release of 4 water molecules, the volume

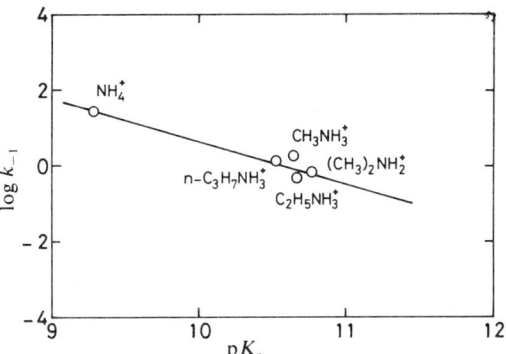

Fig. 9. Plot of log k_{-1} against pK_a.

of the displaced space of the $Na^+ \cdot 4H_2O$ complex would be ca. $0.12\ nm^3$. From a comparison of the estimated volumes of the organic cations and the upper limit in fig. 8(a), one can consider that rejection of a guest ion from the available intracrystalline space of the host cage causes an ion-sieve effect for the larger cations. Furthermore, from the relation found between K' and the cation volume, it can be seen that the extent of exchange of the organic cation for Na^+ in zeolite 4A may be governed mainly by the available intracrystalline space of the cage. This leads to the conclusion that the space requirement of the exchangeable alkylammonium ion is the major factor determining the extent of ion exchange in the present experimental conditions.

DYNAMIC PROPERTIES OF STERIC AND ION-SIEVE FACTORS

Fig. 8(b) shows the dependence of the rate constants obtained on the volume of the cation for NH_4^+, $CH_3NH_3^+$, $C_2H_5NH_3^+$, $n\text{-}C_3H_7NH_3^+$ and $(CH_3)_2NH_2^+$. The values of k_1 and k_2, which are the rate constants for the process of insertion of the alkylammonium ion, decrease linearly with the volume of cation and become zero at a cation volume of nearly $0.12\ nm^3$. This result confirms that the rates of adsorption and intracrystalline diffusion may be controlled by a steric factor of the entering cation. As can be seen from fig. 8(b), however, only the behaviour of the value of k_{-1} differs essentially from those of k_1 and k_2. As is shown in fig. 9, a plot of log k_{-1} against pK_a yields a straight line. This linear relationship indicates that the values of k_{-1} may reflect the difference in the chemical properties of the exchangeable cations rather than the steric factor.

[1] D. W. Breck, *Zeolite Molecular Sieves* (Wiley-Interscience, New York, 1974).
[2] E. M. Flanigen, J. M. Bennet, R. W. Grose, J. P. Cohen, R. L. Patton and R. M. Kirchner, *Nature (London)*, 1978, **271**, 512.
[3] G. T. Kokotailo, S. L. Lawton, D. H. Olson and W. M. Meier, *Nature (London)*, 1978, **272**, 437.
[4] S. T. Wilson, B. M. Lok, C. A. Messian, T. R. Cannan and E. M. Flanigen, *J. Am. Chem. Soc.*, 1982, **104**, 1146.
[5] G. T. Kerr, E. Dempsey and R. J. Mikovsky, *J. Phys. Chem.*, 1965, **69**, 4050.
[6] J. Uytterhoeven, L. G. Christner and W. K. Hall, *J. Phys. Chem.*, 1976, **69**, 2177.
[7] L. V. C. Rees and C. J. Williams, *Trans. Faraday Soc.*, 1965, **61**, 1481.
[8] K. Tanabe, *Solid Acids and Bases* (Kodansha, Tokyo, 1970).
[9] T. Ikeda, M. Sasaki and T. Yasunaga, *J. Phys. Chem.*, 1982, **86**, 1678.

[10] T. Ikeda, M. Sasaki, K. Hachiya, R. D. Astumina, T. Yasunaga and Z. A. Schelly, *J. Phys. Chem.*, 1982, **86**, 3861.
[11] D. W. Breck, W. C. Eversole, R. M. Milton, T. B. Reed and T. L. Thomas, *J. Am. Chem. Soc.*, 1956, **78**, 5963.
[12] H. S. Sherry and H. F. Walton, *J. Phys. Chem.*, 1967, **71**, 1457.
[13] R. M. Barrer and J. Klinowski, *J. Chem. Soc., Faraday Trans. 1*, 1972, **68**, 1956.
[14] L. M. Brown and H. S. Sherry, *J. Phys. Chem.*, 1971, **75**, 3855.
[15] R. M. Barrer and W. M. Meier, *Trans. Faraday Soc.*, 1958, **54**, 1074.
[16] R. M. Barrer, R. Papadopoulos and L. V. C. Rees, *J. Inorg. Nucl. Chem.*, 1967, **29**, 2047.
[17] B. K. G. Theng, E. Vansant and J. B. Uytterhoeven, *Trans. Faraday Soc.*, 1968, **64**, 3370.
[18] K. Hachiya, M. Ashida, M. Sasaki, H. Kan, T. Inoue and T. Yasunaga, *J. Phys. Chem.*, 1979, **83**, 1866.
[19] L. Broussard and D. P. Shoemaker, *J. Am. Chem. Soc.*, 1960, **82**, 1041.
[20] W. Troll and R. K. Cannan, *J. Biol. Chem.*, 1953, **200**, 803.
[21] D. T. Dubin, *J. Biol. Chem.*, 1960, **235**, 783.
[22] C. F. Bernasconi, *Relaxation Kinetics* (Academic Press, New York, 1976).
[23] R. P. Bell, *The Proton in Chemistry* (Cornell University Press, Ithaca, 1959).
[24] R. M. Barrer and R. P. Townsend, *J. Chem. Soc., Faraday Trans. 1*, 1978, **74**, 745.
[25] R. M. Barrer and R. M. Gibbons, *Trans. Faraday Soc.*, 1963, **59**, 2569.

Kinetics of Dissolution of Calcium Hydroxyapatite

By Jørgen Christoffersen* and Margaret R. Christoffersen

Department of Chemistry, Panum Institute, University of Copenhagen, Blegdamsvej 3, DK-2200 Copenhagen N, Denmark

Received 5th December, 1983

The rate of dissolution of calcium hydroxyapatite microcrystals [$Ca_{10}(PO_4)_6(OH)_2$, HAP] in aqueous suspension is controlled by a surface reaction, the diffusion of dissolved substance away from the crystal/liquid interface not influencing the rate. The rate can be described by a polynuclear mechanism. The rate cannot be expressed purely in terms of thermodynamic expressions; kinetic effects are important. The rate constant for the polynuclear mechanism depends on pH. This can be explained by hydrogen-ion catalysis of the linear rate of growth of the dissolution nuclei. Application of the model leads to a value of the Gibbs surface energy of 47 ± 3 mJ m^{-2} for HAP in the range $5.0 \leq pH \leq 7.2$. The acidity constant for the HPO_4^{2-} surface complex is found to be 10^{-6}–10^{-7} mol dm^{-3}, depending on the electrical potential difference between the crystal surface and the solution. Inhibition of the rate of dissolution of calcium hydroxyapatite owing to adsorption of foreign ions or molecules can often be described by a simple Langmuir adsorption isotherm.

SURFACE PROCESSES AND TRANSPORT PROCESSES IN DISSOLUTION KINETICS

When crystals dissolve in an aqueous medium two consecutive reactions take place, a surface process and a transport (diffusion or convective diffusion) process. The surface process involves transfer of substance from the crystalline phase to the solution adjacent to the crystal. This region is for simplicity called the interface region. The transport process involves the transfer of dissolved substance from the interface region to the bulk. The rate may be controlled by either of these two types of processes or by combinations thereof.

For sufficiently large unsuspended crystals, gravity causes the crystals to move relative to the medium, and description of the transport process is complicated.[1] As this complication does not occur for dissolution of microcrystals of HAP, this effect will not be discussed any further here. When the transport process affects the rate of dissolution, a concentration gradient is set up between the interface region and the bulk. In the simple case where the rate is controlled by the diffusion of substance from the interface region to the bulk, the rate of dissolution, J, of a spherical crystal is, in the steady-state approximation, given by

$$J \equiv -dn_{cr}/dt = 4\pi D(C_s - C)r \tag{1}$$

and the rate of change of r, the radius of the crystal, is given by

$$dr/dt = -DV_m(C_s - C)/r \tag{2}$$

in which n_{cr} is the amount of undissolved substance in the crystal, D is the diffusion coefficient, V_m is the molar volume of the crystalline phase, C_s is the solubility of the crystalline phase and C is the bulk concentration of dissolved substance in solution.[2] If the surface processes are not fast enough to keep the concentration,

C', of dissolved substance in the interface layer close to C_s, the rate of transport can still be described by eqn (1) and (2), but with C_s replaced by C'. If the surface process is so slow that C' is close to C, we may neglect the concentration gradient and describe the rate by an expression for the surface process in terms of C.

The natural surroundings of biominerals normally contain inhibitors for growth and dissolution of the mineral. To describe or explain the interaction between the mineral surface and the medium, information about the surface reactions of the pure mineral is most essential. For HAP microcrystals the only method for obtaining such information is to work with a large number of crystals, of the order of 10^{13}. For a polydisperse sample of crystals, all dissolving, one can measure the overall rate of dissolution

$$J = -dn_{cr}/dt = -\sum_i dn_{cr(i)}/dt \tag{3}$$

where index i runs over all crystals, n_{cr} is the amount of undissolved substance and $n_{cr(i)}$ is the amount of substance in the ith crystal.

The aim of studying the overall rate of dissolution of HAP was to find the kinetic law followed by the individual crystals. This is, in principle, impossible unless some auxiliary assumptions are introduced. The linear rate of growth or dissolution perpendicular to the various crystal faces can in general be described by an expression of the type

$$|dr/dt| = f(r)g(C) \tag{4}$$

in which $g(C)$ represents the influence of concentration on the rate and $f(r)$ represents the influence of size and of surface structure, including density of dislocations and types of dislocations.[3,4] With $f(r)$ independent of the concentration history and $g(c)$ being the same for all crystal faces, the overall rate of dissolution or growth of a polydisperse sample of polyhedral crystals may to a good approximation be expressed as

$$J = km_0 F(m/m_0) g(C) \tag{5}$$

in which k is a rate constant, m_0 the initial mass of seed crystals, $F(m/m_0)$ represents the influence of size distribution, $etc.$, of crystals and $g(C)$ represents the influence of concentration on the rate.

A pH-stat technique was used to study the rate of dissolution of HAP microcrystals at constant values of pH in the pH range 5.0–7.2.[4-6] For dissolution into solutions with a calcium-to-phosphate ratio of 1.67, the rate, for a constant value of m/m_0, was found to be proportional to $(1 - C/C_s)^p$, with p in the range 3–6. The rate constant, k, was found to increase with the hydrogen ion concentration (see fig. 1).

From electron micrographs a size distribution of the crystals was obtained. From the size distribution the number of crystals per unit mass, N/m_0, was estimated. Assuming the rate of dissolution to be diffusion controlled, an apparent diffusion coefficient $D_{app} \approx 10^{-9}$ cm^2 s^{-1} was obtained. As diffusion coefficients of small inorganic ions in water are of the order of 10^{-5} cm^2 s^{-1}, the rate of dissolution is $ca.$ 10^{-4} times the rate corresponding to diffusion. Similarly it was shown that the rate of diffusion of H$^+$ from the bulk to the crystal surface cannot be the rate-controlling process.

From the above observations it can be concluded that the rate of dissolution of HAP microcrystals in aqueous solution with $5.0 \leq pH \leq 7.2$ is controlled by a surface reaction. The dependence of the absolute value of the rate constant on pH indicates that the hydrogen ion plays an important role for the dissolution of HAP.

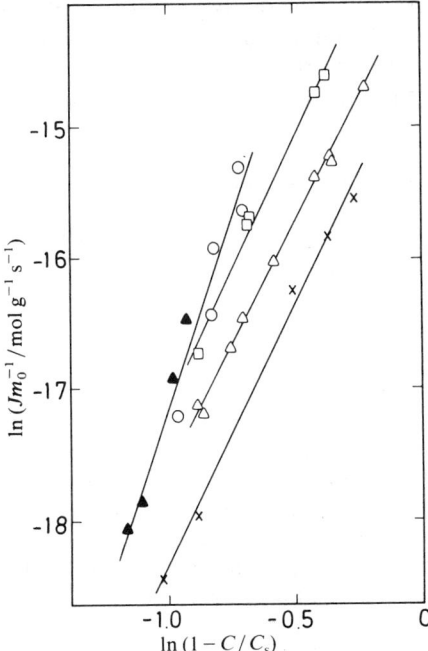

Fig. 1. Plots of $\ln(Jm_0^{-1})$ against $\ln(1 - C/C_s)$ for five constant values of pH: ○, 5.03; ▲, 5.51; □, 6.31; △, 6.76; ×, 7.15. The rates were measured when 30% of the initial mass of crystals had dissolved, i.e. $m/m_0 = 0.7$. The slopes of the lines vary from 4 to 6, indicating that the rate is not controlled by a transport process or by a surface spiral process, for which slopes of 1 and 2, respectively, would be expected.

SURFACE NUCLEATION, A POSSIBLE MECHANISM FOR DISSOLUTION OF HAP

Based on the theory of surface nucleation for growth described by Hillig[7] and on the computer calculation of crystal growth (Gilmer and Bennema,[8,9] De Haan et al.[10] and Bennema and Van der Eerden[11]), we have developed a similar mechanism, but for dissolution of crystals.[4] This mechanism can explain the main features of the rate of dissolution of HAP.

In all nucleation theories for growth and in the present theory for dissolution, nuclei are treated as continuous in size and no distinction is made between the various lattice ions. This is not a severe limitation as the nucleation phenomenon is of a statistical nature.

For dissolution a nucleus is a microscopic hole in the crystal surface, whereas for growth a nucleus is a small hill of material stacked on the crystal surface. Nuclei of different sizes have different energies of formation. The distribution of nuclei according to their energy of formation is assumed to be given by a Boltzmann distribution. ΔG for formation of a cylindrical nucleus, radius r, one mean ionic diameter, a, deep can be expressed as

$$\Delta G = -xkT\beta + 2\pi r a \sigma \qquad (6)$$

in which x is the number of missing ions in the nucleus, $x = \pi r^2/a^2$, kT is Boltzmann's constant times the absolute temperature, β is the dimensionless dissolution affinity for a single ion or growth unit, a is the edge length of the volume-equivalent cube of a growth unit and σ is the Gibbs surface energy, assumed constant, despite the small size of the nuclei. For a critical nucleus $d\Delta G/dr$ is zero. This leads to the following expression for the radius, r^*, of a critical nucleus:

$$r^* = a^3\sigma/kT\beta. \tag{7}$$

Hillig[7] derived the following expression for the rate of surface nucleation

$$I = k''C(1)\beta^{1/2}v_+ \exp[-\pi a^4\sigma^2/\beta(kT)^2] \tag{8}$$

where I is the rate of nucleation per unit surface area and $\beta = -\Delta\mu/kT$, with $\Delta\mu$ equal to the average increase in chemical potential for one growth unit (ion) in the actual reaction. v_+ is the linear rate of growth of a nucleus in the case where no back reaction occurs. For growth, $C(1)$ is the concentration of single growth unit adsorbed in the crystal surface. For dissolution, $C(1)$ is the concentration of holes formed by the loss of a single growth unit. $C(1)$ is assumed to vary relatively little with concentration and may be included in the rate constant, k''.

For a polynuclear mechanism, the rate of nucleation is so fast that nuclei intergrow. Assuming the linear rate of growth, v, of a nucleus to be constant during the lifetime of the nucleus, the nuclei per unit area created in the time interval t_0, $t_0 + dt$ have an area $I\,dt\,\pi v^2(t-t_0)^2$ at a later time t. If the nuclei formed in the time interval t_0, $t_0 + \tau$ cover one unit of area, we have

$$1 = \int_{t_0}^{t_0+\tau} I\pi v^2(t-t_0)^2\,dt = I\pi v^2\tau^3/3. \tag{9}$$

The frequency of removing one atomic layer from the crystal face is thus $\nu = 1/\tau$ and the linear rate of dissolution perpendicular to a crystal face is

$$dr/dt = -\nu a = -a(I\pi v^2/3)^{1/3} \tag{10}$$

which leads to the following expression for the overall rate of dissolution of a polydisperse sample of polyhedral crystals

$$J = k_J m_0 F(m/m_0) v^{2/3} v_+^{1/3} \beta^{1/6} \exp(-\alpha/\beta) \tag{11}$$

with

$$\alpha = \pi a^4\sigma^2/3(kT)^2. \tag{12}$$

The mechanism determining the rate, v, of lateral growth of a nucleus is not known. Gilmer and Bennema[9] expressed v in terms of β; such an expression for v cannot be used to account for specific ion effects. We have suggested[4] a simple kinetic expression for v which has been used in connection with specific ion effects.[6,12] This expression is

$$v = kaC_s|C/C_s - 1| \tag{13}$$

which leads to

$$v^+ = kaC \quad \text{and} \quad v^+ = kaC_s \tag{14}$$

for growth and for dissolution, respectively.

The overall rate of dissolution for the polynuclear mechanism can be expressed as

$$J = k_J m_0 F(m/m_0)(1 - C/C_s)^{2/3}\beta^{1/6} \exp(-\alpha/\beta) \tag{15}$$

with k_J proportional to C_s, and with α given by eqn (12).

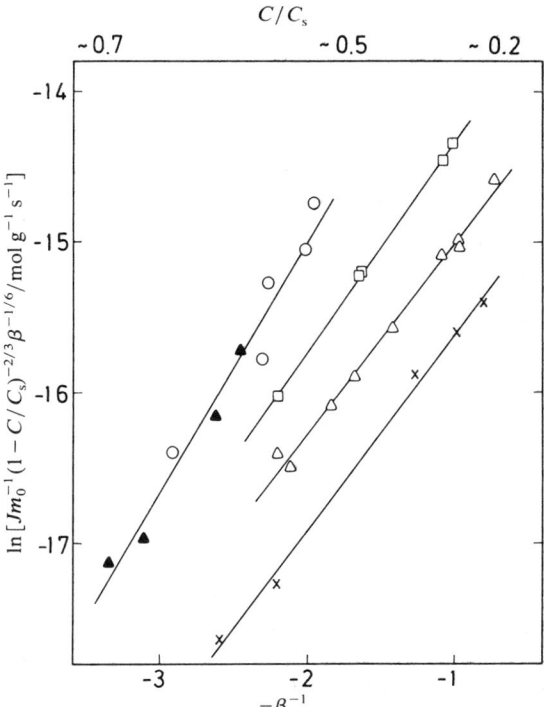

Fig. 2. The experimental results shown in fig. 1 are plotted here according to a polynuclear dissolution mechanism, for which $Jm_0^{-1}(1-C/C_s)^{-2/3}\beta^{-1/6}$ should be proportional to β^{-1}. The slopes of these lines lead to a value of 47 ± 3 mJ m^{-2} for the surface free energy, and the intercepts of the lines give the values of k_J plotted in fig. 3.

In fig. 1 $\ln(Jm_0^{-1})$ is plotted against $\ln(1-C/C_s)$ for dissolution of HAP in solutions with $Ca/P = 1.67$ and for several values of pH. From this plot it is seen that the rate is controlled by a surface process and not by a diffusion process, which would require that the slopes of the lines in fig. 1 should be close to unity. Fig. 2 shows plots of $\ln[Jm_0^{-1}(1-C/C_s)^{-2/3}\beta^{-1/6}]$ against $-\beta^{-1}$ for the same experimental results as shown in fig. 1. The rate constants, k_J, can be determined as the intercepts of the lines with $\beta^{-1} = 0$. k_J is found to increase with decreasing pH. As the slopes of the lines are nearly identical, the values of the Gibbs surface energy, $\sigma = 47 \pm 3$ mJ m^{-2}, determined from these slopes are nearly independent of pH in the range $5 \leqslant pH \leqslant 7$. This indicates that the specific effect of hydrogen ions can be explained by the lateral rate of growth of dissolution nucleus being catalysed by hydrogen ions. The aqueous solution calcium phosphate ion pairs are much more stable than calcium hydrogen phosphate ion pairs. We may similarly expect hydrogen phosphate ions in the crystal surface to be more weakly bonded to surrounding calcium ions than phosphate ions. Assuming the lateral rate of growth of the dissolution nuclei to be proportional to x_{HP}, the mole fraction of the crystal surface phosphate groups in the form of hydrogen phosphate ions, leads to

$$k_J = \tilde{k} x_{HP} = \tilde{k}/[1 + \tilde{K}_{cr}/(H^+)] \tag{16}$$

in which (H^+) is the activity of hydrogen ions in the solution, $\tilde{k}$ is a rate constant, which may depend on $\Delta\phi$, the electrical potential difference between the crystal

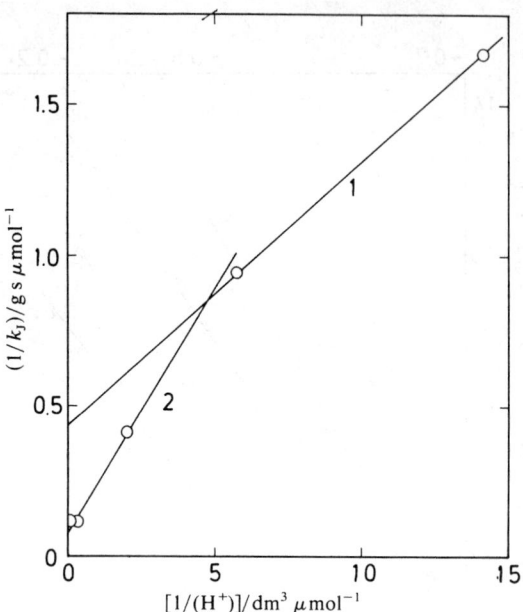

Fig. 3. k_J^{-1} plotted against $(H^+)^{-1}$ for the results shown in fig. 1 and 2. From line 1 we obtain $k = 2.3 \times 10^{-6}$ mol g^{-1} s^{-1} and p$K_{cr} = 6.7$, when $\Delta\phi = 0$. From line 2 we find that $\tilde{k} = 1.3 \times 10^{-5}$ mol g^{-1} s^{-1} and p$\tilde{K}_{cr} = 5.7$, giving $\Delta\phi \cong 60$ mV around pH 5–6.

surface and the solution, and $\tilde{K}_{cr}$ is an acidity constant for the hydrogen phosphate surface complex, defined by

$$\tilde{K}_{cr} = K_{cr} \exp(F\Delta\phi/RT) = (H^+)x_P/x_{HP} \tag{17}$$

with K_{cr} being the acidity constant for $\Delta\phi = 0$ and F Faraday's constant.

In fig. 3 k_J^{-1} is plotted against $(H^+)^{-1}$ for $m/m_0 = 0.7$. If $\tilde{K}_{cr}$ and $\tilde{k}$ were proper constants, a straight line would be obtained. The deviation from a straight line can be explained by $\Delta\phi$ increasing as pH decreases. As an approximation, in fig. 3 we have drawn two straight lines crossing at $1/(H^+) = 4.8$ dm^3 μmol^{-1} (pH 6.7). HAP crystals are not significantly charged at ca. pH 7. From line 1 in fig. 3 we obtain $k = 2.3 \times 10^{-6}$ mol s^{-1} g^{-1} and $K_{cr} = 2 \times 10^{-7}$ mol dm^{-3}, i.e. p$K_{cr} = 6.7$, where k and K_{cr} are the values of $\tilde{k}$ and $\tilde{K}_{cr}$ for $\Delta\phi = 0$. From line 2 in fig. 3 we obtain, for $5 \leq pH \leq 6.3$, $\tilde{k} = 1.3 \times 10^{-5}$ mol s^{-1} g^{-1} and $\tilde{K}_{cr} = 2 \times 10^{-6}$ mol dm^{-3}, i.e. p$\tilde{K}_{cr} = 5.7$. These values of K_{cr} and $\tilde{K}_{cr}$ are only approximate. As the crystals become charged, the acidity constant of HPO$_4^{2-}$ surface ions will gradually change from a value close to 10^{-7} mol dm^{-3} to a value ca. 10 times larger. From eqn (17) we obtain

$$\Delta\phi = (RT/F)\ln(\tilde{K}_{cr}/K_{cr}) \approx 60 \text{ mV}. \tag{18}$$

Assuming that HPO$_4^{2-}$ ions leave the crystal surface relatively easily compared with calcium ions, the activation energy for removing a calcium ion, calculated by inserting the value of k or $\tilde{k}$ in Eyring's formula, is ca. 1.4×10^{-19} J. In HAP, calcium ions are surrounded by 6–8 oxygens from hydroxyl or phosphate groups, half of which are missing in the edge of a nucleus. The activation energy therefore corresponds to breaking 3–4 oxygen–calcium bonds in connection with removal of a calcium ion. The activation energy found is also close to three times the energy required to remove a water molecule from a hydrated calcium ion in solution.[12] Note that knowledge of absolute rate constants for surface-reaction-controlled dissolution processes gives information about the energy needed to break bonds in the crystalline phase, whereas knowledge of absolute rate constants for surface-reaction-controlled crystal-growth processes[13] gives information about the interaction between the solvent and the ions in solution.

INHIBITION OF THE DISSOLUTION PROCESS OF HAP

The effect of an inhibitor for surface nucleation may be described as prevention, or strong retardation, of the nucleation process in areas around the adsorbed inhibitor molecules or ions. Each inhibitor unit in the crystal surface is surrounded by α neighbouring ions, all having the same sign of charge. Owing to interaction with the inhibitor such ions are strongly attached to the crystal surface. These ions, including the inhibitor, occupy an area $\pi r_0^2 \equiv (\alpha + 1)d^2$, in which d is the diameter of a mean ion. The area around an inhibitor unit in which nucleation is strongly retarded is of the order of $\pi(r_0 + r^*)^2$, r^* being the radius of the critical nucleus. This means that the fraction of the crystal surface area in which nucleation is prevented, A_L/A, is larger than the mole fraction, x, of the adsorption sites occupied by the inhibitor, i.e. $A_L/A = \phi x$, with $\phi > 1$. For a Langmuir adsorption isotherm,

$$K_L = x/(1-x)C_L \Leftrightarrow 1/(1-x) = 1 + K_L C_L \qquad (19)$$

we have, for low degrees of surface covering,

$$J_0/J_L \approx A/(A - A_L) = 1/(1 - \phi x) \approx 1 + K_L \phi C_L \equiv 1 + K_{kin} C_L \qquad (20)$$

in which J_0 and J_L are the rates of dissolution without and with the inhibitor present, respectively, both rates being determined for the same values of the parameters influencing J_0. As the solution composition approaches an equilibrium value, r^* increases and so does ϕ. This causes an increase in the effect of the inhibitor. This effect has been demonstrated.[14]

CONCLUSION

The rate of dissolution of HAP can be described by a surface nucleation process, but not purely in thermodynamic terms, kinetic effects being important. This demonstrates the importance of analysing empirical data for growth and dissolution of crystals in terms of models in which chemical reactions in the crystal/solution interface are taken into account.

We are grateful to the Danish Medical Research Council for grants for laboratory assistance (12-0323, 12-2215, 12-3617).

[1] A. E. Nielsen, *J. Phys. Chem.*, 1961, **65**, 46.
[2] A. E. Nielsen, in *Kinetics of Precipitation* (Pergamon Press, Oxford, 1964), p. 34.
[3] J. Christoffersen and M. R. Christoffersen, *J. Cryst. Growth*, 1976, **35**, 79.
[4] J. Christoffersen, *J. Cryst. Growth*, 1980, **49**, 29.
[5] J. Christoffersen, M. R. Christoffersen and N. Kjaergaard, *J. Cryst. Growth*, 1978, **43**, 501.
[6] J. Christoffersen and M. R. Christoffersen, *J. Cryst. Growth*, 1982, **57**, 21.
[7] W. B. Hillig, *Acta Metallurg.*, 1966, **14**, 1868.
[8] G. H. Gilmer and P. Bennema, *J. Cryst. Growth*, 1972, **13/14**, 148.
[9] G. H. Gilmer and P. Bennema, *J. Appl. Phys.*, 1972, **43**, 1347.

[10] S. W. H. de Haan, V. J. A. Meeusen, B. P. Veltman, P. Bennema, C. van Leeuwen and G. H. Gilmer, *J. Cryst. Growth*, 1974, **24/25**, 491.
[11] P. Bennema and J. P. van der Eerden, *J. Cryst. Growth*, 1977, **42**, 201.
[12] S. Petrucci, in *Ionic Interactions, II. Kinetics and Structure*, ed. S. Petrucci (Academic Press, New York, 1971).
[13] A. E. Nielsen, in *Industrial Crystallization 81*, ed. S. J. Jančić and E. J. de Jong (North-Holland, Amsterdam, 1982), pp. 35–44.
[14] J. Christoffersen and M. R. Christoffersen, *J. Cryst. Growth*, 1981, **53**, 42.

Kinetics and Simulation of Dissolution of Barium Sulphate

By Vincent K. Cheng, Bruce A. W. Coller* and John L. Powell†

Department of Chemistry, Monash University, Clayton, 3168 Victoria, Australia

Received 16th December, 1983

Rates of dissolution of a natural barite single crystal mounted in a rotating disc have been found to be proportional to the square of undersaturation in terms of concentration, depressed by added common ion, enhanced by foreign bivalent metal ions and unaffected by monovalent ions.

Activation energies for dissolution of barium sulphate microcrystals of three types show variation from 20 to 80 kJ mol^{-1} and are inversely correlated with energies of dissolution calculated from kinetic solubilities, these energies varying from $\Delta H°$ to near zero.

Monte Carlo simulation of dissolution of crystals with stress fields such as would arise from screw, edge and plane dislocations and from edges and apices gave a corresponding variation of simulated activation energy, depending on the degree of stress and on the initial configuration of the surface.

Effects of foreign ions were simulated by introducing next-nearest-neighbour repulsions in the Monte Carlo treatment of adsorption on an initially flat surface with a 4×4 columnar hole.

The kinetics of growth and dissolution of barium sulphate have been the subject of a number of previous studies.[1-6] The processes are surface controlled and provide a test case for theories of growth and dissolution of ionic crystals. The rates of growth and dissolution of microcrystalline suspensions were reported to be proportional to the square of supersaturation or undersaturation in terms of concentration

$$R = k(A)(C_{eq} - C)^2. \qquad (1)$$

Similar behaviour was reported for strontium and lead sulphates[7,8] but runs counter to the naive expectation that rates of dissolution or growth of solid salts might be proportional to the distance from equilibrium in terms of ionic product, P_I:

$$R = k(A)(K_s - P_I) \qquad (2)$$

which for stoichiometric solution would take the form

$$R = k(A)(C_{eq}^2 - C^2) \qquad (3)$$

reducing, when close to equilibrium, to the linear form

$$R \approx 2k(A)C_{eq}(C_{eq} - C). \qquad (4)$$

Davies and Jones[9] intuitively identified the observed proportionality of the rate of growth of silver chloride to the square of supersaturation with the binary ionic nature of the salt in an adsorbed monolayer of mobile hydrated ions, but they had no sound physical basis for their model. However, later workers have reported rates of dissolution depending on the cube of undersaturation for the ternary ionic compounds Ba(IO$_3$)$_2$[10] and Ag$_2$(CrO$_4$).[11]

---

† Present address: Oil and Gas Division, BHP Ltd, 35 Collins Street, Melbourne, 3000 Victoria, Australia.

The kinetic solubilities of certain barium sulphate preparations have been found to be significantly greater than the accepted equilibrium solubility.[3] The extent of solubility enhancement is dependent on the conditions of nucleation and growth and thus on the regularity of the crystals. Successive dissolution runs using one batch of crushed crystals showed progressive decreases in both kinetic solubility and dissolution rate constant.[3]

Our earlier finding of the elevation of kinetic and saturation solubilities for barium sulphate depending on crystal type and on radioactive labelling were attributed to submicrometre crystallites on the surfaces of the crystal and as grains within the bodies of the crystals.[3] We would now also give weight to the effect of preferential dissolution at dislocation centres and lines of dislocation (grain boundaries).

Nancollas et al.[4] found that rate constants for growth and dissolution of different crystal types are not proportional to their specific surface areas and suggested the involvement of active sites yet to be identified. Similar observations were recently reported by Christoffersen in which the rate constants for growth and dissolution of calcium sulphate dihydrate were found to depend on the initial growth conditions and on the washing and aging treatment.[12] The cause of such variation in rate constant was given in terms of the dislocation density and grain boundary structure at the surface.

The topography of a crystal surface can be modified as a result of preferential dissolution at stressed regions. The extent of changes in surface activity will therefore be determined by the perfection of the crystals, the solvent undersaturation and also by the initial mass and the duration of dissolution. The initial perfection is affected by the method of preparation; microscopic and X-ray examination of barium sulphate crystals[13,14] reveals that crystals grown under vigorous conditions contain extensive imperfections, such as dislocations and mosaic and grain boundaries within both the surface and bulk of the crystal. Well aged crystals appear to have smooth surfaces.[4]

In accordance with the Kossel–Stranski model for crystal growth,[15] the majority of growth events may be considered in terms of a three-dimensional lattice (B_n) with steps terminated by kink units (K) to which free units (F) from the solution become attached after one or two stages of diffusion, desolvation and adsorption. Each such event adds one unit to the lattice:

$$B_n K + F \rightleftarrows \rightleftarrows \rightleftarrows B_{n+1} K. \tag{5}$$

The rate of growth is often interpreted in terms of equilibrium between free units (F) and terrace units (T), with terrace diffusion-controlled rates of attachment of units to steps (ledges, L) in equilibrium with kinks. The rate of growth should then depend on the number of available kinks (N_K) and the oversaturation in terms of concentration:

$$\text{growth rate} = k N_K (C - C_{eq}). \tag{6}$$

Screw dislocations are widely held to be necessary for the growth of crystals, both ionic and molecular. Burton, Cabrera and Frank (B.C.F)[16] showed that the screw-dislocation mechanism can provide a continuous supply of step and kink sites with a steady-state area density such that

$$\text{growth rate/area} = k \ln (C_{eq}/C)^{-1}(C - C_{eq}). \tag{7}$$

A similar expression would apply to dissolution if this were dominated by reverse spirals developing around screw dislocations, and would tend toward proportionality to the square of undersaturation in terms of concentration when close to equilibrium for simple crystals. Such an approximation may also apply for ionic crystals dissolving into stoichiometric solutions by the mechanism controlled by terrace diffusion and screw dislocation. With a more generalised rate law, van Rosmalen[17] showed that the rate of growth of calcium sulphate dihydrate close to equilibrium is consistent with the B.C.F. process.

We have studied dissolution rates with a cleaved surface of barite which was mounted in a rotating disc to compare the applicability of the B.C.F. equation with the square-of-undersaturation equation given above. The surfaces of similar barite samples were examined by scanning electron microscopy before and after dissolution.

The activation energy for the dissolution of barium sulphate, reported by Jones ($E_a = 24$ kJ mol^{-1}),[2] is comparable to the activation energy of diffusion, despite the clear demonstration of surface-controlled kinetics. Our studies are now extended to values of the activation energy determined from the temperature dependence of the initial rate using microcrystalline suspensions prepared by different methods. Apparent enthalpies of dissolution are also evaluated from kinetic solubilities. The dependence of the activation energy on crystal surface stress was simulated by calculating rates at different simulated temperatures using Bennema's Monte Carlo model incorporating a cylindrical stress field.[18]

Reductions of crystal growth and dissolution rates in the presence of added impurities are commonly reported. Small quantities of polyelectrolytes, or other high-molecular-weight material, are commonly inhibitors of the growth and dissolution of barium sulphate[5,19,20] and calcium sulphate.[21] Simon reported that addition of cadmium chloride retarded the dissolution of sodium chloride.[22] The blocking of movement of surface steps by the adsorption of high-molecular-weight impurities has been used to account for the inhibition effect.[20,21,23] However, traces of sodium chloride enhanced the growth of calcium sulphate,[24] and potassium dichromate and potassium ferrocyanide were found to enhance the growth of barium sulphate.[5]

Computer simulation studies have shown[25,26] that traces of impurities which form strong impurity–host bonds at the surface may cause enhancement of the growth rate through nucleation. Impurities that form surface clusters cause enhancement of the growth rate by providing a large number of step sites. Thus the effectiveness of impurities in altering crystal growth rates will depend on their adsorption characteristics.[26] The dissolution rate can be expected to be enhanced if adsorbed impurities can reduce the binding energy of the nearby surface lattice ions. At low solution concentrations simple ionic compounds are adsorbed on the surface of barium sulphate[27] in electroneutral amounts. Their characteristic adsorption isotherms indicate that the adsorbate ions are well isolated from one another. Under these conditions lattice ions which are next-nearest neighbours to an adsorbed ion will experience repulsion. Such weakening of the binding energy of surface ions provides the potential for enhancement of the dissolution rate. We have added trace quantities ($<10^{-4}$ mol dm^{-3}) of simple ionic additives such as chlorides of potassium, strontium, calcium and barium and sodium sulphate to the dissolving barite to investigate the alteration of the dissolution rate. The results are compared with those obtained from a simulation model which incorporates next-nearest-neighbour lattice-adsorbed ion repulsions in the manner described by Bennema *et al.*[28]

EXPERIMENTAL DISSOLUTION STUDIES

Three types of barium sulphate microcrystals were prepared as previously described.[3] Type I crystals were made by slow mixing (with paddle stirring) of 10^{-3} mol dm^{-3} reactant solutions. Type II crystals were prepared by rapid mixing of 0.1 mol dm^{-3} reactant solutions. Type III crystals were prepared from solutions in methanol, which were mixed rapidly at room temperature. The colloid was allowed to coagulate and settle before washing and storage. The precipitates were aged for an initial period of 4 h at 90 °C. At weekly intervals during subsequent storage the overlying solutions were gently decanted and replenished with doubly distilled water. The crystals were crushed and mounted onto polythene plates as described previously.[3]

Dissolution of barium sulphate was monitored conductimetrically. Rate coefficients and kinetic solubilities were obtained from a plot of (rate)$^{1/2}$ against concentration and initial rates calculated therefrom. Arrhenius activation energies were calculated from ratios of rate coefficients determined before and after changing the temperature from 25 to 35 °C in mid run. Plates with a given type of crystals were re-used in several runs.

The composition and crystallographic characteristics of the natural barite used were determined by AAS, X-ray fluorescence and X-ray diffraction. The barite contains over 96% barium sulphate and the other impurity detectable by X-ray fluorescence was strontium. Its unit-cell dimensions (5.3 ×7.2 ×8.4 Å^3) were compatible with the literature values.[29] A single crystal was cleaved at an existing plane of weakness (100) and mounted in an Araldite disc to enable rotation in a dissolution medium. The barite surface was not polished. Conductimetric measurements were used to determine the rate of dissolution into doubly distilled water and into diluted samples of solution that had become saturated (12.4 μmol dm^{-3}) in contact with a microcrystal suspension. Effects of added salts were determined by adding measured amounts of the salts after a period of dissolution (ca. 3 h) into distilled water and continuing the run for a similar period of time. Linear conductance–time relations were observed and each dissolution run corresponded to an increase of ca. 10^{-7} mol dm^{-3} in the concentration of barium sulphate.

MONTE CARLO SIMULATION METHODS

Monte Carlo simulation of the dissolution of a stressed surface was carried out by means of a computer program kindly supplied by P. Bennema.[18] The model incorporates an effective nearest-neighbour attraction energy, ϕ. The cylindrical stress field located around the centre of the surface is given by

$$U(r) = \frac{U(0)}{1+(r/r_h)^2} \qquad (8)$$

where r is the distance from the dislocation centre, $U(0)$ is the stress energy density at the dislocation centre, r_h is the Hooke radius $[r_f \phi / U(0)]$ and r_f is the Frank radius, which effectively measures the size of the stress field or dislocation region.

Surface diffusion is not included, and the microscopic reversibility condition for the ratio of deposition and detachment probability at each site depends on the number of lateral nearest neighbours, n:

$$\frac{k^-}{k^+} = \frac{\exp\{[(4-2n)\phi + U(r)]/kT\}}{C/C_{eq}}. \qquad (9)$$

The program has been extended to allow initial surface configurations that are not flat.

On an unstressed lattice surface the absolute energy required to remove one kink unit and thus create one less bulk unit is 6ϕ. This corresponds to the enthalpy ($\Delta H°$) of dissolution of the stress-free crystal. In a real crystal activation barriers in the elementary deposition and detachment events are to be expected. It has been shown that activation barriers can be factorised into pre-exponential frequency factors in the attachment and detachment probability expressions.[30]

The reference conditions chosen were $\phi/kT = 2$ and $C/C_{eq} = 0.135$. Rate calculations were then carried out assuming constant ϕ and C with selected values of $U(0)$ and initial configurations. Changes of temperature of 2.5%, up and down, were imposed. The new C/C_{eq} inputs were calculated by allowing for the change in equilibrium concentration due to temperature change

$$\frac{d \ln (C_{eq})}{d(1/kT)} = -6\phi. \tag{10}$$

Simulations were carried out for an initially stress-free surface, for one starting with a 4×4 columnar hole[25] at its centre and for surfaces with different stress fields and initial configurations. The rate of dissolution was calculated from the change in the number of lattice units per unit area over a period of time sufficient to obtain a steady average rate and root-mean-square deviation.

Monte Carlo simulation of the effects of the foreign bivalent metal ions on the kinetics of dissolution of barium sulphate was achieved by means of a newly developed program based on Bennema's previous work referring to a two-component solid.[28] The initial crystal lattice was assigned a flat surface with a 4×4 columnar hole without specific recognition of its binary nature. Foreign ions were assumed to attach to the surface, on top of lattice ions of opposite charge, with the attractive interaction energy ϕ_{IL} equal to the lattice-ion interaction energy ϕ_L. Next-nearest neighbours in the lattice surface adjacent to the adsorbed ion, being of like charge, were then assigned an interaction energy $\alpha\phi_{IL}$, to replace the stress function in eqn (9) above, using negative values of α to represent the repulsive character of the interaction. The foreign ions in solution were assigned a potential μ_I (solution) such that the average fractional coverage (θ_I) of the surface by adsorbed ions would tend with time to an equilibrium value such that $kT \ln [\theta_I/(1-\theta_I)] \approx \mu_I$ (solution).

RESULTS

KINETIC DATA

Rates of dissolution of barium sulphate into partly saturated solutions from a single crystal barite surface are represented in fig. 1. The observations conform to the square of distance from a kinetic solubility C_∞ [eqn(1)]:

$$R = k(A)C_\infty^2(1 - C/C_\infty)^2 \tag{11}$$

rather than to the B.C.F. theory [eqn (7)]. The corresponding dissolution rate constant $k(A)$ was found to be 190 dm^3 mol^{-1} min^{-1}. If this is divided by the area it is compatible with that reported for the type A crystals used by Liu et al.[4] Extrapolation of fig. 1 to zero of (rate)$^{1/2}$ gives the kinetic solubility of barite as 11.5 ± 0.9 μmol dm^{-3}, intermediate between the saturation solubility of the barium sulphate suspension (12.4 μmol dm^{-3}) and the accepted equilibrium solubility (10.4 μmol dm^{-3}).[3]

SEM STUDIES

Plate 1 shows the barite disc used in the kinetic studies. Slight dulling of the surface as a result of the dissolution is evident. However, it appears that dissolution has occurred to a greater degree near natural cracks and edges. Plate 2 shows a low-magnification micrograph of a similar barite surface obtained by cleavage along planes of weakness. Flat and smooth areas were separated by multiple step features (plate 3) originating from fracture between mismatched planes of weakness. A third

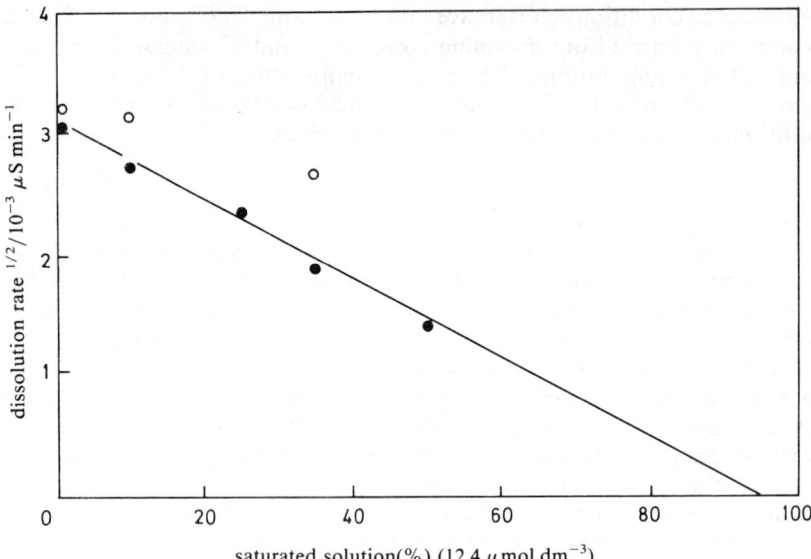

Fig. 1. Square root of rate of dissolution from rotated single barite crystal face plotted against percentage of solution saturated by microcrystals (12.4 μmol dm^{-3}) ●, Stoichiometric solution; ○, with added SrCl$_2$ (10 μmol dm^{-3}).

sample (plate 4), after standing in water for several days, showed smooth areas that were largely unaffected, together with clear indications of dissolution and roughening along fracture steps and in other areas, presumably of surface imperfections such as emergent dislocations and grain structure. Plate 5, at high magnification, shows the erratic etching that occurs in such areas. Thus it is evident that dissolution occurs preferentially in the vicinity of steps, edges, cracks and other lattice defects.

EXPERIMENTAL ACTIVATION ENERGIES

Table 1 shows the ranges of kinetic solubilities at 25 and 35 °C, and the corresponding energy changes (enthalpies) of dissolution, together with activation energies based on calculated initial rates $[k(A)C_\infty^2]$, for several microcrystalline barium sulphate preparations mounted on polythene plates. An inverse correlation (fig.2) exists between the activation energies and the energy changes (enthalpies) of dissolution, while both of these properties are roughly correlated with kinetic solubilities at 25 °C, the more soluble crystals showing less positive energies of dissolution, as would normally be expected. However, the variation in Gibbs free energy of dissolution (see table 1) is small while the energy change varies from approximately the calorimetric value (19.2 kJ mol^{-1})[31] to near zero. The activation energies vary from 20 up to 80 kJ mol^{-1}, the crystals having the less positive energy changes requiring the *higher* activation energies for dissolution. Variability is also seen for crystals of each type. This is likely to be due to use of each plated specimen in several successive dissolution-and-storage cycles. Such variation may also be expected to occur in wash-and-age cycles during storage of seed crystals.[3]

Specimens prepared with radioactive tracer (^{35}S, 20 and 50 mCi)[3] showed behaviour patterns similar to those of the corresponding inactive preparations and will not be discussed separately.

Plate 1. Barite disc (area 131 mm²) used in the experiment, showing surface texture after dissolution.

Plate 2. Scanning electron micrograph of a mechanically cleaved surface (100) before dissolution. Regular microsteps are evident. 50×magnification. Bar represents 200 μm.

Plate 3. Specimen of plate 2 at 1000 × magnification, showing the topography of fractured microsteps. Bar represents 5 μm.

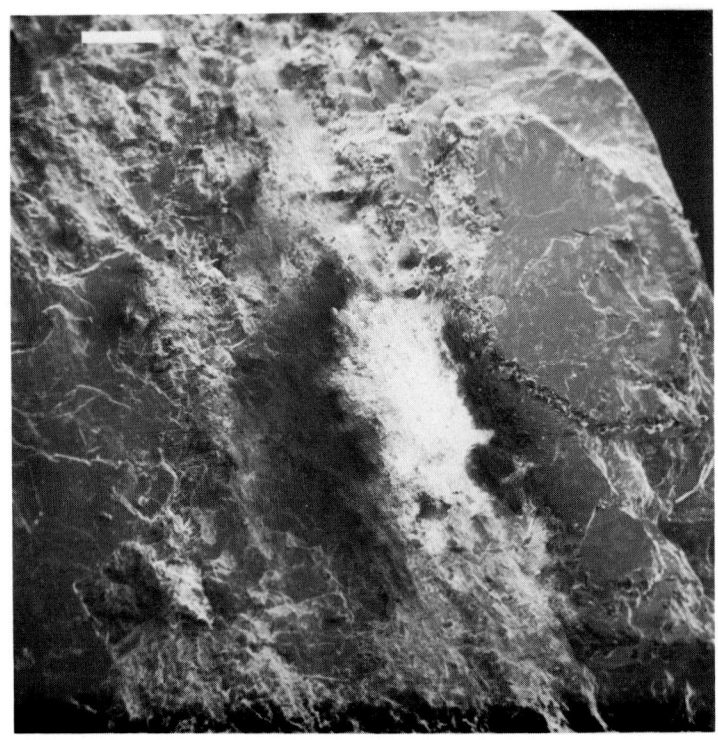

Plate 4. Scanning electron micrograph of a similarly cleaved surface (100) after dissolution. Roughening of edges, steps and patches is evident. 50 × magnification. Bar represents 200 μm.

Plate 5. Specimen of plate 4 at 5000×magnification, showing the irregular pits in the roughened patch. Bar represents 2 μm.

Table 1. Characteristics and dissolution behaviour of three types of microcrystalline barium sulphate[a]

parameter	crystal type		
	I	II	III
size^3/μm	5–15	<5	<0.2
shape	rectangular [4]	irregular stars [4]	aggregated colloids
precipitation	slow, 10^{-3} mol dm^{-3}	rapid, 10^{-1} mol dm^{-3}	from methanol
nucleation (inferred)	heterogeneous	homogeneous	homogeneous
defect structure (inferred)	central region of stress/dislocation [4,13]	high surface energy near edges and apices	grain boundaries, high surface energy
kinetic solubility μmol dm^{-3} (suspension)	10.4[b]	11.0	13.5
saturation solubility^3 /μmol dm^{-3} (suspension)	12.5	10.9	13.0
kinetic solubility /μmol dm^{-3} (from plates)[c]	10.5–11.1[d]	12.5–15.2	14.6–17.0
$-\frac{1}{2}\Delta\bar{G}$/kJ mol^{-1}	0.03–0.16	0.42–0.94	0.85–1.20
$-\frac{1}{2}\Delta\bar{H}$/kJ mol^{-1} (kinetic solubilities)	19–11	16–0	11–2
	$\Delta H° \approx \Delta\bar{H}$	$\Delta H° \geqslant \Delta\bar{H}$	$\Delta H° > \Delta\bar{H} > 0$
E_a/kJ mol^{-1}	20–50	40–80	40–70
dissolution rate law	$k(C_\infty - C)^2$	$k'(C_\infty - C)^2$	$k''(C_\infty - C)^2$

[a] Accepted equilibrium solubility, 10.4 μmol dm^{-3}.^{3} Calorimetric energy (enthalpy) of dissolution, $\Delta H° = 19.2$ kJ mol^{-1}.^{31} [b] 11.0 for radioactive specimen. [c] This work. [d] 11.1 to 13.0 μmol dm^{-3} for radioactive specimens.

SIMULATED DISSOLUTION OF CRYSTALS WITH STRESS FIELD

Simulated dissolution of an unstressed initial flat surface showed the establishment of a more or less steady surface configuration with a small number of pits (1 or 2) of depth one lattice unit. The rates of dissolution showed oscillations corresponding to initiation, propagation and completion in the removal of surface layers. Table 2 shows that with increasing intensity of stress at the centre [$U(0)$] and with greater radius over which stress is present (r_r), average rates of dissolution become higher and root-mean-square deviations in the oscillations become smaller. The steady-state surface configurations are marked by deeper pits centred on the stressed region. Arrhenius activation energies, based on rates obtained at three simulated temperatures for a fixed small concentration (ca. $0.135 C_{eq}$) of dissolved units, were seen to decrease with increasing intensity and increasing extension of the stress field, as intuitively expected.

For a given stress field, an initial surface configuration, which had previously been obtained from extensive simulated etching, led to a higher steady average dissolution rate and a lower activation energy. This result may correspond to the

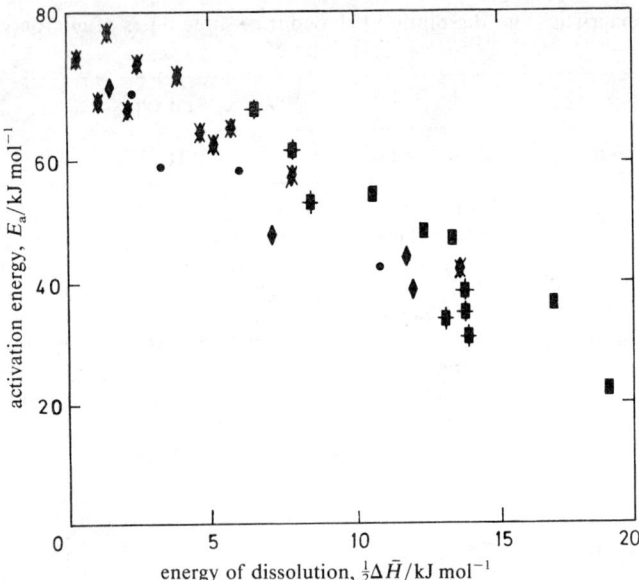

Fig. 2. Activation energies and energy changes for dissolution for three types of microcrystalline barium sulphate: ■, type I; ◆, type II; ●, type III. Star points on symbols indicate radioactive labelling (20 or 50 mCi).

Table 2. Simulated activation energy at various stress field strengths and initial surface configurations[a]

$U(0)/kT$	rate and r.m.s. dev. at $\phi/kT = 2.0$	E_a/ϕ	pit depth[b] initial → final
$r_f = 10.0$			
0	−0.0069 (±0.0035)	9.0 ± 0.1	0 → 1
2.5	−0.0337 (±0.0080)	9.0 ± 0.3	0 → 4
	−0.0326 (±0.0075)	4.7 ± 0.1	29 → 5[c]
4.5	−0.1102 (±0.0039)	3.4 ± 0.2	0 → 29[d]
open core	−0.0689 (±0.0030)	3.8 ± 0.4	0 → 14
$r_f = $	20.0		
2.5	−0.0489 (±0.0103)	5.3 ± 0.1	0 → 5
4.5	−0.1574 (±0.0058)	2.4 ± 0.4	0 → 50[d]

[a] $\phi/kT = 2.05, 2.00, 1.95.$ $C/C_{eq} = 0.182, 0.135, 0.100$ ($\Delta H° = 6\phi$). [b] Pit depth is defined as the difference between the maximum and minimum height of the steady-state surface configuration. [c] Initial pit depth 29 units (reduced by dissolution of upper layers). [d] Pit depth increases with temperature. The value given is at $\phi/kT = 2.0$.

Table 3. Relative initial rate of dissolution of barite with added salts in solution

salt added	final solution concentration/μmol dm^{-3}	relative initial dissolution rate[a]
KCl	100	1.01
Na$_2$SO$_4$	10	0.60
	20	0.34
BaCl$_2$	9	0.68
CaCl$_2$	10	1.18
	54	1.25
	100	1.45
SrCl$_2$	10	1.13
	20	1.41
	10	1.33[b]
	10	2.03[c]

[a] Relative rate = rate in salt solution/rate in distilled water. [b] When SrCl$_2$ was added to a 10% saturated BaSO$_4$ solution. [c] When SrCl$_2$ was added to a 35% saturated BaSO$_4$ solution.

variation in activation energy of dissolution for successive runs with a given type of barium sulphate crystals, as described above.

For the system with a 4×4 open core, simulating a case where deep etch-pit formation has led to removal of a stressed region around a dislocation, the dissolution rate and activation energy were intermediate between those for high stress [$U(0)/kT = 4.5$] and low stress [$U(0)/kT = 2.5$]. This again shows how variability of behaviour can arise from differences in the initial surface configuration. An otherwise perfect lattice can be expected to dissolve more rapidly when an etch pit is present. Simulation, with a columnar core present at the start, also models the effects of outer crystal edges, but not apices, on rates of dissolution.

EFFECTS OF ADDED SALTS ON RATES OF DISSOLUTION

The effects of added salts on the rate of dissolution of barium sulphate from the single crystal are shown in fig. 1 and table 3. The simple univalent ions K$^+$ and Cl$^-$ had no significant effect on the rate of dissolution and should have a negligible effect on the position of equilibrium. On the other hand, addition of either barium chloride or sodium sulphate reduced the rate of dissolution. Since these salts involve the common ions Ba^{2+} and SO$_4^{2-}$, the distance from equilibrium in terms of concentration and ionic product is also reduced. Addition of foreign alkaline-earth ions to the solution, either Sr^{2+} or Ca^{2+}, enhanced the dissolution rate but will hardly have altered the distance from equilibrium.

SIMULATED EFFECTS OF ADSORBED IONS

Results obtained by Monte Carlo simulation of dissolution in the presence of foreign ions which can be adsorbed with various strengths of the next-nearest-neighbour (NNN) repulsions are represented in fig 3. Rates of simulated dissolution were found to increase with the magnitude of the NNN repulsion.

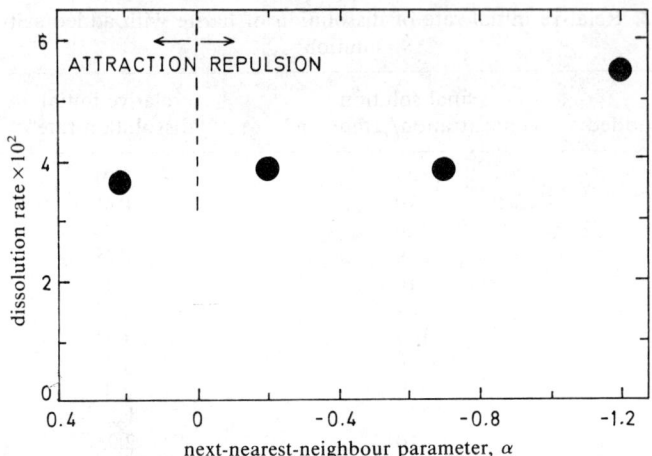

Fig. 3. Dependence of simulated dissolution rate (dimensionless) on the next-nearest-neighbour (NNN) parameter, α. $\phi/kT = 2.0$, $\Delta\mu_L/kT = -1.0$, $\mu_I/kT = -6.0$.

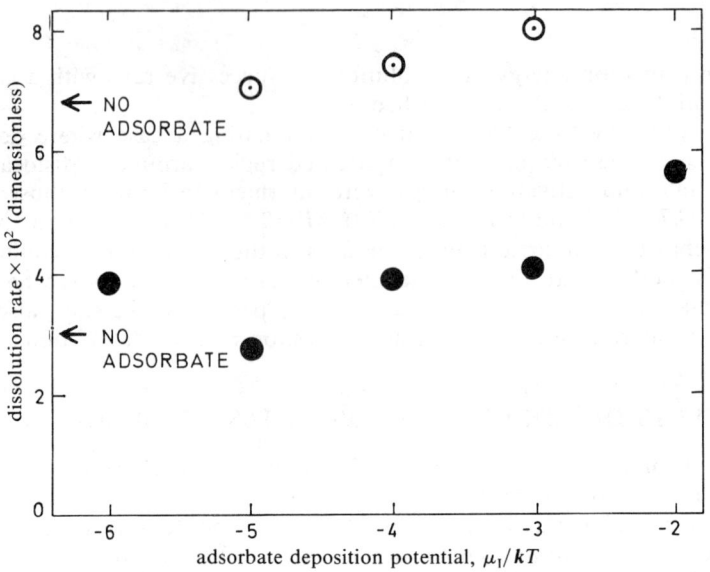

Fig. 4. Simulated dissolution rate (dimensionless) at various adsorbate deposition potentials, $\bigcirc$, $\Delta\mu_L/kT = -2.0$; $\bullet$, $\Delta\mu_L/kT = -1.0$. $\phi/LT = 2$, $\alpha = -0.7$.

The effectiveness of the NNN repulsion in the enhancement of dissolution rate is evident even at low fractional coverage of foreign ions on the surface ($\ln \theta_I \approx \mu_I/kT = -6.0$). Extensive clustering of the adsorbate was evident at high μ_I/kT ($\theta_I \approx 0.5$), but this will only be realistic for ions if counter-ions are taken into account. None of the foreign entities was found to be incorporated into the lattice.

The dissolution rates at various simulated undersaturations and foreign-ion concentrations are plotted in fig. 4. At low impurity deposition potential ($\mu_I/kT < -3.0$, $\theta_I < 0.05$) and moderate undersaturation the isolated adsorbate ions did not affect the dissolution rate to any considerable degree. Otherwise the presence of adsorbed ions tended to increase the dissolution rate. For every 2.7-fold increase in adsorbate-ion concentration in the solution the rate enhancement at high undersaturation ($C/C_{eq} = 0.135$, fig. 4, $\Delta\mu_L = -2.0$) was found to be ca. 10%, whereas at moderate undersaturation ($C/C_{eq} = 0.37$) the higher value of ca. 30% was found. Thus we can explain why the observed degree of enhancement of dissolution rate for barium sulphate was greater when strontium chloride was added to a partially saturated solution.

Overall, the simulation results show that non-nearest-neighbour repulsive effects of adsorbed ions can lead to enhancement of dissolution rate of the kind found when Sr^{2+} or Ca^{2+} ions are added to a solution into which barium sulphate is dissolving.

DISCUSSION OF RESULTS

The present results show that the rate of dissolution of barium sulphate from a single-crystal face is proportional to the square of distance from equilibrium in terms of concentration. Such a rate law can be expected if the mechanism is one in which the rate of reaction of certain active units (in this case kink units) is proportional to the distance from equilibrium while the number of active units [eqn (6)] is also proportional to the distance from equilibrium.

Preferentially dissolved regions were found on the barite surface at steps, edges and cracks and other regions of structural imperfection. The pits generated as a result of dissolution at imperfect regions may serve as a ready-made supply of layer openings, so that only step openings are required to sustain dissolution. This can be expected to give rise to a higher order than simple proportionality to undersaturation. However, the consistent proportionality of rates to the square of undersaturation in terms of concentration remains to be explained quantitatively in terms of mechanism.

Table 1 lists observed and inferred characteristics of the three types of microcrystalline barium sulphate used in the present study. For microcrystalline barium sulphate of type I, with kinetic solubility close to the accepted equilibrium solubility, Arrhenius activation energies based on rates of dissolution were found to be similar to, or more than, the calorimetric energy of dissolution. The energies (enthalpies) of dissolution based on temperature coefficients of kinetic solubilities were similar to or less than this quantity. Crystals of types II and III showed considerably higher activation energies and considerably lower energy changes for dissolution.

The above observations are explicable insofar as the conditions of nucleation and growth of the crystals determine their defect structure and dictate the mechanism of dissolution. The inverse correlation between energy of dissolution and activation energy can be explained at least partly by reference to surface topographies and inferred internal stresses as indicated below.

Type I crystals grown from dilute neutral solutions after heterogeneous nucleation appear to have large flat faces with line defects (possibly screw dislocations) at their centres, as evidenced by the flower-like growths[4] and dissolution pits[13] that develop around these points. The energy of dissolution would be high and the activation energy low if the dissolution process was dominated by detachment of kink units from steps between the flat terrace areas with the steps emanating from an etch pit formed in a more labile stressed area around the central line defect.

The above proposals can also be used to explain why the saturation solubility of type I crystals[3] exceeds both the kinetic and equilibrium solubilities. From the rate law, which involves higher order than simple proportionality to undersaturation, it is clear that the number of active dissolution units (kink or step units) decreases as saturation is approached. The units exposed at the central etch pit will be subjected to residual stress and will continue to dissolve, while those on the outer unstressed areas have passed their point of equilibrium and may slowly take up units from the solution.

Type II crystals, grown from more concentrated media with homogeneous nucleation, show more erratic or dendritic surface topography and are likely to be affected by plane dislocations (grain or crystallite or dendrite boundaries) as well as by axes of stress (screw or edge dislocations). Plane dislocations, being more extensive in their areas of influence, may thus be expected to affect both solubilities and rates of dissolution. The dissolution process, on a crystal lattice plane without emergent screw dislocations, requires initiation of layer opening for each new layer dissolved. Smaller crystallites, having more edge or apex units and thus having higher average molar surface energies, may have higher solubilities and lower energies of dissolution, and yet require higher energies of activation to initiate layer opening by detachment of units from edges or apices, because of the absence of adjacent steps.

Kinetic and saturation solubilities for type II crystals are equal to one another, but significantly higher than the accepted equilibrium solubility of barium sulphate. This can be explained if detachment of units from apices, edges and corners has a major role in determining rates and remains important in the exchange between surfaces and solution when saturation is reached.

Type III crystals, being aggregates of crystallites of colloidal dimensions and thus having a high proportion of edges and apices, can also be expected to display high solubilities, low energies of dissolution and high activation energies for dissolution. The saturation solubility for type III crystals appears to be less than the kinetic solubility.

Variability for a given type of barium sulphate crystal appears to arise from the crushing of samples to increase their surface areas before mounting them on the polythene plates. The initial sample will thus involve many crystals of smaller sizes which give higher kinetic solubilities, lower energies of dissolution and higher activation energies. After several trials of dissolution the smaller crystals will be dissolved away and the size distribution restored more nearly to the normal for the particular crystal type.

Variation of activation energy and rate of dissolution with surface stress has been demonstrated by the present simulation studies. Both the real and the simulated dissolution must consist of the coupling of initiation of dissolution (nucleation) and its propagation by detachment of kink units in succession along steps. Thus the activation energy is a combination of the energies required to execute the two types of events. At low stress-field strength the energy for initiation is large and layer opening is less frequent. High activation energy is a result of this factor in the rate of dissolution. At high stress field, initiation of layer opening at the stress centres becomes energetically favourable and step opening becomes more frequent. The activation energy may thus approach a lower value, corresponding more nearly to that for kink-unit detachment. If the stress field is of larger radius the simulated activation energy is found to be smaller because the region of easy step opening is more extended.

The weekly age-and-wash cycle previously used in extended storage of seed crystals is likely to lead to periodic removal of stressed material to an extent determined by the fraction of mass dissolved and the strength of the stress field.

Although topographically imperfect, such crystals will be energetically more stable. Ageing in saturated solution will partially refill pits and smoothen surfaces but the stress is also likely to be restored insofar as it arises from long-range and permanent mismatching of crystal lattice regions. Further dissolution of already pitted surfaces may give rise to progressively higher energies of dissolution because of the reduction in stress, while the development of more steps around the pit may lead to lowering of the activation energy.

The effectiveness of foreign bivalent ions, Ca^{2+} and Sr^{2+}, in the enhancement of the dissolution rate of barium sulphate is attributed to their double ionic charge. The monovalent K^+ and Cl^-, which would exert weaker repulsions on next-nearest-neighbour ions in $BaSO_4$, did not perceptibly change the dissolution rate. Simulation indicates that ionic adsorbates with weak NNN repulsions should have little effect on the dissolution rate, while those with stronger NNN repulsions lead to marked enhancement of simulated dissolution rates even at average coverage as little as 0.1% (fig. 3).

Low surface coverage by adsorbate ensures that the impurities are well isolated and the NNN repulsions exerted by the adsorbed ions are not cancelled by the presence of nearby ions of opposite charge. A reduction rather than an enhancement of dissolution rate when sodium chloride was dissolved into 0.07 mol dm^{-3} cadmium chloride, reported by Simon,[22] may be associated with clustering of ions at the higher surface coverage to be expected with adsorption from this more concentrated solution of bivalent metal ion.

Dissolution-rate enhancement in the presence of added Sr^{2+} was found to be greater when closer to equilibrium. The simulation study gives a similar result. This appears to arise from the frequency of initiation of layer opening around adsorbed ions being higher relative to the rate of kink-unit detachment at lower undersaturation. Such an explanation is similar to that given by Gilmer[26] for impurity-enhanced nucleation of crystal growth due to strong impurity–host interaction.

CONCLUSIONS

This work shows that the sensitivity of rate coefficient, activation energy and solubility to the origin and history of crystals may cause confusion in understanding the mechanisms of growth and dissolution, although concentration-dependent terms in the rate laws may remain unaffected. The surface-activity hysteresis found in successive dissolution trials with a given batch of microcrystals limits their usefulness for studying the effects of concentration variables, as when common ions or foreign species are added. The reproducible results obtained from the barite rotating-disc assembly simplifies this problem.

The experimental results on the variation of the activation energy of dissolution and enhancement of dissolution rate by foreign divalent metal ions indicates that the common cause for such observations is the weakening of the binding energy of the surface ions. Dislocations displace ions from their equilibrium positions while the NNN repulsion exerted by adsorbed but isolated simple ionic species is a manifestation of the long-range electrostatic repulsion between ions of like charge. Because of the higher ionic charge in these sparingly soluble salts, a given perturbation to the ion–ion distances inside a dislocation is likely to cause a large change in binding energy of lattice ions and also in the dissolution activation energy. The specific effectiveness of dissolved impurities on crystal growth and dissolution rates will be related to their surface coverage by adsorption and consequently to their solution concentrations.

The asymmetry between crystal growth and dissolution[32] suggests that the activation energy for growth will behave differently. We have not found significant variations of activation energies reported in studies of growth rates of ionic crystals.[33]

We thank Prof. P. Bennema of the University of Nijmegen for his generosity in supplying a copy of the stress-field simulation program.

[1] G. H. Nancollas and N. Purdie, *Trans. Faraday Soc.*, 1963, **59**, 735.
[2] C. H. Bovington and A. L. Jones, *Trans. Faraday Soc.*, 1970, **66**, 764.
[3] J. L. Powell, B. A. W. Coller and A. L. Jones, *J. Cryst. Growth*, 1978, **43**, 185.
[4] S. T. Liu, G. H. Nancollas and E. A. Gasiecki, *J. Cryst. Growth*, 1976, **33**, 11.
[5] E. N. Rizkalla, *J. Chem. Soc., Faraday Trans. 1*, 1983, **79**, 1857.
[6] G. M. van Rosmalen, M. C. van der Leeden and J. Gouman, *Krist. Tech*, 1980, **15**, 1213.
[7] D. M. S. Little and G. H. Nancollas, *Trans. Faraday Soc.*, 1970, **66**, 3103.
[8] J. R. Campbell and G. H. Nancollas, *J. Phys. Chem.*, 1969, **73**, 1375.
[9] C. W. Davies and A. L. Jones, *Trans. Faraday Soc.*, 1955, **51**, 812.
[10] A. L. Jones, G. A. Madigan and I. R. Wilson, *J. Cryst. Growth*, 1973, **20**, 93; 99.
[11] A. L. Jones, H. G. Linge and I. R. Wilson, *J. Cryst. Growth*, 1974, **26**, 37; 1975, **28**, 254.
[12] M. R. Christoffersen, J. Christoffersen, M. P. C. Weijnen and G. M. van Rosmalen, *J. Cryst. Growth*, 1982, **58**, 585.
[13] K. Takiyama, *Bull. Chem. Soc. Jpn*, 1959, **32**, 68.
[14] I. V. Melikhov and Z. Vukovic, *J. Chem. Soc., Faraday Trans. 1*, 1975, **71**, 2017
[15] I. N. Stranski, *Z. Phys. Chem.*, 1928, **136**, 259.
[16] W. K. Burton, N. Cabrera and F. C. Frank, *Philos. Trans. R. Soc. London, Ser. A*, 1951, **243**, 299.
[17] G. M. van Rosmalen, P. J. Daudey and M. G. J. Marchee, *J. Cryst. Growth*, 1981, **52**, 801.
[18] G. Z. Liu, J. P. van der Eerden and P. Bennema, *J. Cryst. Growth*, 1982, **58**, 152.
[19] G. M. van Rosmalen, M. C. van der Leeden and J. Gouman, *Krist. Tech.*, 1980, **15**, 1269; 1982, **17**, 627.
[20] S. T. Liu and G. H. Nancollas, *J. Colloid Interface Sci.*, 1975, **52**, 602.
[21] S. T. Liu and G. H. Nancollas, *J. Colloid Interface Sci.*, 1975, **52**, 593.
[22] B. Simon, *J. Cryst. Growth*, **52**, 1981, 789.
[23] R. J. Davey, *J. Cryst. Growth*, 1976, **34**, 109.
[24] W. P. Brandse, G. M. van Rosmalen and G. Brouwer, *J. Inorg. Nucl. Chem.*, 1977, **39**, 2007.
[25] G. H. Gilmer, *J. Cryst. Growth*, 1977, **42**, 3.
[26] G. H. Gilmer, *Science*, 1980, **208**, 355.
[27] L. de Brouckere, *Ann. Chim.*, 1933, **XIX**, 86.
[28] T. A. Cherepanova, J. P. van der Eerden and P. Bennema, *J. Cryst. Growth*, 1978, **44**, 537.
[29] *DANA Handbook of Minerology* (1954), p. 407.
[30] T. A. Cherepanova, in *Industrial Crystallisation*, ed. J. W. Mullin (Plenum Press, New York, 1976), p. 113.
[31] *Natl Bur. Stand. (U.S.) Circ.*, 1952, **500**.
[32] J. W. Mullin, *Crystallisation* (Butterworths, London, 1972) p. 199.
[33] G. E. Cassford, W. A. House and A. D. Pethybridge, *J. Chem. Soc., Faraday Trans. 1*, 1983, **79**, 1617.

Study of the Dynamic Equilibrium in the CaF_2/Aqueous Solution System using $^{45}Ca^{2+}$ as Radiotracer

By Johannes J. M. Binsma* and Zvonimir Kolar

Interuniversity Reactor Institute, Mekelweg 15, 2629 JB Delft, The Netherlands

Received 9th January, 1984

The transport of Ca^{2+} ions at the interface between CaF_2 crystals and an aqueous solution of CaF_2 has been studied under equilibrium conditions with the aid of $^{45}Ca^{2+}$ as a radioactive tracer. The kinetics of the isotopic exchange between CaF_2 and its saturated solution appears to consist of a fast and a slow component with rate constants of ca. 2×10^{-5} and 1×10^{-6} s^{-1}, respectively. Total fluxes of Ca^{2+} from the solution to individual crystal faces of the order of 10^{-10} mol m^{-2} s^{-1} were found. The exchange was retarded by the presence of some foreign ions in the solution, but eventually the same number of Ca^{2+} ions were transported as for pure solutions. The crystallographic orientation ({100} or {111}) of the faces did not have a large influence on the exchange rate, but the relative amount of transported Ca^{2+} ions (compared with the amount present in the first crystal layer) was lower for the {100} faces than for the {111} faces. Also, it has been demonstrated that some 30% of the $^{45}Ca^{2+}$ remains inside the crystal when re-exchange with a saturated CaF_2 solution which initially does not contain tracer ions is attempted, probably by diffusion into the crystal.

In recent years many studies have been undertaken concerning the kinetics at crystal/solution interfaces under non-equilibrium conditions, *i.e.* during growth or dissolution [see ref. (1)–(3) for example]. Under such circumstances a driving force is present which results in a net transport of material towards or from the crystal. This means that attention is focussed on the dependence of the net transport during growth or dissolution on the various relevant parameters (*e.g.* supersaturation or undersaturation, temperature, interface properties). In performing this kind of investigation, however, it should be kept in mind that the net transport is the result of oppositely directed fluxes of material. The absolute value of these fluxes is a very important parameter for the phenomenon of 'interfacial instability'. In addition, the fluxes provide for the attainment of the dynamic equilibrium between crystal and solution under equilibrium conditions. In the latter case the interfacial instability will be entirely determined by the inward and outward fluxes. In practice, interfacial instability is found in a wide variety of systems, *e.g.* at the surface of ion-selective electrodes, in ion exchange in soils and in bone formation.

Transport kinetics (exchange) in equilibrium have been the subject of many experimental studies in which radiotracers are used to follow the material in one of the subsystems (solution or crystal).[4–8] With the aid of so-called tracer kinetics[9] the tracer experiments may lead to a description of the exchange in terms of compartments and transport rates between these compartments. The term compartment has also been used in a more physical sense in the description of the crystal/solution interface, namely as a region with a specific environment for the building unit(s) (*e.g.* ions or molecules) of the crystal. For instance, the Gouy–Chapman layer, the Stern layer, the outer crystal layer and the inner (bulk) crystal might be considered as such compartments.[10] Also, the different positions (surface, step and

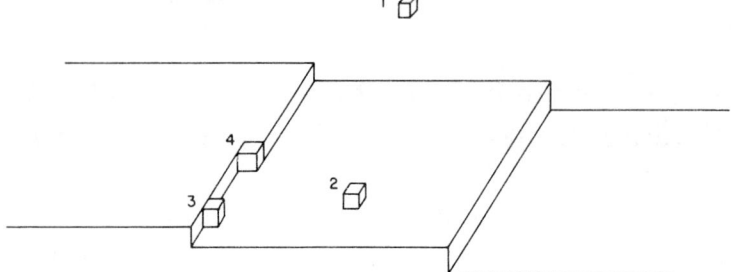

Fig. 1. Crystal/solution interface with various possible positions for a building unit: (1) in solution, (2) at a flat surface, (3) at a step and (4) at a kink site.

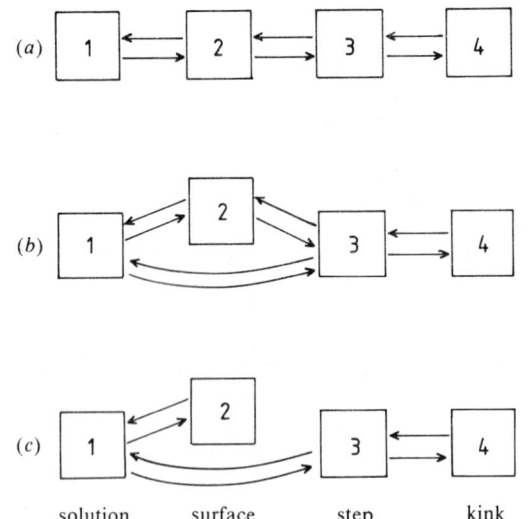

Fig. 2. Different possible routes for transport from solution to the crystal in terms of compartment models.

kink) for building units at a crystal interface (see fig. 1), often used in describing crystal growth processes, may be thought to constitute different compartments. Such an interpretation of the results obtained for tracer experiments in a crystal/solution system has not yet been attempted.

Different possible routes for transport from the solution to surface, step or kink sites are shown in fig. 2. The compartments are distinguished by the different amounts of building units which they contain and by their specific bond strengths for the building units, and will thus have different exchange rates.[9]

The building units may diffuse from their position at the interface to the inner crystal layers. These layers can also be included in the compartment models and then form a fifth compartment. Whether a compartment really comes to the fore in an experiment depends on the magnitude of the exchange rate constants and on the amount of material in each compartment (the compartment size).

The exchange rates and thus the interfacial instability may be influenced by various factors such as the crystallographic structure of the outer solid layer, the surface morphology (roughness) and the presence of impurities which can be preferentially adsorbed at the interface. The aim of this study is to gain more insight into the relevant influencing factors in the system CaF_2/aqueous solution. For the solid CaF_2 phase, larger single crystals (centimetre size) were taken because the characterization of the surface is much more easily performed on such crystals than for a population of small crystals in suspension. In addition, recrystallization effects have to be expected because of Ostwald ripening for suspensions which are not completely monodisperse.[11] CaF_2 has been chosen as a model compound because it possesses a simple crystal structure (fluorite), making the surface structures simple too, and because a suitable radiotracer is available ($^{45}Ca^{2+}$). Moreover, from a comparison with the behaviour of SrF_2 and BaF_2 (which have the same structure) the influence of the solid phase on the exchange rate might be deduced. In the present study the exchange between CaF_2 crystals and the saturated solution is followed by measuring the uptake of ^{45}Ca from the tracer-containing solution onto the individual CaF_2 crystal faces.

EXPERIMENTAL

Synthetic CaF_2 single crystals were obtained from Dr K. Korth (Monokristalle–Kristalloptik oHG, 'K-crystals') and from Dr H. W. den Hartog of the University of Groningen ('H-crystals'). Natural CaF_2 crystals were also employed. The synthetic crystals were colourless, while the natural ones had slight yellow colouring. By neutron activation analysis it was found that the most important impurities were Sr (3.1×10^2 ppm) and Ba (1.9×10^2 ppm) in the K-crystals and Sr (1.2×10^2 ppm) in the H-crystals. The dominant impurities in the natural crystals were found to be Ti (7.4×10^3 ppm) and As (2.7×10^2 ppm). CaF_2 crystals of typical size $10 \times 10 \times 5$ mm^3 which were to be used in the exchange experiments were cleaved ({111} faces) or cut ({100} faces) from the specimens mentioned above. The {100} faces were polished up to optical quality. The orientation of the faces was checked by X-ray diffraction.

Some of the synthetic H-crystals and of the natural crystals were used to prepare saturated CaF_2 solutions. The concentration of impurities introduced by the solid was ca. 2×10^{-8} and 2×10^{-6} mol dm^{-3} for the two types of solutions, respectively, as can be calculated on the basis of the results of the chemical analysis of the crystals.

After equilibration at 25 °C for some days in doubly distilled water, the concentrations of Ca^{2+} and F^- were checked using calcium and fluoride ion-selective electrodes, respectively. If the concentrations corresponded to the equilibrium values, the liquid was separated by filtration. The crystal was then added to this liquid in a polyethylene vessel, and after several days of equilibration a few mm^3 of an aqueous solution of $^{45}CaCl_2$ (Amersham, 2.16 mCi cm^{-3}, 2.9×10^{-3} mol dm^{-3} Ca^{2+}) was added. The crystal was taken out of the solution after a time t (varying from 1 min at the beginning of the experiment to several days at the end of the experiment). All experiments were carried out at 25 °C. In order to remove the adherent solution, the crystal was washed once with water and twice with ethanol.

The activity present at the various faces was counted by means of a Geiger–Müller counter. Because of the rather low β energy of ^{45}Ca ($E_{max} = 0.252$ MeV) it is possible to measure the activity at one face without contributions of other faces if the crystal is not too thin (thicker than ca. 0.3 mm) and the edges of the crystal are not included in the counting. This is achieved by using a stainless-steel diaphragm with an opening of 2×5 mm^2 and by counting only the central part of a face. Calibration was carried out by measuring the activity of the solution in the same geometry as that used for the crystals. This was achieved by attaching a drop of 1 mm^3 of the solution to a glass plate and measuring the activity after all water had evaporated. In this way it was possible to determine the fraction of the activity present at the crystal faces.

By autoradiography using electron-sensitive film (Agfa-Gevaert Scientia) we checked whether the activity was distributed homogeneously over the crystal faces.

RESULTS AND DISCUSSION

For some natural crystals which manifested small cracks, preferred accumulation of activity along the intersecting lines of the cracks with the surface was observed by means of autoradiography. Also, relatively high activities were found at the crystal edges and for faces which had no specific crystallographic orientation (rounded faces). These inhomogeneous distributions of activity point to recrystallization processes. For quantitative determinations of the activity only those crystals were taken into consideration which showed no inhomogeneous distributions of activity. In addition, the edges and rounded (side) faces were excluded from the measurements (see previous section).

Fig. 3 shows the relative activity (the ratio of the count rate at time t, R_t, and the final equilibrium count rate, R_∞) as a function of time for different crystal faces of the CaF_2 crystals S_1 and N_1 in solutions obtained from synthetic and natural CaF_2 crystals ('pure' and 'impure' solutions, see previous section). The influence of impurities present in the solution on the exchange rate is obvious. The impurities retard the exchange processes appreciably, possibly by preferential adsorption at the CaF_2/solution interface. The influence of crystallographic orientation ($\{100\}$ or $\{111\}$) is much smaller under the present conditions. In contrast to the exchange rate the final activity does not show significant differences for crystals in pure and impure solutions.

The total flux of Ca^{2+} ions from solution to a crystal face can be evaluated from the initial uptake of activity, Δq, by that face after a very short time Δt, according to

$$F = \frac{\Delta q}{q_s \Delta t} Q_s \quad (1)$$

where q_s and Q_s are the activity and the amount of Ca^{2+} in the solution, respectively. Fluxes of 1.1×10^{-10} mol m^{-2} s^{-1} ($\{100\}$) and 3.7×10^{-10} mol m^{-2} s^{-1} ($\{111\}$) were calculated for crystal N_1 and 7.3×10^{-11} mol m^{-2} s^{-1} ($\{100\}$) and 5.6×10^{-11} mol m^{-2} s^{-1} ($\{111\}$) for crystal S_1. Thus, the fluxes appear to be smaller in the impure solution than in the pure solution.

If $\ln(R_\infty - R_t)$ is plotted against t, a curve is obtained in which two linear parts with slopes g_1 and g_2 can be discerned, as illustrated in fig. 4 for crystal S_1 (impure solution). This means that three different compartments are involved, one of which must be the saturated solution. Because only the activity incorporated in the crystal is measured, the two other compartments should correspond to Ca^{2+} ions at different types of position in the crystal. Only if these two compartments have no exchange between each other and are both connected to the very large central solution compartment individually can the slopes g_1 and g_2 be assumed to be equal to the rate constants (k_1 and k_2) for transport between the central compartment and the two peripheral ones. The residence times of the Ca^{2+} ions in the peripheral compartments can then be calculated according to $\tau_i = k_i^{-1} = g_i^{-1}$ ($i = 1, 2$). Other compartment models do not give such a simple relation between τ_i and g_i, and require a different type of evaluation which is still in progress. In table 1 the g values and the residence times calculated on the basis of the assumption given above are listed for different natural (N) and synthetic (S) CaF_2 crystals in pure and impure solutions. The residence times appear to lie in the range $(3.0–6.7) \times 10^4$ s for the fast exchange and in the range $(0.37–1.7) \times 10^6$ s for the slow exchange.

The number of exchangeable Ca^{2+} ions per cm^2 in the crystal can be determined from the final count rate, R_∞, because the specific activity of Ca in the compartments of the crystal which are taking part in the exchange will be equal then to the specific

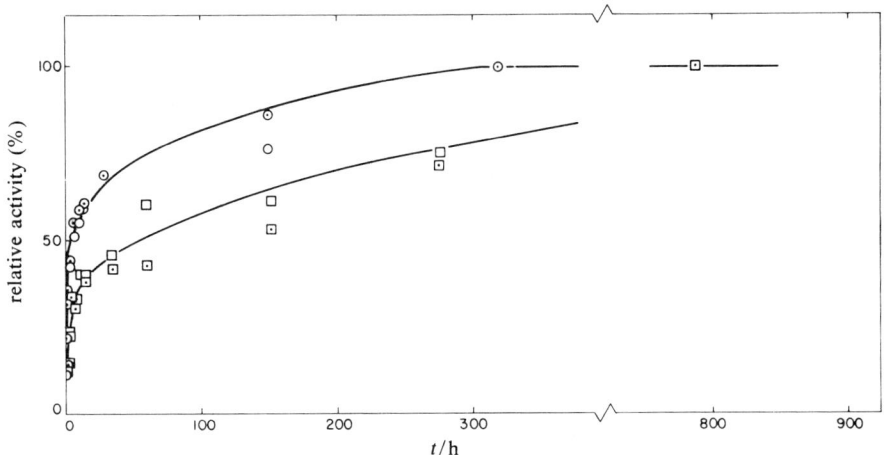

Fig. 3. Relative activity plotted as a function of immersion time for CaF_2 crystals in saturated solutions of different degrees of purity (see text). Crystal N_1 (pure solution): ○, {100} face; ⊙, {111} face; crystal S_1 (impure solution): □, {100} face; ⊡, {111} face.

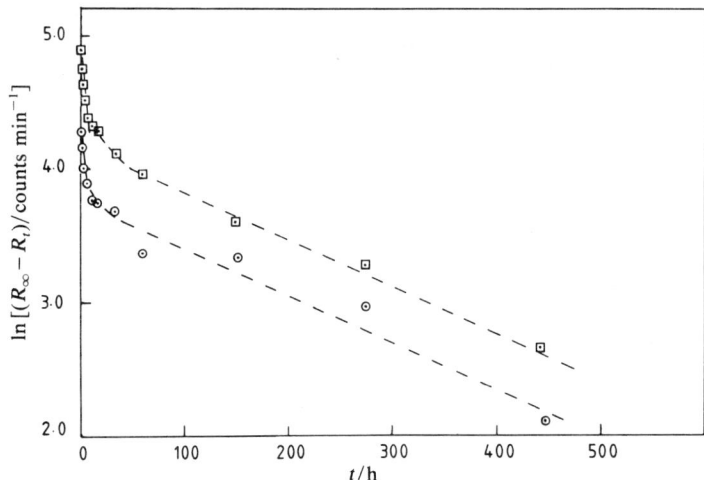

Fig. 4. Plot of $\ln(R_\infty - R_t)$ against t for CaF_2 crystal S_1: □, {111} face; ⊙, {100} face.

Table 1. g_i values and residence times τ_i ($\tau_i = 1/g_i$) for different crystal faces of various CaF_2 crystals under equilibrium conditions

crystal	solution	face	$g_1/10^{-5}\,s^{-1}$	$g_2/10^{-6}\,s^{-1}$	$\tau_1/10^4\,s$	$\tau_2/10^6\,s$
N_1	pure	{100}	1.5	0.58	6.7	1.7
		{111}	1.7	0.67	5.9	1.5
N_2	impure	{100}	3.3	1.81	3.0	0.55
		{111}	2.8	2.69	3.5	0.37
S_1	impure	{100}	2.0	0.97	5.0	1.0
		{111}	1.6	0.97	6.3	1.0
S_2	pure	{111}	2.3	1.92	4.3	0.52
		{111}	1.9	1.00	5.3	1.0

Table 2. Number of exchangeable Ca^{2+} ions Q_c and f values for different crystal faces of various CaF_2 crystals under equilibrium conditions

crystal	solution	face	Q_c/nmol cm^{-2}	f
N_1	pure	{100}	0.35	0.63
		{111}	0.40	0.93
N_2	impure	{100}	0.23	0.41
		{111}	0.30	0.70
S_1	impure	{100}	0.32	0.57
		{111}	0.40	0.93
S_2	pure	{111}	0.24	0.56
		{111}	0.21	0.49

activity of Ca^{2+} in the solution. We can then write:

$$\frac{q_s}{Q_s} = \frac{q_c}{Q_c} \quad (2)$$

where q_c and Q_c are, respectively, the activity of the crystal face under consideration and the amount of Ca^{2+} ions present at the crystal face. The activity q is related to the measured count rate R by $q = R/\varepsilon$, where ε is the overall efficiency of the counting technique. Because the activity of the solution and of the crystal faces is measured with the same efficiency, we can substitute count rates for activities in eqn (2), which leads to

$$Q_c = \frac{R_\infty}{R_s} Q_s. \quad (3)$$

The numbers of exchangeable ions Q_c which are obtained on the basis of eqn (3) for the {100} and {111} faces of the same crystals as listed in table 1 are given in table 2, together with the ratio f, which is defined as $f = Q_c/Q_t$ where Q_t is the theoretical number of Ca^{2+} ions present at the crystal face as calculated from the crystallographic structure. For the {100} and {111} faces Q_t takes the values 0.557 and 0.429 nmol cm^{-2}, respectively. It appears that only a very small amount of material (0.2 to 0.4 nmol cm^{-2}) takes part in the exchange. In fact less than one lattice layer is involved in the exchange as f ranges from 0.41 to 0.63 for the {100} face and from 0.49 to 0.93 for the {111} face. For each crystal the value of f for the {100} face is smaller than that for the {111} face, but whether the differences found are significant is not yet clear. For the growth of CaF_2 crystals (and also SrF_2 and BaF_2) from aqueous solutions it is known[12,13] that the {100} faces are the slowest growing. This might be in accordance with the rather small f values found for the {100} faces in this study; however, one should be careful in comparing the observations made for non-equilibrium systems (growth) with those of an equilibrium system in which no net transport of material takes place.

The so-called 're-exchange' was studied by measuring the decrease in activity at the faces of a crystal which had reached equilibrium activity and which was immersed in an initially inactive saturated aqueous solution of CaF_2. It appears that the exchange is not entirely reversible; some 30% of the $^{45}Ca^{2+}$ ions remain in the crystal. A similar effect was found for $CaCO_3$ by Möller and Sastri,[6] who ascribed it to diffusion into the inner parts of the crystal. These Ca^{2+} ions are then no longer available for the exchange processes and do not play a role in maintaining the dynamic equilibrium.

The present study has shown that there are at least two different types of bonding for Ca^{2+} ions within the crystal or at its surface. Different rate constants are found for these two types of Ca^{2+} ions, which if they are considered to belong to compartments that are only connected with the solution compartment correspond to residence times of ca. 5×10^4 and ca. 1×10^6 s, respectively. This different behaviour of one type of ion could be explained by the fact that the Ca^{2+} ions differ in type of bonding at the crystal surface, e.g. Ca^{2+} in kink- and step-like positions. The real {100} and {111} faces of CaF_2 differ from the face represented in fig. 1, but in analogy with the latter, ions in the kink positions possess half the number of bonds of ions inside the crystal and ions in step positions one bond less. Another explanation of the existence of two different compartments for the Ca^{2+} ions would be that the ions having the larger rate constants are situated in the first crystal layer and those having the smaller rate constants correspond to ions in the bulk of the crystal. The first explanation is sustained by the fact that the equivalent of approximately one monolayer (probably the first layer at the surface) takes part in the exchange processes. The second explanation is sustained, however, by the re-exchange experiment, which showed that some (ca. 30%) of the $^{45}Ca^{2+}$ ions remain inside the crystal. Further studies will be performed in order to identify the correct explanation.

The influence of impurities present in the solution is reflected in the exchange rates, which are reduced if impurities are present. The impurities do not influence the number of exchangeable ions, however.

By means of tracer experiments useful information can thus be obtained with respect to rate constants and the attainment of dynamic equilibrium in a crystal/aqueous solution system. From the rate constants residence times may be deduced; these are a measure of interfacial instability. This information can be obtained on the basis of very small fluxes of material by means of tracer experiments. Further studies on the influence of impurities and surface morphology on the exchange are under way.

We thank Mr J. F. van Lent and Dr Th. H. de Keyser (both of the Metallurgy Laboratory, University of Technology, Delft) for performing the X-ray diffraction experiments.

[1] *Proc. 7th Symp. on Industrial Crystallization*, ed. E. J. de Jong and S. J. Jančić (North Holland, Amsterdam, 1979).
[2] *Proc. 8th Symp. on Industrial Crystallization*, ed. S. J. Jančić and E. J. de Jong (North Holland, Amsterdam, 1982).
[3] W. J. P. van Enckevort, Thesis (Nijmegen, 1982).
[4] K. E. Zimens, *Ark. Kemi, Mineral. Geol.*, 1945, **21A**, no. 16.
[5] G. Lang and K. H. Lieser, *Z. Phys. Chem. N.F.*, 1973, **86**, 143.
[6] P. Möller and C. S. Sastri, *Inorg. Nucl. Chem. Lett.*, 1973, **9**, 759.
[7] T. C. Huang, K. Y. Li and S. C. Hoo, *J. Inorg. Nucl. Chem.*, 1972, **34**, 47.
[8] Y. Inoue and Y. Yamada, *Bull. Chem. Soc. Jpn*, 1983, **56**, 705.
[9] R. A. Shipley and R. E. Clark, *Tracer Methods for In Vivo Kinetics* (Academic Press, New York, 1972).
[10] W. E. Brown and L. C. Chow, *Colloids Surf.*, 1983, **7**, 67.
[11] T. Sugimoto and G. Yamaguchi, *J. Cryst. Growth*, 1976, **34**, 253.
[12] P. Hartman, *Mineral Genesis* (Bulgarian Academy of Sciences, Sofia, 1974), p.111.
[13] Z. Kolar, J. J. M. Binsma and B. Subotić, *J. Cryst. Growth*, accepted for publication.

High-temperature Dissolution of Nickel Chromium Ferrites by Oxalic Acid and Nitrilotriacetic Acid

BY ROBIN M. SELLERS* AND WILLIAM J. WILLIAMS

Central Electricity Generating Board, Berkeley Nuclear Laboratories, Berkeley, Gloucestershire GL13 9PB

Received 14th December, 1983

A study of the dissolution of a number of spinel-type oxides containing iron(III) ions by oxalic acid and nitrilotriacetic acid (NTA) is reported. Increasing the chromium content of oxides of general composition $Ni_{0.6}Cr_xFe_{2.4-x}O_4$ ($x = 0.3$–1.5) brought about a marked reduction in dissolution rate, and it is suggested that this arises by a mechanism involving a change from kink-site attack to ledge-site attack as the former become 'blocked' by less reactive chromium(III) ions. The $Ni_{0.6}Cr_{0.6}Fe_{1.8}O_4$ oxide was investigated in more detail. The concentration dependences determined suggest that both oxalic acid and NTA are adsorbed at surface sites prior to dissolution. NTA brings about dissolution simply by 'complexing' attack, but with oxalic acid it was not possible to distinguish directly between complexing and reductive attack. The divalent cation also plays a role, for in the dissolution of a number of ferrites, AFe_2O_4 (A = Co, Fe, Mn, Ni), in oxalic acid appreciable differences in dissolution rate were found, the order of reactivity being Fe > Mn > Co > Ni. The reasons for this are discussed.

Oxalic acid and strong complexants such as nitrilotriacetic acid (NTA) have been widely used to remove oxide deposits in power plants or to decontaminate pipework surfaces in water-cooled nuclear reactors.[1-5] Despite the considerable practical experience with these reagents there is still only a poor understanding of how they bring about dissolution and the factors which influence their reactivity.[1,6] In practice the oxides to be dissolved are often of the spinel type (AB_2O_4), usually either magnetite or magnetite substituted by nickel(II), zinc(II), chromium(III) etc., although goethite or haematite may be significant under certain circumstances.[7,8] Complexants bring about the dissolution of the oxide lattice at the site of adsorption. Such processes tend to be slow, but if the oxide contains iron(II) *and* iron(III) ions an autocatalytic process may occur as a result of a reductive dissolution reaction:[9,10]

$$Fe^{II}L_n + >Fe^{3+} \rightarrow Fe^{III}L_n + Fe^{2+} \quad (1)$$

$$Fe^{2+} + nL \rightarrow Fe^{II}L_n. \quad (2)$$

Oxalic acid may also bring about dissolution by complexing attack (certainly this is what happens with materials such as chrysotile which contain no variable-valency metal ions[11]), but reductive dissolution is possible with oxides rich in iron(III). Here again autocatalytic processes have been observed due to the formation of ferrous oxalate.[6,12,13] Although oxalic acid tends to be more reactive than strong complexants such as NTA, even more rapid dissolution can be achieved with one-electron reductants such as tris(picolinato)vanadium(II).[14,15] Detailed studies of this reagent have been made in our laboratory; amongst other things, these have shown that increasing the chromium content of a series of nickel chromium ferrites dramatically reduced the dissolution rate.[6,13] In this paper we describe what happens when dissolution of the same oxides is induced by oxalic acid and NTA.

EXPERIMENTAL

MATERIALS

All solutions were prepared using triply distilled water and analytical-reagent-grade chemicals, except nitrilotriacetic acid, which was B.D.H. general-purpose-reagent grade. The pH was adjusted by addition of H_2SO_4 or NaOH. The oxides were prepared by coprecipitation of the appropriate hydroxides by addition of Na_2CO_3 to solutions of the metal nitrates (the 'carbonate' method[15]), washing, drying and calcining under argon for 6 h at 1400 °C. Finally the oxides were ground and sieved, the fraction passing through a 53 μm sieve being retained for use. Mean particle diameters were measured using a model TAII Coulter counter and were typically ca. 6 μm. The composition of the oxides was checked by chemical analysis, infrared spectroscopy and X-ray crystallography. Further details of these measurements will be given in a separate publication.

One of the oxides used, $Ni_{0.6}Cr_{0.6}Fe_{1.8}O_4$, showed some ageing between the initial series of experiments (carried out in late 1981/early 1982) and the final part of the work (summer 1983). Several of the earlier runs were repeated in the latter period, and there appeared to be a constant relation between the two sets of data. For convenience we treat the oxide before and after ageing as two different batches, referred to here as batches A and B, respectively.

APPARATUS AND PROCEDURE

Neither oxalic acid nor nitrilotriacetic acid dissolve the spinel oxides under consideration here sufficiently rapidly at temperatures below 100 °C for convenient measurement. We therefore carried out most of our measurements at 140 °C using an apparatus consisting of a 50 cm^3 round-bottomed flask of specially thickened glass, a Gyrolok adapter, a Pierce and Warriner septum swinger and a frame to restrain the flask. A seal was obtained by compression of a rubber O-ring onto the glass neck of the flask. Solutions could be injected into the flask at temperature through the septum (using a Precision Sampling Corp. hypodermic syringe), but for reasons of safety we did not attempt to remove solution under these conditions. Heating was achieved by immersion of the flask (but not the septum assembly) in a bath of water-soluble oil. Solutions were stirred continuously by magnetic stirrer. Various safety devices were incorporated into the design to prevent overheating and overpressurisation of the apparatus. Only modest pressure rises were required to keep the bulk of the water in the liquid state, and in our experience the apparatus could be operated quite safely at temperatures up to ca. 150 °C.

Solutions plus ca. 3 mg oxide were degassed at room temperature by bubbling with high-purity argon for ca. $\frac{1}{2}$ h. The septum swinger was then closed and the apparatus heated. The pressure was relieved periodically at temperatures up to ca. 80 °C by opening the swinger. When the apparatus had reached the desired temperature the reaction was initiated by injection of oxalic acid and/or nitrilotriacetic acid as appropriate. The solution injected was at room temperature, so some small decrease in temperature in the first few minutes of reaction was inevitable, but we do not believe that this will have materially affected the results obtained. Reaction was terminated by removal of the whole assembly from the oil bath; the temperature fell rapidly, and so the dissolution reaction was quickly inhibited. When sufficiently cool the solution was removed, filtered through a 0.22 μm filter (Millipore Millex GS) to remove unreacted oxide, diluted ×10 with 0.1 mol dm^{-3} HCl (B.D.H., C.V.S. reagent) and analysed for iron, nickel, chromium etc. by atomic absorption using a Baird A5100 instrument. Ferrous ion was estimated spectrophotometrically as the 1,10-phenanthioline complex.[16] Oxalic acid interferes strongly with the estimation of Fe^{3+} by this method, so this was calculated as the difference between the total iron (from atomic absorption measurements) and iron(II). Oxalate reduces the tris(1,10-phenanthroline)iron(III) complex over a period of hours, so all spectrometric measurements were made on freshly prepared solutions. The pH was measured on the cooled, but undiluted, solution at the end of each run using an EIL type 7050 meter.

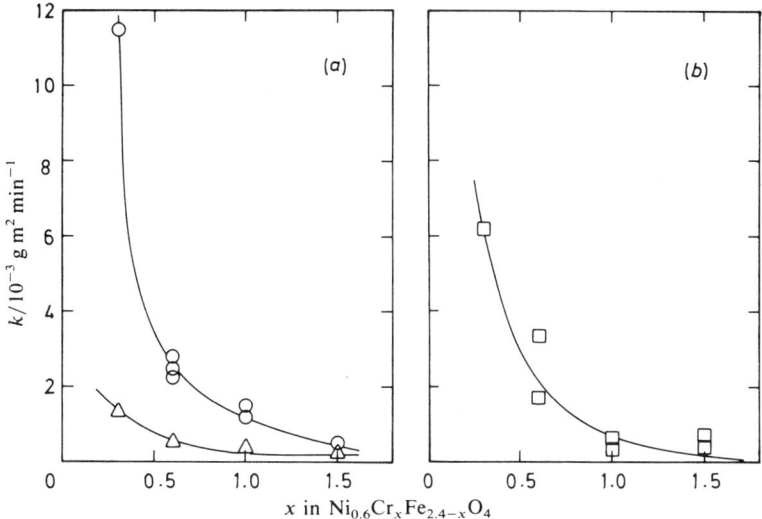

Fig. 1. Effect of chromium content on the dissolution of $Ni_{0.6}Cr_xFe_{2.4-x}O_4$ by (a) oxalic acid (○) and NTA (△) and (b) oxalic acid + NTA (□). The measurements were made in solutions containing: ○, 5.3×10^{-3} mol dm^{-3} oxalic acid, pH_{25} 2.4, 140 °C; △, 5.2×10^{-3} mol dm^{-3} NTA, pH_{25} 5.0, 140 °C and □, 2.6×10^{-3} mol dm^{-3} oxalic acid + 5.2×10^{-3} mol dm^{-3} NTA, pH_{25} 3.0, 140 °C.

Some difficulty was experienced in preliminary runs with leaks. We found that this could readily be overcome by careful attention to detail during assembly, but as a cross-check the flask and its contents were weighed before and after each experiment. If >3 cm³ (10% of the solution) had been lost the run was rejected; if smaller than this, appropriate corrections were made to the concentrations of species in solution deduced from the atomic absorption measurements.

RESULTS

EFFECT OF CHROMIUM CONTENT ON THE DISSOLUTION OF NICKEL CHROMIUM FERRITES BY OXALIC ACID AND NITRILOTRIACETIC ACID

The effect of increasing chromium content on the dissolution of nickel chromium ferrites of general formula $Ni_{0.6}Cr_xFe_{2.4-x}O_4$ by oxalic acid, NTA and their mixtures is shown in fig. 1. Four different oxides were employed in the experiments, with chromium contents corresponding to $x = 0.3, 0.6, 1.0$ and 1.5. With a fifth oxide, $NiCr_2O_4$, there was <1% dissolution in 4 h in oxalic acid + NTA mixtures. Values of α were calculated from the amounts of iron dissolved after 4 h at 140 °C. The corresponding figures for nickel dissolution were a few percent smaller, whilst those based on chromium were consistently 15% or more smaller, indicating some specific leaching of both iron and nickel by both reagents. The concentrations of the reagents in the three series of experiments were, respectively, (a) 5.3×10^{-3} mol dm^{-3} oxalic acid, pH_{25} 2.4, (b) 5.2×10^{-3} mol dm^{-3} NTA, pH_{25} 5.0 and (c) 2.7×10^{-3} mol dm^{-3} oxalic acid + 5.2×10^{-3} mol dm^{-3} NTA, pH_{25} 3.0.

DISSOLUTION OF $Ni_{0.6}Cr_{0.6}Fe_{1.8}O_4$ BY OXALIC ACID

An appreciable fraction of the $Ni_{0.6}Cr_{0.6}Fe_{1.8}O_4$ oxide dissolved in the 4 h of the experiments described above, and so this oxide was selected for more detailed study.

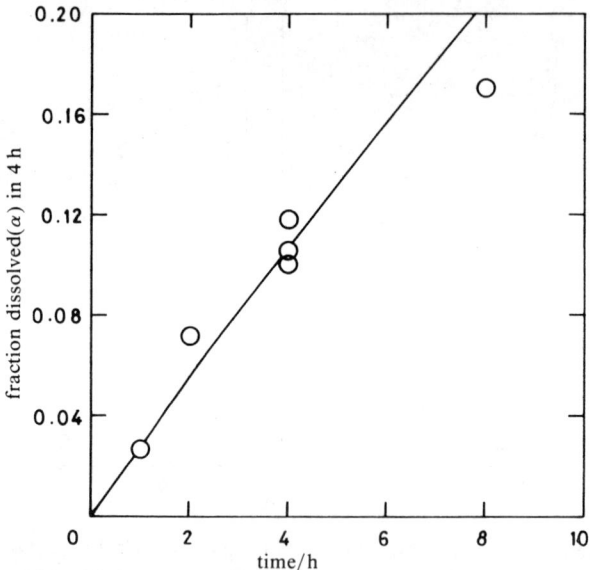

Fig. 2. Time dependence of the dissolution of $Ni_{0.6}Cr_{0.6}Fe_{1.8}O_4$ by 5.3×10^{-3} mol dm^{-3} oxalic acid at pH_{25} 2.5 and 140 °C. Line calculated from eqn (3) taking $k_{obs} = 1.25 \times 10^{-4}$ min^{-1}.

The rate law for its dissolution by oxalic acid at 140 °C was determined by carrying out a series of runs of varying duration. The fraction of the oxide dissolved varied with time as shown in fig. 2. Some curvature is apparent, but when $(1-\alpha)^{1/3}$ was plotted against time a linear dependence was obtained. This corresponds to the shrinking-core model of dissolution,[6,15] for which the rate law is of the form

$$(1-\alpha)^{1/3} = 1 - k_{obs} t \qquad (3)$$

where $k_{obs} = k/r_0\rho$ with r_0 the initial particle radius and ρ the oxide density. The curve in fig. 2 has been calculated according to eqn (3). There was no evidence for an autocatalytic pathway. Even when ferrous ions were injected at the commencement of a run to give a solution containing 9×10^{-5} mol dm^{-3} ferrous oxalate only a slight increase in dissolution rate could be detected. Since this ferrous ion concentration is at least a factor of five higher than produced in oxalic acid solutions at $\alpha < 0.2$, it is clear that the ferrous oxalate pathway plays only a minor part under our conditions. In the remaining work described in this section the validity of eqn (3) has been assumed, and it has been used to calculate values of k_{obs} from 'single-point' dissolution measurements.

The rate of dissolution of $Ni_{0.6}Cr_{0.6}Fe_{1.8}O_4$ at 140 °C increased as the oxalic acid concentration increased. The dependence was not linear, but showed a tendency to level off at oxalic concentrations above ca. 0.1 mol dm^{-3}, as shown in fig. 3. Decreasing the pH from 6 to 1.5 brought about a marked increase in dissolution rate (fig. 4). At very low pH it seemed probable that some dissolution was being brought about by proton attack at the oxide lattice. This was checked using sulphuric acid solutions of varying concentration, and was shown to be significant below pH ca. 2.5. Increasing the temperature also brought about an increase in dissolution rate. A plot of log (k_{obs}) against T^{-1} was linear (temperature range 105–160 °C), and from this the value for the activation energy given in table 1 was calculated.

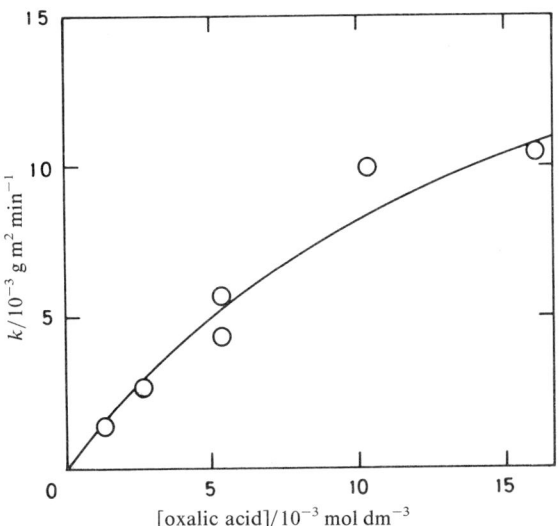

Fig. 3. Effect of oxalic acid concentration on the dissolution of $Ni_{0.6}Cr_{0.6}Fe_{1.8}O_4$ at 140 °C. Measurements made at pH_{25} 2.5 and using oxide batch A.

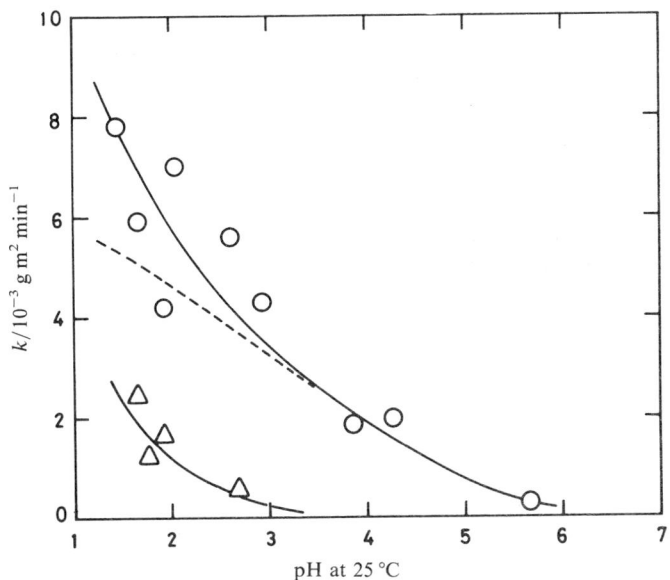

Fig. 4. Effect of pH on the dissolution of $Ni_{0.6}Cr_{0.6}Fe_{1.8}O_4$ by oxalic acid at 140 °C (○); dashed line after correction for proton-induced dissolution. Measurements made in 5.3×10^{-3} mol dm^{-3} oxalic acid and using oxide batch A; △, dissolution by sulphuric acid.

Practically all the iron dissolved by oxalic acid with this oxide was found to be iron(II). Other experiments showed that iron(III) was rapidly reduced by oxalic acid, so any iron dissolved as iron(III) would be rapidly reduced in the bulk solution to iron(II) and be indistinguishable from iron dissolved reductively.

Table 1. Activation energies for the dissolution of $Ni_{0.6}Cr_{0.6}Fe_{1.8}O_4$ by oxalic acid and nitrilotriacetic acid

system	E_a/kJ mol^{-1}
5.3×10^{-3} mol dm^{-3} oxalic acid, pH$_{25}$ 2.5, oxide batch A	42 ± 6
5.2×10^{-3} mol dm^{-3} NTA, pH$_{25}$ 4.9, oxide batch B	18 ± 3

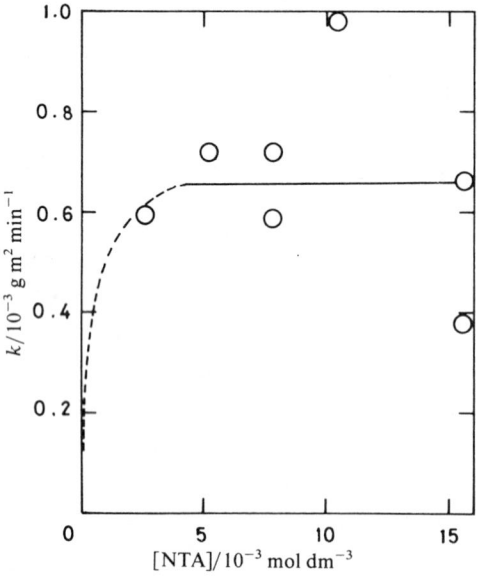

Fig. 5. Effect of nitrilotriacetic acid concentration on the dissolution of $Ni_{0.6}Cr_{0.6}Fe_{1.8}O_4$ at 140 °C. Measurements made at pH$_{25}$ 3.1 and using oxide batch B.

DISSOLUTION OF $Ni_{0.6}Cr_{0.6}Fe_{1.8}O_4$ BY NITRILOTRIACETIC ACID

Dissolution of $Ni_{0.6}Cr_{0.6}Fe_{1.8}O_4$ by NTA at 140 °C was in general considerably slower than with oxalic acid. Unfortunately this resulted in large errors in the estimated rate constants (particularly with the batch B oxide), and in part masked the details of the dependences. For NTA concentrations above *ca.* 2×10^{-3} mol dm^{-4} at pH 3.0 the dissolution rate was practically independent of complex concentration (fig. 5) and increased with decreasing pH at 5.0×10^{-3} NTA (fig. 6). Increasing temperature caused an increase in rate consistent with the Arrhenius law and from which the activation energy shown in table 1 was calculated. In an experiment in which ferrous ions were added at the start of the reaction to give 9×10^{-5} mol dm^{-3} nitrilotriacetatoiron(II) a factor of *ca.* 3 increase in dissolution rate was found. At the concentrations of iron(II) formed in solutions containing initially only NTA this pathway will not have been significant.

DISSOLUTION OF OTHER FERRITES BY OXALIC ACID

To investigate the effect of the divalent cation on the dissolution kinetics, the ability of oxalic acid to dissolve four different ferrites (AFe_2O_4 with A = Co, Fe,

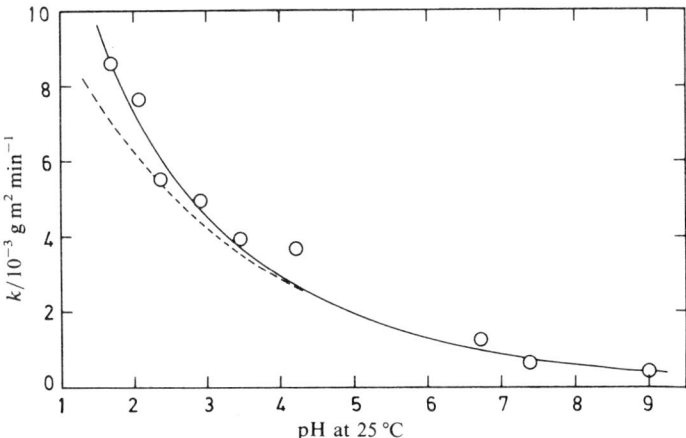

Fig. 6. Effect of pH on the dissolution of $Ni_{0.6}Cr_{0.6}Fe_{1.8}O_4$ by nitrilotriacetic acid at 140 °C. Measurements made in 5.2×10^{-3} mol dm^{-3} NTA and using oxide batch B (○); dashed line after correction for acid-induced dissolution (cf. fig. 4).

Table 2. Rate constants for the dissolution of some ferrites by oxalic acid at 140 °C[a]

oxide	k/gm^2 min^{-1}	log K_1[b]	solubility[c]/g dm^{-3}
$NiFe_2O_4$	0.011	5.2	i
$CoFe_2O_4$	ca. 0.04[d]	4.7	i
$MnFe_2O_4$	0.087	3.7	0.031
Fe_3O_4	0.35	3.0	0.022

[a] 5.3×10^{-3} mol dm^{-3} oxalic acid, pH$_{25}$ 2.5. [b] K_1 (in units of dm^3 mol^{-1}) is the equilibrium constant for $M^{2+} + C_2O_4^{2-} \rightleftharpoons MC_2H_4O_2$ [values taken from ref. (18)]. [c] Solubility of the oxalate complexes ($MC_2O_4 \cdot 2H_2O$) in water at 25 °C (i = insoluble) [from ref. (23)]. [d] $r_0 = 2.5$ μm (assumed).

Mn, Ni) was studied under otherwise identical conditions. The measurements were made in solutions containing 5.3×10^{-3} mol dm^{-3} oxalic acid at pH 2.5 and 140 °C. The runs were terminated after $\frac{1}{2}$ or 1 h depending on the reactivity of the oxide. The results obtained are shown in table 2. The rate constants have been corrected for both differences in initial particle size and density (calculated from X-ray crystallographic data) and are therefore directly comparable. They show a marked dependence on the nature of the divalent cation, the order of reactivity being Fe > Mn > Co > Ni.

DISCUSSION

DISSOLUTION BY OXALIC ACID

The cubic rate law, eqn (3), clearly establishes that the dissolution process involves reaction at the oxide particle surface as the rate-determining step. The oxalic acid concentration dependence we interpret in terms of adsorption of the oxalic acid at the surface prior to reaction according to the Langmuir adsorption isotherm, reaction (4):

$$C_2O_4^{2-} + >S \rightleftharpoons >S\cdots C_2O_4^{2-}. \tag{4}$$

We have no direct information on the nature of the adsorbed complex, but there is evidence from infrared measurements that adsorption of oxalic acid on goethite gives a binuclear complex,[17] and this may well occur with the ferrites under consideration here. The rate-determining step may involve either electron transfer from the adsorbed complex to Fe^{3+} ions at the surface:

$$>S\cdots C_2O_4^{2-} \rightarrow >S^-\cdots C_2O_4^- \tag{5}$$

followed by dissolution of the site as Fe^{2+}_{aq}:

$$>S^-\cdots C_2O_4^- \rightarrow S^-_{aq} + CO_2 + \tfrac{1}{2}C_2O_4^{2-} \tag{6}$$

or desorption of the adsorbed complex without change in valency state:

$$>S\cdots C_2O_4^{2-} \rightarrow S(C_2O_4^{2-})_{aq}. \tag{7}$$

The very rapid reduction of Fe^{3+}_{aq} by oxalic acid under our experimental conditions, and hence the failure to detect any iron(III) in the solutions following dissolution, prevents us from distinguishing directly between the reductive and complexing pathways.

A mechanism consisting of reaction (4) followed by either reaction (5) or (7) as the rate-determining step predicts an oxalic acid concentration dependence of the form

$$k_{obs} = \frac{kK_4[C_2O_4^{2-}]_T}{1 + K_4[C_2O_4^{2-}]_T} \tag{8}$$

where $k = k_5$ or k_7, and $[C_2O_4^{2-}]_T$ represents the total oxalic acid concentration. This describes the data in fig. 2(a) well. The solid line in fig. 2(a) is calculated from eqn (8) taking $k = 0.020$ min^{-1} and $K_4 = 70$ dm^3 mol^{-1}.

The pH dependence (fig. 3) arises from both the acid-base properties of oxalic acid ($pK_1 = 3.6$, $pK_2 = 1.2$ at 25 °C and $I \approx 1$ mol dm^{-3} [18]) and the oxide ($pK \approx 3.5$ for NiFe$_2$O$_4$ at 80 °C [15]). There are no acid dissociation constants available for oxalic acid at high temperatures. Taking into account the six possible parallel pathways, a very complicated rate expression can be derived. In view of the large number of adjustable parameters (6 rate constants and 3 equilibrium constants) and the uncertainties in the experimental data ($\pm 20\%$) we have not attempted to obtain a best fit to the data.

Nickel, chromium and ferrous ions are presumably dissolved by a complexing mechanism, or pass into solution as the hexa-aquo ions. The specific leaching of iron and nickel suggest that the surface becomes slightly enriched in chromium, but this did not appear to limit the reaction rate under our conditions (typically <20% dissolution). Surface enrichment has also been observed in the reductive dissolution of magnetite[19] [enriched in iron(II)] and franklinite[20] [enriched in zinc(II)] by tris(picolinato)vanadium(II), and in the oxalic-acid-induced dissolution of chrysotile[11] (enriched in silicon). It may be that the chromium(III) ions only dissolve after the neighbouring ions have been removed, i.e. by an 'undercutting' mechanism. A similar suggestion has been made for the dissolution of the nickel ion in NiFe$_2$O$_4$ induced by tris(picolinato)vanadium(II).

The marked effect of the chromium content on the dissolution rate parallels the situation with NTA, bis(histidinato)vanadium(II)[13] and tris(picolinato)-

vanadium(II),[13] and it seems likely that some common factor is responsible. Since these reagents unquestionably bring about dissolution in different ways [complexing attack by NTA, reductive attack by the vanadium(II) complexes] it is probable that some physical, rather than chemical, characteristic of the surface is involved. We suggest that this can be understood in terms of a 'surface-blocking' mechanism. We envisage that dissolution occurs primarily by the sequential dissolution of iron(III) ions in kink sites[6,21] (the dissolution of a kink site will in general result in the formation of a new kink site). When a chromium(III) ion occupies a kink site dissolution can only proceed *via* the dissolution of an iron(III) in a ledge site (which in general gives rise to two kink sites). Thus as the chromium content of the oxide increases the dissolution of iron(III) in ledge sites becomes progressively more important. If the ledge sites are more difficult to dissolve than the kink sites, as seems reasonable since they have more nearest neighbours, the overall dissolution rate will fall as the chromium content increases. A diagrammatic representation of this in terms of a simple 'building-block' model has been given by Segal and Sellers.[6] These arguments neglect some of the complexities of the spinel structure (for instance only half the octahedral holes and a quarter of the tetrahedral holes are occupied), and it has not proved possible to account quantitatively for the results obtained. Nevertheless these suggestions do, we believe, provide a framework within which the observations can be rationalised.

The effect of changing the divalent cation in simple ferrites (AFe_2O_4) is in marked contrast to the behaviour of the same oxides with tris(picolinato)vanadium(II),[13] where only a factor of two or so separate the fastest and the slowest. The dissolution rate constants are inversely correlated with the stability constants for the formation of the 1:1 metal-ion–oxalate complexes in aqueous solutions (table 2). This probably arises because of binding by oxalate at divalent cation sites (or by bridging between a divalent site and a neighbouring ferric ion), which may inhibit dissolution either by slowing down the rate at which the site passes into solution or by hindering the mobility of the divalent cations on the surface. Dissolution is presumably favoured by adsorption at iron(III) sites alone. The values may also reflect (at least in part) the differences in the solubility of the oxalate complexes (*cf.* table 2).

Two other factors may also play some part in the varying reactivity of these oxides. First, small differences in the fault density, surface morphology *etc.* due to slight variations in the preparation and calcining history of the samples may affect the reactivity. [Every attempt was made to keep these to a minimum; the results of dissolution in tris(picolinato)vanadium(II) suggest that such differences are quite minor.] Secondly, we note that the divalent cations in these oxides do not occupy identical sites in the oxide lattice. Magnetite, nickel ferrite and cobalt ferrite are inverse spinels, in which the divalent cations occupy octahedral holes [the iron(III) ions are equally divided between octahedral and tetrahedral holes], whilst manganese ferrite (which has very nearly a normal spinel structure) has 80% of the manganese(II) ions in tetrahedral holes.[22]

DISSOLUTION BY NITRILOTRIACETIC ACID

The concentration dependences in the dissolution of $Ni_{0.6}Cr_{0.6}Fe_{1.8}O_4$ by NTA parallel those in the oxalate system and can be understood in similar terms to those described above. Not surprisingly NTA appears to be much more strongly adsorbed than oxalate, and at the concentrations employed in this work the oxide particles appear to be more or less completely covered. This accounts for the much smaller activation energy measured in these solutions in comparison with those containing oxalic acid (table 1). In NTA solutions eqn (8) simplifies to $k_{obs} = k$, and so the activation energy refers only to reaction (5) or (7), whereas with oxalic acid the activation energy relates to reaction (5) or (7) *and* reaction (4). The acid dependence again reflects both protonation of surface sites and the acid-base behaviour of the complex itself, for which $pK_1 = 9.7$, $pK_2 = 2.5$ and $pK_3 = 1.9$ at 20 °C.[18] The effect of increasing the chromium content of the oxides can be understood in terms of the surface blocking mechanism described above. That iron and nickel are leached in preference to chromium is no doubt related to the substitution inertness of the chromium(III) ion.

We are indebted to Dr D. Bradbury, who first suggested and set up the high-temperature reaction cell used in this work. We also thank Drs T. Swan and M. G. Segal for their advice and encouragement. This paper is published by permission of the Central Electricity Generating Board.

[1] M. A. Blesa and A. J. G. Maroto, in *Decontamination of Nuclear Facilities, Keynote Addresses* (American Nuclear Society), p. 1.
[2] *Decontamination of Nuclear Reactors and Equipment*, ed. J. A. Ayres (Ronald Press, New York, 1970).
[3] G. R. Choppin, R. L. Dillon, B. Griggs, A. B. Johnson, J. F. Remark and A. E. Martell, *Electric Power Research Institute Report*, no. EPRI NP-1033 (1979).
[4] A. B. Johnson, B. Griggs, R. L. Dillon and R. A. Shaw, in *Decontamination and Decommissioning of Nuclear Facilities*, ed. M. Osterhout (Plenum Press, New York, 1980), p. 65.
[5] C. S. Lacy and B. Montford, in *Decontamination and Decommissioning of Nuclear Facilities*, ed. M. Osterhout (Plenum Press, New York, 1980), p. 93.
[6] M. G. Segal and R. M. Sellers, *Adv. Inorg. Bioinorg. Mech.*, in press.
[7] Y. L. Sandler, *Corrosion*, 1979, **35**, 205.
[8] A. B. Johnson, B. Griggs, F. M. Kustas and R. A. Shaw, in *Water Chemistry of Nuclear Reactor Systems 2* (British Nuclear Energy Society, London, 1981), p. 389.
[9] G. V. Buxton, T. Rhodes and R. M. Sellers, *Nature (London)*, 1982, **295**, 538.
[10] G. V. Buxton, T. Rhodes and R. M. Sellers, *J. Chem. Soc., Faraday Trans. 1*, 1983, **79**, 2961.
[11] J. H. Thomassin, J. Goni, P. Baillif, J. C. Touray and M. C. Jaurand, *Phys. Chem. Miner.*, 1977, **1**, 385.
[12] E. Baumgartner, M. A. Blesa, H. A. Marinovich and A. J. G. Maroto, *Inorg. Chem.* 1983, **22**, 2226.
[13] R. M. Sellers, unpublished data.
[14] M. G. Segal and R. M. Sellers, *J. Chem. Soc., Chem. Commun.*, 1980, 991.
[15] M. G. Segal and R. M. Sellers, *J. Chem. Soc., Faraday Trans. 1*, 1982, **78**, 1149.
[16] A. E. Harvey, J. A. Smart and E. S. Amis, *Anal. Chem.*, 1955, **27**, 26.
[17] R. L. Parfitt, V. C. Farmer and J. D. Russell, *J. Soil Sci.*, 1977, **28**, 29.
[18] (a) *Chemical Society Special Publication No. 17* (The Chemical Society, London, 1964); (b) *Chemical Society Special Publication No. 25* (The Chemical Society, London, 1971).
[19] G. C. Allen, R. M. Sellers and P. M. Tucker, *Philos. Mag.*, 1983, **48**, L5.
[20] G. C. Allen, C. Kirby, R. M. Sellers and P. M. Tucker, unpublished work.
[21] G. M. Rosenblatt, in *Treatise on Solid State Chemistry*, ed. N. B. Hannay (Plenum Press, New York, 1976), vol. 6A, p. 165.
[22] A. F. Wells, *Structural Inorganic Chemistry* (Clarendon Press, Oxford, 4th edn, 1976), p. 492.
[23] *Handbook of Chemistry and Physics*, ed. R. C. Weast (C.R.C. Press, Boca Raton, Florida, 62nd ed, 1981/82), pp. B95, B109, B118 and B124.

Interfacial Kinetics in Solution

Linear Free-energy Relationships Applicable to Heterogeneously Catalysed Reactions in Solution

BY MICHAEL SPIRO

Department of Chemistry, Imperial College of Science and Technology, London SW7 2AY

Received 6th December, 1983

The Brönsted and Hammett linear free-energy relations (LFER), widely used for homogeneous reactions, are applicable also to heterogeneously catalysed reactions in solution. When the reaction at the surface is sufficiently fast the kinetics become diffusion controlled. An analysis of this situation reveals a new LFER in which $\alpha (=\mathrm{d} \ln k/\mathrm{d} \ln K$, where k is the rate constant and K the equilibrium constant) depends only on the stoichiometric coefficients of the reaction, as do the kinetic orders. Redox reactions catalysed by metals exhibit a different kind of diffusion control when the formal electrode potentials are far apart. If the two interacting redox couples are electrochemically reversible, the $\ln k$ against $\ln K$ diagram resembles the shape of a Brönsted-Eigen plot, although both the rising section and the plateau region are now controlled by diffusion processes.

THE BRÖNSTED AND HAMMETT RELATIONS

It has long been recognised that relationships exist between the rate constants (k) and the equilibrium constants (K) of groups of closely related reactions. Sixty years ago Brönsted and Pedersen[1] proposed an equation of the type

$$\log k = \alpha \log K + \text{constant} \quad (1)$$

to describe the way in which the rate of a base-catalysed reaction depends on the dissociation constant K of the conjugate acid of the catalysing base; α here is negative. A similar Brönsted relation with positive α values connects the rate constants of acid-catalysed reactions with the dissociation constants of the catalysing acids.[2] The correlation is improved by taking statistical factors into account.[2] Brönsted and Pedersen already recognised in this first paper that the rates of simple base- (acid-) catalysed reactions would become independent of the dissociation constant of the relevant acid for bases (acids) so strong that every collision with the substrate resulted in reaction. Plots of $\ln k$ against $\ln K$ were therefore expected to take the form of the curve in fig. 1. Their prediction was confirmed after experimental techniques had been developed for measuring fast reaction rates. Eigen[3] then showed that many plots of $\ln k$ against $\ln (K_{\text{donor}}/K_{\text{acceptor}})$ were in fact gentle curves that tended to diffusion-controlled plateau values ($\alpha = 0$). The initial value of the slope α was unity because here the back reaction became sufficiently fast to be diffusion controlled. In between, and over a range of several pK units, data points of $\ln k$ against $\ln K$ scatter about a reasonable straight line. Such a plot is equivalent to the existence of a linear relation between the Gibbs free energies of activation and the 'standard' Gibbs free-energy changes for the acid dissociations

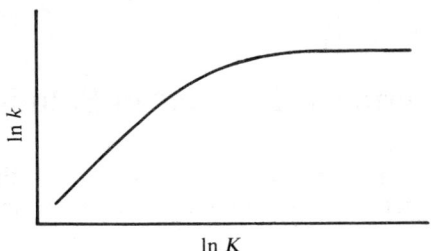

Fig. 1. Schematic Brönsted–Eigen plot for homogeneous acid–base reactions.

(in the prevailing medium), as follows from the introduction of transition-state theory:

$$\Delta G^{\ddagger} = \alpha \Delta G^{\ominus} + \text{constant}'. \qquad (2)$$

A low value of the Brönsted exponent α indicates a similarity in structure between the transition state and the reactants while a value approaching unity suggests structural similarity between the transition state and the products.[4] A more explicit analysis of α has been given by Marcus.[5] For fairly strong overlap reactions such as proton transfers he showed that

$$\alpha = \tfrac{1}{2}(1 + \Delta G^{\ominus}/4\Delta G_0^{\ddagger}) \qquad (|\Delta G^{\ominus}| \le 4\Delta G_0^{\ddagger}) \qquad (3)$$

where $\Delta G_0^{\ddagger}$ is the Gibbs free-energy barrier for proton transfer within the encounter complex when $\Delta G^{\ominus} = 0$. The effects of two work terms have been omitted in eqn (3). The same equation has been derived by a different route by Murdoch.[6] For a series of structurally similar acids $\Delta G_0^{\ddagger}$ is probably constant[4] and its value can be calculated from the curvature of the Brönsted plot since

$$d\alpha/d\Delta G^{\ominus} = 1/8\Delta G_0^{\ddagger}. \qquad (4)$$

Thus proton transfers to and from O and N atoms, which are intrinsically fast because $\Delta G_0^{\ddagger}$ is small, will show significant curvature in their Brönsted plots. Proton transfers to and from most carbon atoms, on the other hand, involve large energy barriers and therefore display good LFER over a wide range of $\Delta G^{\ominus}$.[7,8]

In the diffusion-controlled plateau region the Marcus analysis is no longer appropriate. Here α will be zero and independent of $\Delta G^{\ominus}$. Since the rate constant is now given by the Smoluchowski equation,[2] $\Delta G^{\ddagger}$ will be a function of the activation energies of diffusion of the two reacting species.[9] One finds similar diffusion-limited plateaux in homogeneous reactions that involve fast electron transfer in the encounter complex, as in certain fluorescence quenching reactions.[10]

Fendler et al.[11] have correlated the rates of hydrolysis of 2,4-dinitrophenyl sulphate with various amines in the presence of micellar catalysts by means of the Brönsted relation, and they found that α varied according to the surfactant used. This example illustrates the value of linear free-energy relationships in microheterogeneous catalysis. Catalyses of this kind exhibit many features in common with true heterogeneous catalysis in solution.[12] The Brönsted relation has also been introduced by Gold et al.[13] to help in understanding the catalytic effect of weak-acid ion-exchange resins on a hydrolysis reaction subject to general acid catalysis.

The most numerous LFER are forms of the Hammett equation[14,15]

$$\log(k_{ij}/k_{0j}) = \rho_j \sigma_i \qquad (5)$$

where k_{ij} is the rate (or equilibrium) constant of a given reaction j when one of the reactants bears a substituent i and k_{0j} is the corresponding constant in the absence of any substituent. The primary value of the substituent parameter σ_i, characteristic of the type and position of the substituent i and independent of the reaction, is given by

$$\sigma_i = \log(K_i/K_0) \qquad (6)$$

where K_0 and K_i are the dissociation constants of the unsubstituted and of the correspondingly substituted benzoic acids, respectively. The reaction constant ρ_j, on the other hand, depends only upon the reaction and the experimental conditions (medium, temperature). Combination of eqn (5) and (6) and introduction of transition-state theory shows that for any reaction

$$\Delta G^\ddagger = \rho_j \Delta G^\ominus + \text{constant}. \qquad (7)$$

The Hammett relations work well for a wide range of *meta*- and *para*-substituted aromatic compounds.[16] To make this type of equation applicable also to *ortho*-substituted aromatics and to substituted aliphatic compounds, Taft[17] proposed the modified form

$$\log(k_{ij}/k_{0j}) = \rho_j^* \sigma_i^* + s_j E_{si} \qquad (8)$$

where E_{si} is a steric substituent parameter[18] and σ_i^* is a polar substituent parameter calculated from the experimental rates of acidic and alkaline hydrolysis of the appropriate esters. Its use has recently been criticised, however.[19] Several other variants of eqn (5) are known.[15,20] The reaction parameters ρ and ρ^* are not infrequently >1 or <-1, in contrast to Brönsted α values which rarely stray outside the bounds of 0 and 1.[7]

Literature discussions about the Hammett and Taft equations regularly omit mention of diffusion control. This is easily understood. The homogeneous organic reactions involved are relatively slow: on the Marcus model, they involve sizeable intrinsic energy barriers.[8] These lead to the long linear free-energy relations observed and would require huge values of $\log(K_i/K_0)$ before a diffusion-limited plateau region can be reached.[7]

Kraus has shown in two reviews[21,22] that the rates of many gas-phase and some liquid-phase heterogeneously catalysed reactions can also be correlated by means of σ, E_s or, most frequently, σ^* parameters. It is quite surprising to find that these parameters, derived from homogeneous liquid-phase equilibria at 25 °C, are capable of relating the catalytic rates of gas reactions at temperatures from 100 to 500 °C. The slopes ρ or ρ^*, respectively, are generally high and in a few cases lie outside the limits ± 10. The processes studied were predominantly eliminations and hydrogenations, as well as some dehydrogenations, hydrogenolyses, esterifications and miscellaneous organic reactions; the catalysts employed were metals or metal oxides. It should be borne in mind, however, that several of these correlations refer not to relative rate constants but to relative rates. These are the products of heterogeneous rate constants and (Langmuir) adsorption coefficients which do not always change in the same direction on substitution. Nor has it been established in every case that the catalytic mechanism remains the same when one of the reagents is progressively substituted. Nevertheless, the fact that so many correlations have been observed points to the usefulness of LFER in heterogeneous as in homogeneous reactions. To allay any remaining doubts it might be as well to test the applicability of the LFER with a simpler reaction that can be more thoroughly studied. A good candidate would be the unimolecular solvolysis of a series of substituted t-butyl halides. These reactions are heterogeneously catalysed by silver salts, silver metal and other solids incorporating soft acid sites on which the halide end of the substrate adsorbs.[23,24]

Because the interaction here is of the soft-acid–soft-base type, the Brönsted relation is inapplicable. More appropriate would be the Hammett equation or a variant of it such as that suggested by Swain and Scott[25] for nucleophile–electrophile interactions and subsequently applied to inorganic as well as organic reactions in solution.[26]

In view of the way in which Hammett or Taft parameters have been found to fit both homogeneous and heterogeneous rates, one would expect that catalytic rates at micellar surfaces in solution would also lend themselves to this type of correlation. Little seems to have been done along these lines. However, LFER of the Hammett type have been used for some time[27,28] to correlate the biological activity of drugs and are known as QSAR (quantitative structure–activity relationships).[29]

DIFFUSION-CONTROLLED REACTIONS

We have seen that in homogeneous solution, fast proton transfers with small intrinsic energy barriers $\Delta G_0^\ddagger$ show curved $\ln k$ against $\ln K$ plots which also display diffusion-limited plateaux. In contrast, slow proton transfers to and from carbons that obey the Brönsted relation, as well as organic reactions that follow the Hammett or Taft equations, involve quite large intrinsic energy barriers. These barriers will be lower when the reactions are heterogeneously catalysed. If they are speeded up sufficiently, one would again expect diffusion to play a role in the kinetics. In this section we shall derive the kinetics for a catalysed reaction that is fully diffusion controlled.

Consider a general chemical reaction

$$\nu_A A + \nu_B B + \cdots \underset{v_{-1s}}{\overset{v_{1s}}{\rightleftharpoons}} \nu_X X + \nu_Y Y + \cdots \tag{I}$$

taking place at the surface of a catalyst with a net velocity v_s (mol m^{-2} s^{-1}) given by

$$v_s = v_{1s} - v_{-1s}. \tag{9}$$

These velocities are functions of the concentrations of the species i (mol m^{-3}) at the surface, c_{is}. We may visualise these species as present in a thin layer at the surface[30] or as physically or chemically adsorbed, but no assumption about its exact state is needed in the present treatment.

If the catalyst is sufficiently powerful $v_{1s} \approx v_{-1s} \gg v_s$ so that the surface reaction is virtually at equilibrium. Of course the species in the bulk solution will not be in equilibrium for a long time because the reactants must diffuse to the catalyst surface before they can react there and the products must diffuse away. The disappearance of A, B etc. from the bulk solution and/or the gradual appearance of X, Y etc. are the very phenomena monitored by the experimenter, who will interpret them in terms of the stoichiometric reaction (I). He will therefore write for the observed rate of reaction v_{obs} (mol m^{-3} s^{-1})

$$v_{\text{obs}} = -\frac{1}{\nu_A}\frac{\partial c_A}{\partial t} = \cdots = \frac{1}{\nu_X}\frac{\partial c_X}{\partial t} = \cdots. \tag{10}$$

It follows from dimensional analysis that

$$v_{\text{obs}} = v_s S / V \tag{11}$$

where S is the area of the catalyst surface and V is the volume of the solution. Should reaction (I) also proceed to an appreciable extent in the bulk solution, the

v_{obs} value in the present treatment must be corrected for the homogeneous contribution.[31]

In the steady state, the flux of reactant A (mol s^{-1}) diffusing towards unit area of catalyst must equal the net rate at which A disappears by chemical reaction at the surface. By Fick's first law

$$D_A\left(\frac{\partial c_A}{\partial x}\right) = D_A\left(\frac{c_A - c_{As}}{\delta_A}\right) = v_s \nu_A \qquad (12)$$

where D_A is the diffusion coefficient of A and δ_A is the effective thickness of the (Nernst) diffusion layer for this species. Hence

$$c_{As} = c_A - v_s \nu_A \delta_A / D_A. \qquad (13)$$

In the case of a product such as X, its net steady-state rate of formation by the catalytic reaction equals the rate at which it diffuses away into the bulk solution. Thus

$$D_X(c_{Xs} - c_X)/\delta_X = v_s \nu_X \qquad (14)$$

$$c_{Xs} = c_X + v_s \nu_X \delta_X / D_X. \qquad (15)$$

When the catalytic reaction is so fast that the surface reaction is effectively at equilibrium

$$K_s = \frac{c_{Xs}^{\nu_X} c_{Ys}^{\nu_Y} \cdots}{c_{As}^{\nu_A} c_{Bs}^{\nu_B} \cdots} = \frac{\prod_{\text{prod}} (c_X + v_s \nu_X \delta_X / D_X)^{\nu_X}}{\prod_{\text{react}} (c_A - v_s \nu_A \delta_A / D_A)^{\nu_A}}. \qquad (16)$$

Initially $c_X = c_Y = \cdots = 0$ and $c_A, c_B, \ldots$ are large. Hence

$$K_s \approx \frac{v_s^{(\nu_X + \nu_Y + \cdots)} \prod_{\text{prod}} (\nu_X \delta_X / D_X)^{\nu_X}}{\left(1 - v_s \sum_{\text{react}} (\nu_A^2 \delta_A / D_A c_A)\right) \prod_{\text{react}} c_A^{\nu_A}} \qquad (17)$$

whence

$$\frac{1}{v_s} = \frac{1}{v_s'} + \sum_{\text{react}} \left(\frac{\nu_A^2 \delta_A}{D_A c_A (\nu_X + \nu_Y + \cdots)}\right) \qquad (18)$$

$$v_s' = K_s^{[1/(\nu_X + \nu_Y + \cdots)]} \prod_{\text{prod}} (D_X/\nu_X \delta_X)^{[\nu_X/(\nu_X + \nu_Y + \cdots)]} \prod_{\text{react}} c_A^{[\nu_A/(\nu_X + \nu_Y + \cdots)]}. \qquad (19)$$

The summation function in eqn (18) is likely to be of the order of $1/10^{-4}$ mol m^{-2} s^{-1}, whereas typical values of v_s are of the order[31] of 10^{-6} to 10^{-5} mol m^{-2} s^{-1}. We may therefore reasonably adopt the more severe approximation

$$v_s = v_s'. \qquad (20)$$

An experimental test of these equations is discussed in the final section.

Certain conclusions can now be drawn about these heterogeneously catalysed reactions. First, they are sufficiently fast to become wholly transport controlled. The product of the (D_X/δ_X) terms in eqn (19) is essentially a mean (D/δ) raised to the first power, and the summation term in eqn (18) contains a corresponding parameter. Thus if the catalyst is present in the form of a rotating disc (see below), the overall rate will be directly proportional to the square root of its rotation speed. Secondly, the catalytic rate is of fractional order in the various reactants A, B *etc.* but independent of the concentrations of the products. The magnitudes of these orders depend solely on the stoichiometry of the reaction. Because of the general model used, the fact that the orders are fractional does not imply Freundlich or any other particular kind of adsorption of the reactants on the catalyst surface. It also follows from eqn (16) that addition of either product to the initial reaction mixture will considerably decrease the catalytic rate and also change the kinetic orders of the reactants. Thirdly, the catalytic rate constant k_{cat} is given by the terms preceding the final product sign in eqn (19). Taking logarithms

$$(\nu_X + \nu_Y + \cdots) \ln k_{cat} = \ln K_s + \sum_{prod} [\nu_X \ln D_X - \nu_X \ln (\nu_X \delta_X)]. \qquad (21)$$

The relationship between the catalytic rate constant and the equilibrium constant is therefore given by

$$\alpha = \frac{\partial \ln k_{cat}}{\partial \ln K} = \frac{1}{\nu_X + \nu_Y + \cdots} \qquad (22)$$

and also depends only on the stoichiometry of the reaction.

Two comments are required here. First, K in eqn (22) is the equilibrium constant of the reaction under study and not that of some related process such as the dissociation of an acid. Secondly, the equation shows that a significant degree of thermodynamic control is retained even in the diffusion region. If $\nu_X = \nu_Y = 1$, an LFER with a slope α of $\frac{1}{2}$ is obtained. This stands in strong contrast to homogeneous reactions in which diffusion control manifests itself by a horizontal plateau region in the $\ln k$ against $\ln K$ plots,[3] *i.e.* by $\alpha = 0$. It is possible, therefore, that experimenters have studied heterogeneous reactions in this region and not been aware of it, especially if the reactions were carried out under constant-stirring conditions. It is by changing the conditions of hydrodynamic flow, and thus the thickness of the Nernst layer, that one can best test whether diffusion is playing a role in heterogeneous processes at smooth surfaces or even at moderately rough ones.[32]

Application of the transition-state equations[9] for both k and D leads to

$$(\nu_X + \nu_Y + \cdots)\Delta G^{\ddagger}_{cat} = \Delta G^{\ominus} + \sum_{prod} \nu_X \Delta G^{\ddagger}_{X\,diffn} + RT \sum_{prod} \nu_X \ln (\nu_X \delta_X / \lambda^2) \qquad (23)$$

where R is the gas constant, T is the absolute temperature and λ is the distance between successive equilibrium positions in diffusion. The first term on the right-hand side of eqn (23) represents the thermodynamic contribution to the Gibbs free energy of activation of the catalysed reaction while the subsequent terms constitute the kinetic or flow contributions. The corresponding enthalpy relationship can be derived by inserting the transition-state equations into eqn (21) and differentiating with respect to $1/T$:

$$(\nu_X + \nu_Y + \cdots)\Delta H^{\ddagger}_{cat} = \Delta H^{\ominus} + \sum_{prod} \nu_X \Delta H^{\ddagger}_{X\,diffn}. \qquad (24)$$

In deriving eqn (24) it was assumed that λ and δ were independent of the temperature. The value of δ is essentially fixed by the hydrodynamic regime in the reaction vessel, whether this is streamline or turbulent, and will depend upon the size and shape of the catalyst, the diffusion coefficient of the species involved and the properties of the solvent. If, for example, the catalyst is introduced in the form of a rotating disc, the Levich equation tells us that under conditions of streamline flow[33a]

$$\delta_i = 1.613 D_i^{1/3} \nu^{1/6} \omega^{-1/2} \tag{25}$$

where ν is the kinematic viscosity of the medium and ω is the angular velocity of the disc. The situation is quite different for a rapidly swirled suspension of spherical catalyst particles. If their radius r exceeds an eddy length given approximately by ν/U, where U is the velocity of the liquid in the reaction vessel, then it follows from an analysis by Levich[33b] that

$$\delta_i \approx D_i L^{1/3}/(2r)^{1/3} U \tag{26}$$

where L is the scale of motion of the agitated fluid as a whole. Values of δ_i calculated from eqn (26) are typically 100–1000 times smaller than those given by eqn (25).

Diffusion through the Nernst layer can become rate-determining in other interfacial processes, as the papers on liquid–liquid extraction in this Discussion testify. It also plays an important role in interfacial drug transfer.[29]

REDOX REACTIONS CATALYSED BY METALS

Metals and carbons often catalyse oxidation–reduction reactions of the type

$$\nu_{ox_2} ox_2 + \nu_{red_1} red_1 \rightarrow \nu_{red_2} red_2 + \nu_{ox_1} ox_1 \tag{II}$$

where the stoichiometric coefficients ν are such as to refer to the transfer of one electron. Particular attention has been paid to the aqueous redox reaction

$$Fe(CN)_6^{3-} + \tfrac{3}{2} I^- \rightarrow Fe(CN)_6^{4-} + \tfrac{1}{2} I_3^- \tag{III}$$

which is strongly catalysed at the surfaces of several noble metals[34] as well as by carbons.[35] By combining kinetic and electrochemical measurements at the same anodically pretreated platinum surface it has been demonstrated[36,37] that the catalysis takes place by a purely electrochemical mechanism. In the reacting solution the platinum automatically adopts a so-called mixture (or mixed) potential E_{mix} such that no net current passes, the anodic current caused by the oxidation of iodide at its surface being exactly equal in magnitude to the cathodic current produced by the surface reduction of ferricyanide. This is illustrated in fig. 2 and 3. These show the current–potential curves of the individual redox couples ox_1/red_1 (e.g. I_3^-/I^-) and ox_2/red_2 [e.g. $Fe(CN)_6^{3-}/Fe(CN)_6^{4-}$]. When both couples are present together, their curves are usually treated as additive according to a principle enunciated by Wagner and Traud.[38] At any given potential, therefore, the net current in the mixture equals the algebraic sum of the contributing currents. At the mixture potential the net current is zero. The two contributing currents, equal in magnitude, are designated I_{mix}, and by Faraday's law are proportional to the catalytic rate

$$v_s = I_{mix}/FS \tag{27}$$

where F is the Faraday constant. The values of v_s calculated by eqn (27) from electrochemical experiments on the separate I_3^-/I^- and $Fe(CN)_6^{3-}/Fe(CN)_6^{4-}$ couples were found to be equal, within experimental error, to the catalytic rates obtained from kinetic runs with reaction (III).[36,37] Also equal were the electrochemically determined mixture potentials E_{mix} and the potentials E_{cat} taken up by the catalysing platinum disc in the reaction mixtures. These sets of experiments clearly establish

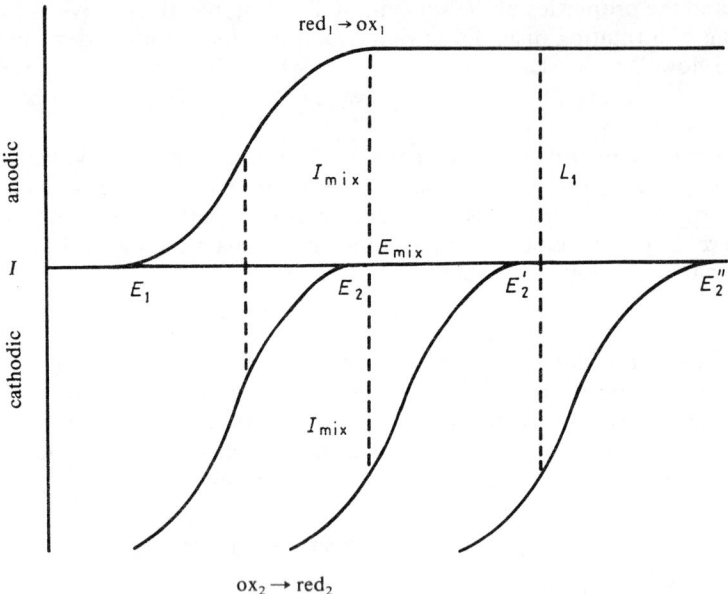

Fig. 2. Schematic plots of current against potential for sets of electrochemically irreversible couples. Couple 1 is paired successively with three different couples 2 of increasing formal potential.

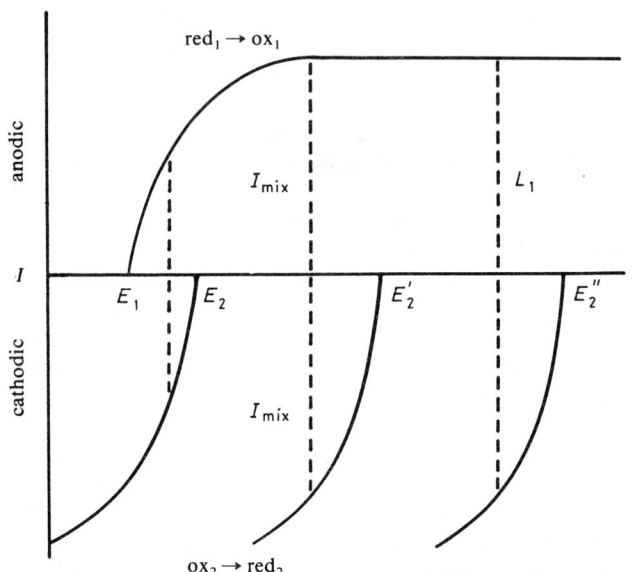

Fig. 3. Schematic plots of current against potential for sets of electrochemically reversible couples. Couple 1 is paired successively with three different couples 2 of increasing formal potential.

the electrochemical mechanism. Evidence in the literature makes it likely that numerous other redox reactions of type (II) are catalysed at the surfaces of noble metals[39] and carbons[35] by the same mechanism of electron transfer through the solid.

We shall defer to the next section the equations that apply to the regions of the gradually rising curves in fig. 2 (electrochemically irreversible couples) or the steeply rising curves in fig. 3 (electrochemically reversible couples).[40] Here we shall treat the situations that arise in both cases when the formal electrode potentials are sufficiently far apart. This is shown in both diagrams where couple 1 is successively paired with three other couples of increasing formal potential. In each case conditions are so arranged as to make the limiting current L_1 for the oxidation of red_1 less than the limiting currents L_2 for the reduction of the various oxidants ox_2. It can be seen that above a certain value of E_{mix}, the mixture current for the two couples is simply equal to L_1, so that[41]

$$v_s = \frac{I_{mix}}{FS} = \frac{D_{red_1} c_{red_1}}{\nu_{red_1} \delta_{red_1}}. \tag{28}$$

The reaction therefore remains wholly diffusion controlled through the inverse dependence on δ, and has become first order in one of the reactants (red_1) and zero order in the other (ox_2). This kind of diffusion control can occur even with couples whose formal potentials are close together if the concentration of one of the reactants (red_1) is very low, for its limiting current region will then begin very near its formal potential. Two very early catalytic studies in the literature appear to fall into this category: that of Jablczynski[42] of the reaction

$$Cr^{II} + H^+ \rightarrow Cr^{III} + \tfrac{1}{2}H_2 \tag{IV}$$

and that of Denham[43] of the reaction

$$Ti^{3+} + H^+ \rightarrow Ti^{4+} + \tfrac{1}{2}H_2 \tag{V}$$

both catalysed by platinised platinum. More recently the same interpretation has been applied to the catalysis by noble-metal colloids of the reaction

$$MV^+ + H^+ \rightarrow MV^{2+} + \tfrac{1}{2}H_2 \quad (MV = \text{methyl viologen}) \tag{VI}$$

in a scheme for photocatalytic water reduction.[44] The electrochemical model underlying eqn (28) has also been used[45] to interpret previously puzzling corrosion data in the literature and is generally accepted in metal-dissolution theory.[46]

Certain corollaries follow from eqn (28). If one inserts the transition-state equations for the rate constant v_s/c_{red_1} and for D_{red_1}, and assumes as before that δ remains constant, then one obtains

$$\Delta G^{\ddagger}_{cat} = \Delta G^{\ddagger}_{red_1(diffn)} + RT \ln(\nu_{red_1} \delta_{red_1}/\lambda^2). \tag{29}$$

Thermodynamic control has thus vanished completely. The same point is made by the equation

$$\alpha_{el} = \frac{\partial \ln I_{mix}}{\partial (E_2^\circ - E_1^\circ)} = \frac{\partial \ln L_{red_1}}{\partial (E_2^\circ - E_1^\circ)} = 0 \tag{30}$$

where E_i° is the formal electrode potential of couple i in the medium employed. This kind of diffusion control therefore differs radically from that given in eqn (18)–(23). Indeed, it bears a resemblance to the type of diffusion control found in homogeneous reactions except that $\Delta G^{\ddagger}_{cat}$ in eqn (29) depends on the activation energy of diffusion of only one of the reacting species instead of on both and also depends on the stirring conditions.

METAL-CATALYSED REDOX REACTIONS INVOLVING TWO IRREVERSIBLE COUPLES OR TWO REVERSIBLE COUPLES

We now return to the value of the catalytic rate when E_{mix} lies within the rising sections of the current–potential curves. Consider fig. 2, in which the curves for both couples possess the shapes typical of redox systems that are electrochemically irreversible,[40] i.e. couples whose electrochemical rate constants and exchange current densities are small. The exponentially rising parts of the curves are described by Tafel equations, and from these and the additivity principle[38] it is possible to calculate the potential E_{mix} at which the anodic and cathodic currents are equal, as well as the currents themselves (I_{mix}) and hence v_s. When the anodic process of couple 1 is just first order in red_1 and the cathodic reaction of couple 2 first order in ox_2, the resulting equation is[41]

$$v_s = k_1^{r_2} k_2^{r_1} \exp[\alpha_2 z_2 r_1 F(E_2^\circ - E_1^\circ)/RT] c_{red_1}^{r_2} c_{ox_2}^{r_1} \tag{31}$$

where k_i is the electrochemical rate constant of the couple ox_i/red_i, α_i is its cathodic transfer coefficient, z_i is the charge-transfer valence and the r_i values are given by

$$r_1 = 1 - r_2 = \frac{(1-\alpha_1)z_1}{(1-\alpha_1)z_1 + \alpha_2 z_2}. \tag{32}$$

If the electrode kinetics are not both first order, the concentration terms are more complex.[41] The fractional kinetic orders of red_1 and ox_2 are a natural consequence of the electrochemical mechanism and do not signal Freundlich adsorption. By collecting all the non-concentration terms into a catalytic rate constant k_{cat}, and remembering that the equilibrium constant of reaction (II) is given by

$$K = \exp[F(E_2^\circ - E_1^\circ)/RT] \tag{33}$$

we obtain another LFER

$$\ln k_{cat} = r_2 \ln k_1 + r_1 \ln k_2 + \alpha_2 z_2 r_1 \ln K. \tag{34}$$

Since k_1 and k_2 vary with the couples in question, and depend upon the respective free-energy changes of solvent reorganisation[8,47] as well as the energies needed to break chemical bonds, the linearity condition should be tested by plotting $\ln(k_{cat}/k_1^{r_2} k_2^{r_1})$ against $\ln K$. Commonly $\alpha \approx \frac{1}{2}$ and $z = 1$, which makes $r_1 \approx \frac{1}{2}$ and the slope $\alpha_2 z_2 r_1 = (1-\alpha_1)z_1 r_2 \approx \frac{1}{4}$.

Let us now turn to pairs of electrochemically reversible couples whose electrochemical rate constants and exchange current densities are large. Fig. 3 shows the shapes of their current–potential curves. From the known equations of the steeply rising sections and the additivity principle, Freund and Spiro[48] have calculated the mixture potentials and currents when the two couples are present together at the metal surface. The resulting equation [eqn (33) in ref. (48)] is the exact electrochemical analogue of eqn (18) and (19). This equation was rigorously tested in the laboratory for reaction (III) at an anodically pretreated platinum rotating-disc catalyst.[31] All the theoretical predictions were verified including the proportionality of the rate to the square root of the disc rotation speed, the kinetic orders of I^- (1) and of $Fe(CN)_6^{3-}$ (2/3), the effect on those orders of adding one of the products to the mixture, the value of the initial rate constant and the value of the activation energy. One may therefore have confidence in the validity of eqn (18) and corollaries based on it. This applies specifically to the resulting linear free-energy relations (21) and (22). For reaction (III) the value of the slope α equals $1/(1+\frac{1}{2}) = \frac{2}{3}$.

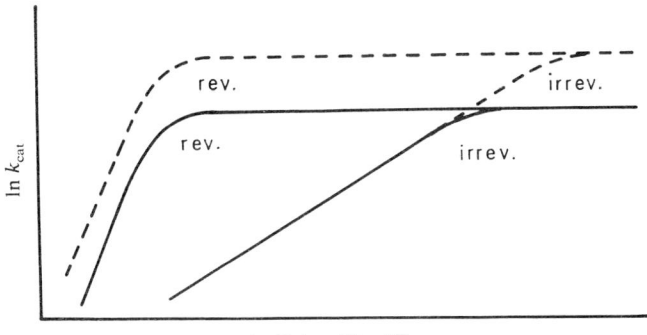

Fig. 4. Schematic plots of ln k_{cat} against ln K (or $E_2^\circ - E_1^\circ$). Full curves represent conditions of slow stirring, dashed curves conditions of fast stirring. See the text for other conditions.

We are now in a position to compare the behaviour of different metal-catalysed redox reactions when the various reactants are present in similar concentrations. Plots of ln k_{cat} against ln K (or $E_2^\circ - E_1^\circ$) will exhibit the shapes sketched in fig. 4. The initial linear portion possesses a slope of $\alpha_2 z_2 r_1$ for pairs of irreversible couples [if we plot ln $(k_{cat}/k_1^{r_2} k_2^{r_1})$] and of $1/(\nu_X + \nu_Y + \cdots)$ for pairs of reversible couples. For the latter type, we should compare only reactions of the same stoichiometry or else make an appropriate correction to the data according to eqn (19). With irreversible couples the linear section will be unaffected by faster stirring, with reversible ones it will rise to a parallel position. When $(E_2^\circ - E_1^\circ)$ is sufficiently large the rate constants of both types of catalysed reaction will reach a diffusion-controlled limit. According to eqn (28), this limit too will increase with faster stirring because this decreases the thickness of the Nernst layer. It is apparent that the shapes of all these curves resemble that of the Brönsted–Eigen plot in fig. 1. Whereas, however, only the plateau region is diffusion controlled in homogeneous reactions and in catalysed reactions between irreversible redox couples, the whole curve is diffusion controlled for metal-catalysed reactions between reversible redox couples. In the linearly rising region the rate of the latter depends upon the diffusion of the reactants towards the catalyst and, particularly in the early stages, on the diffusion away of the products. In the plateau region only the diffusion towards the catalyst of one of the reactants (generally the one of smallest concentration) is of kinetic significance. Over the whole potential range, then, k_{cat} for reversible couples will be a function of the hydrodynamic regime and will be directly proportional to the square root of the rotation speed with a rotating-disc catalyst. This kind of plot represents an extension of the traditional Brönsted–Eigen or Hammett picture.

---

[1] J. H. Brönsted and K. Pedersen, *Z. Phys. Chem.*, 1924, **108**, 185.
[2] R. P. Bell, *The Proton in Chemistry* (Chapman and Hall, London, 2nd edn, 1973), chap. 7 and 10.
[3] M. Eigen, *Angew. Chem., Int. Ed.*, 1964, **3**, 1.
[4] J. E. Crooks, in *Proton-transfer Reactions*, ed. E. F. Caldin and V. Gold (Chapman and Hall, London, 1975), chap. 6.
[5] R. A. Marcus, *J. Phys. Chem.*, 1968, **72**, 891; A. O. Cohen and R. A. Marcus, *J. Phys. Chem.*, 1968, **72**, 4249.
[6] J. R. Murdoch, *J. Am. Chem. Soc.*, 1972, **94**, 4410.
[7] A. J. Kresge, *Chem. Soc. Rev.*, 1973, **2**, 475.
[8] W. J. Albery, *Annu. Rev. Phys. Chem.*, 1980, **31**, 227.

9 S. Glasstone, K. J. Laidler and H. Eyring, *The Theory of Rate Processes* (McGraw-Hill, New York, 1941), chap. IX.
10 D. Rehm and A. Weller, *Isr. J. Chem.*, 1970, **8**, 259; F. Scandola, V. Balzani and G. B. Schuster, *J. Am. Chem. Soc.*, 1981, **103**, 2519.
11 J. H. Fendler, E. J. Fendler and L. W. Smith, *J. Chem. Soc., Perkin Trans. 2*, 1972, 2097.
12 M. Spiro, in *Essays in Chemistry*, ed. J. N. Bradley, R. D. Gillard and R. F. Hudson (Academic Press, London, 1973), vol. 5, p. 63.
13 V. Gold, C. J. Liddiard with J. L. Martin, *J. Chem. Soc., Faraday Trans. 1*, 1977, **73**, 1128.
14 L. P. Hammett, *J. Am. Chem. Soc.*, 1937, **59**, 96.
15 L. P. Hammett, *Physical Organic Chemistry* (McGraw-Hill Kogakusha, Tokyo, 2nd edn, 1970), chap. 11.
16 P. R. Wells, *Chem. Rev.*, 1963, **63**, 171.
17 R. W. Taft Jr, *J. Am. Chem. Soc.*, 1952, **74**, 3120; 1953, **75**, 4231.
18 R. Gallo, in *Progress in Physical Organic Chemistry*, ed. R. W. Taft (Wiley-Interscience, New York, 1983), vol. 14, p. 115.
19 M. Charton, in *Progress in Physical Organic Chemistry*, ed. R. W. Taft (Wiley-Interscience, New York, 1981), vol. 13, p. 119.
20 J. Shorter, in *Advances in Linear Free Energy Relationships*, ed. N. B. Chapman and J. Shorter (Plenum Press, London, 1972), chap. 2.
21 M. Kraus, in *Advances in Catalysis*, ed. D. D. Eley, H. Pines and P. B. Weisz (Academic Press, New York, 1967), vol. 17, p. 75.
22 M. Kraus, in *Advances in Catalysis*, ed. D. D. Eley, H. Pines and P. B. Weisz (Academic Press, New York, 1980), vol. 29, p. 151.
23 E. F. G. Barbosa, R. J. Mortimer and M. Spiro, *J. Chem. Soc., Faraday Trans. 1*, 1981, **77**, 111.
24 R. J. Mortimer and M. Spiro, *J. Chem. Soc., Perkin Trans. 2*, 1982, 1031.
25 C. G. Swain and C. B. Scott, *J. Am. Chem. Soc.*, 1953, **75**, 141.
26 R. G. Pearson, in *Advances in Linear Free Energy Relationships*, ed. N. B. Chapman and J. Shorter (Plenum Press, London, 1972), chap. 6.
27 C. Hansch, R. M. Muir, T. Fujita, P. P. Maloney, F. Geiger and M. Streich, *J. Am. Chem. Soc.*, 1963, **85**, 2817.
28 A. Cammarata and K. S. Rogers, in *Advances in Linear Free Energy Relationships*, ed. N. B. Chapman and J. Shorter (Plenum Press, London, 1972), chap. 9.
29 J. T. M. van de Waterbeemd, *Doctoral Thesis* (University of Leiden, 1980); H. van de Waterbeemd, in *Quantitative Approaches to Drug Design*, ed. J. C. Dearden (Elsevier, Amsterdam, 1983), p. 183.
30 P. D. Totterdell and M. Spiro, *J. Chem. Soc., Faraday Trans. 1*, 1976, **72**, 1477.
31 P. L. Freund and M. Spiro, *J. Chem. Soc., Faraday Trans. 1*, 1983, **79**, 491.
32 M. Spiro and D. S. Jago, *J. Chem. Soc., Faraday Trans. 1*, 1982, **78**, 295.
33 V. G. Levich, *Physicochemical Hydrodynamics* (Prentice-Hall, Englewood Cliffs, N.J., 1962), (a) p. 69, (b) sections 4, 25 and 33.
34 M. Spiro, *J. Chem. Soc.*, 1960, 3678.
35 J. M. Austin, T. Groenewald, and M. Spiro, *J. Chem. Soc., Dalton Trans.*, 1980, 854.
36 M. Spiro and P. W. Griffin, *Chem. Commun.*, 1969, 262.
37 M. Spiro and P. L. Freund, *J. Electroanal. Chem.*, 1983, **144**, 293.
38 C. Wagner and W. Traud, *Z. Elektrochem.*, 1938, **44**, 391.
39 M. Spiro and A. B. Ravnö, *J. Chem. Soc.*, 1965, 78.
40 A. J. Bard and L. R. Faulkner, *Electrochemical Methods* (Wiley, New York, 1980), chap. 3.
41 M. Spiro, *J. Chem. Soc., Faraday Trans. 1*, 1979, **75**, 1507.
42 K. Jablczynski, *Z. Phys. Chem.*, 1908, **64**, 748.
43 H. G. Denham, *Z. Phys. Chem.*, 1910, **72**, 641.
44 D. S. Miller, A. J. Bard, G. McLendon and J. Ferguson, *J. Am. Chem. Soc.*, 1981, **103**, 5336.
45 T. P. Hoar, in *Modern Aspects of Electrochemistry*, ed. J. O'M. Bockris (Butterworths, London, 1959), no. 2, chap. 4.
46 M. Spiro, in *The Physical Chemistry of Solutions*, ed. D. V. Fenby and I. D. Watson (Massey University Press, New Zealand, in press).
47 R. A. Marcus, *J. Chem. Phys.* 1965, **43**, 679; *Electrochim. Acta*, 1968, **13**, 995.
48 P. L. Freund and M. Spiro, *J. Chem. Soc., Faraday Trans. 1*, 1983, **79**, 481.

GENERAL DISCUSSION

Dr. B. H. Robinson (*University of Kent*) said: I wish to put three questions to Prof. Yasunaga. (1) The rate constant of the hydrolysis of hydroxyl groups at the zeolite/solution interface in table 2 of your paper was determined to be the order of $10^2 \, \text{dm}^3 \, \text{mol}^{-1} \, \text{s}^{-1}$. If this rate is governed by a diffusion-controlled reaction, why is the value of the rate constant not of the same order (10^8–$10^{10} \, \text{dm}^3 \, \text{mol}^{-1} \, \text{s}^{-1}$) as in homogeneous systems? (2) Did you determine the temperature dependence for the rate constants shown in table 3 of your paper? (3) How did the relaxation amplitudes vary for the series of alkylammonium ions?

Prof. T. Yasunaga (*Hiroshima University, Japan*) said: For the first question, the following two reasons may be proposed to explain the smaller value of the rate constant. First, in the adsorption–desorption of ions at the solid/liquid interface, since the reaction occurs at a two-dimensional interface, it may be argued that the reduction of a degree of freedom lowers the rate constant by *ca.* 10^2–$10^4 \, \text{dm}^3 \, \text{mol}^{-1} \, \text{s}^{-1}$.[1] The extent of this reduction is thought to be affected by differences in the site density and particle size. Another possible explanation is that there is a higher activation energy for surface hydrolysis owing to the strongly bound water molecule formed during hydrolysis. It is well known that metal oxides strongly adsorb water molecules on particle surfaces. Taking this fact into account, mechanism (I) in our paper can be rewritten as

$$\text{SOH} + \text{OH}^- \rightleftharpoons \text{SO}^- \cdots \overset{H}{\underset{H}{\diagup}} \negthickspace\text{O} \rightleftharpoons \text{SO}^- + \text{H}_2\text{O} \qquad (\text{I}')$$

$$\text{steady state}$$

$$\begin{pmatrix} \text{diffusion-controlled} \\ \text{step} \end{pmatrix} \begin{pmatrix} \text{rate-determining} \\ \text{step} \end{pmatrix}$$

where $\text{SO}^- \cdots \overset{H}{\underset{H}{\diagup}}\negthickspace\text{O}$ is in the steady state. If the second step, *i.e.* the step involving the release of the bound water molecule, is the rate-determining step in the above mechanism, then the reduction of the rate constant can reasonably be explained by the higher activation energy of the rate-determining desorption of water from the steady-state intermediate.

With regard to the second question, we did not carry out additional experiments on the temperature dependence for the rate constants obtained. In our paper the effect of molecular size of the adsorbing molecule in the ion-exchange reaction of the alkylammonium ion for Na^+ in the cage was studied only at 25 °C. A temperature-dependence experiment and the determination of activation parameters would be useful and are under consideration for future work.

For the third question, the amplitude of the relaxation signal decreased and diminished with increasing alkyl chain length (the volume of the cation).

---

[1] R. D. Astumian and Z. A. Schelly, *J. Am. Chem. Soc.*, 1984, **106**, 304.

Dr. G. Sartori (*Exxon, Annandale, N.J.*) said: I would like to address a question to Prof. Yasunaga regarding the molecular size of the ion-exchanging molecules. You have shown that a bulky alkyl group, *e.g.* isopropyl, directly attached to the nitrogen atom greatly reduces the ability of the ammonium ion to exchange with Na^+ in the cage of zeolite 4A. What would happen if the bulky substituents were not directly attached to the nitrogen atom, but removed from it by one carbon, such as in

$$\text{C}-\underset{|}{\overset{\overset{\text{C}}{|}}{\text{C}}}-\text{C}-NH_3^+ \quad \text{and} \quad \text{C}-\underset{\underset{\text{C}}{|}}{\overset{\overset{\text{C}}{|}}{\text{C}}}-\text{C}-NH_3^+ ?$$

Prof. T. Yasunaga (*Hiroshima University, Japan*) said: We did not argue the basis of heterogeneities of the hydrocarbon and the charge density of the penetrating alkylammonium ions. We have studied the ion-exchange kinetics of the alkylammonium ion, in which a bulky alkyl group is attached directly to the nitrogen atom in order to avoid the hydrophobic effect due to the alkyl chain length of the entering alkylammonium ion. Several investigators have reported that larger molecules having long hydrocarbon chains are difficult to exchange in zeolite 4A[1,2] and that for higher concentrations of the larger cations ($\geqslant 5$ mol dm^{-3}) some cations identified by Dr. Sartori can exchange slightly. Under the present circumstances, as pointed out by Dr. Sartori, it is considered that the conclusion in our paper may be applicable for the alkylammonium ions having homogeneous distributions of both the charge density and the hydrophobicity of the cation itself. Kinetic studies of the hydrophobic effect in the ion-exchange reaction are now in progress[3] and the results will be reported in the near future.

[1] D. W. Breck, *Zeolite Molecular Sieves* (Wiley-Interscience, New York, 1974).
[2] R. M. Barrer and W. M. Meier, *Trans. Faraday Soc.*, 1958, **54**, 1074.
[3] T. Ikeda and T. Yasunaga, *J. Phys. Chem.*, in press.

Dr. B. A. W. Coller (*Monash University, Australia*) said: I address my remarks to Prof. Christoffersen.

(1) How was β, the scaled potential, calculated in this work?

(2) The formulation of the polynuclear mechanism, leading to eqn (15) of the paper, does not explicitly deal with the components of the ionic product. Have you investigated the dependence of rate of dissolution on the concentrations of calcium and phosphate ions varied separately?

(3) Added inert electrolyte can be expected to alter the electric potential at the interface. Has the effect of a salt such as potassium chloride been investigated?

Prof. J. and Dr. M. R. Christoffersen (*University of Copenhagen, Denmark*) replied: We define the dimensionless affinity of dissolution, β, as the difference between the chemical potential of one mole of substance in the solid phase, μ_{cr}, and the chemical potential of one mole of substance in solution, μ_{aq}, divided by the number of ions in one mole of substance and divided by kT, *i.e.*

$$\beta = (\mu_{cr} - \mu_{aq})/18kT = \ln(K_s/Y)/18$$

where $Y = [Ca^{2+}]^{10}[PO_4^{3-}]^6[OH^-]^2$ and K_s is the corresponding ionic product at equilibrium, for which we have used the value $10^{-116.4}$ mol^{18} dm^{-54}. More details of the calculation are given in Appendix 2, ref. (4) of our paper.

It is correct that eqn (15) in our paper does not deal specifically with the components of the ionic product. This equation should only be applied for solutions with a Ca/P ratio of 1.67. We have investigated the dependence of the rate of dissolution on variations in the Ca/P ratio. Far from equilibrium a constant rate is obtained for constant values of the product $[Ca]^\alpha[P]$, with $\alpha \approx 3$; as equilibrium is approached α approaches 1.67. For further details see ref. (1).

We have not found any large change in the rate due to increased concentrations of inert electrolytes, such as potassium nitrate. We have avoided the presence of chloride ions in our experiments because these ions may enter the crystal lattice. Increasing the concentration of inert electrolyte will reduce the magnitude of the zeta potential, but is not expected to change the charge on the crystal surface.

[1] J. Christoffersen and M. R. Christoffersen, *J. Cryst. Growth*, 1979, **47**, 671

Dr. W. A. House (*Freshwater Biological Association, Dorset*) said: I also turn to Prof. Christoffersen.

(1) Could you explain in a little more detail how you obtained an apparent diffusion coefficient, $D_{app} \approx 10^{-9}$ cm^2 s^{-1}, from your dissolution data? Does your calculation involve an implicit assumption about the effective diffusion layer boundary thickness for the particles.

(2) Do you know of any other evidence that is available concerning the proposed change in $\Delta\phi$ with pH? The microelectrophoresis work of Foxall et al.[1] for a precipitated $Ca_3(PO_4)_2$ solid indicated that the mobility of these particles was essentially independent of pH in the region pH 8–11.

[1] T. Foxall, G. C. Peterson, H. M. Rendall and A. L. Smith, *J. Chem. Soc., Faraday Trans. 1*, 1979, **75**, 1034.

Prof. J. and Dr. M. R. Christoffersen (*University of Copenhagen, Denmark*) answered: For the estimation of an apparent diffusion coefficient we have assumed the crystals to be spherical and so small that they can be treated as stagnant in the medium. Assuming the amount of substance, K, which per unit time diffuses through a set of concentric spheres, with the crystal situated in the centre, to be independent of the radius of these spheres, r, leads to

$$4\pi r^2 D \left(\frac{dC}{dr}\right)_{r=r_1} = 4\pi r^2 D \left(\frac{dC}{dr}\right)_{r=r_2} \cdots = K \tag{1}$$

in which D is a diffusion coefficient and $r_1, r_2, \ldots$ are the radii of the spheres. Integrating this equation gives

$$K \int_{r=r_{cr}}^{\infty} \frac{dr}{r^2} = 4\pi D \int_{C_s}^{C} dC \tag{2}$$

from which we obtain

$$K = 4\pi D r (C_s - C). \tag{3}$$

Combining eqn (1) and (3) gives

$$\left(\frac{dC}{dr}\right)_{r=r_{cr}} = \frac{C_s - C}{r_{cr}}. \tag{4}$$

The overall rate per unit mass, J/m, can thus be expressed as

$$J/m = A_{sp}D_{app}(C_s - C)/r_{cr} \qquad (5)$$

with D_{app} being an apparent diffusion coefficient. For pH ca. 7 the solubility of HAP is ca. 10^{-5} mol dm^{-3}, the experimentally determined rate is $<10^{-6}$ mol s^{-1} g^{-1},[1] the specific surface area of our HAP crystals is ca. 30 m^2 g^{-1} and the linear dimension of the crystals, r_{cr}, is ca. 0.03 μm. Inserting these values in eqn (5) leads to $D_{app} \approx 10^{-9}$ cm^2 s^{-1}, which is of the order of 10^{-4} times the diffusion coefficient of small ions in aqueous solution. The above derivation of eqn (4) is not original.[2,3]

The electrophoretic mobility of HAP is pH-dependent and is zero around pH 7.[4-6]

[1] J. Christoffersen, M. R. Christoffersen and N. Kjaergaard, *J. Cryst. Growth*, 1978, **43**, 501.
[2] A. E. Nielsen, *J. Colloid Sci.*, 1955, **10**, 576.
[3] M. V. Smoluchowski, *Z. Phys. Chem.*, 1918, **92**, 129.
[4] F. Z. Saleeb and P. L. de Bruyn, *J. Electroanal. Chem.*, 1972, **37**, 99.
[5] P. Somasundaran and G. E. Agar, *Trans. Soc. Mining Eng.*, 1972, **252**, 348.
[6] S. K. Doss, *J. Dental Res.*, 1976, **55**, 1067.

Mr. V. K. Cheng (*Monash University, Australia*) said: Nucleation mechanisms, in general, are well known for their ineffectiveness close to equilibrium. In fact the B.C.F. theory for spiral growth was proposed to overcome such difficulties.[1] Does the polynuclear model show such a property and how does the dissolution rate of HAP compare with theory close to saturation?

The factor $C(1)$ in eqn (8) of Prof. Christoffersen's paper for the rate of surface nucleation is described, for dissolution, as the concentration of holes formed by the loss of a single growth unit, earlier indicated as being one ion.

Electroneutrality must apply unless the numbers of missing charges are small, and so also should the principles of equilibrium. Therefore $C(1)_{Ca^{2+}}^{10} C(1)_{PO_4^{3-}}^{6} C(1)_{OH^-}^{2}$ should be proportional to the corresponding ionic product, $[Ca^{2+}]^{10}[PO_4^{3-}]^6[OH^-]^2$, for the solution. The values of $C(1)$ for each ion should thus vary more or less in proportion to the lattice ions in solution and also with pH.

In the polynuclear mechanism for dissolution, what would be the effects of undersaturation and pH on $C(1)$?

The removal of the first dissolution unit during nucleation of holes requires a large activation energy, and consequently there is a slower dissolution rate. Crystal edges and apices have been proposed to be viable alternative sites for the initiation of dissolution.[2] What are the roles of crystal edges and apices and perhaps, spiral steps which may be present on the surface, in determining the dissolution rate of HAP?

The activation energy for solvent exchange in the vicinity of alkaline-earth ions has not been reported.[3] How can the proposal that the calculated activation energy for the detachment of a calcium ion during nucleation is three times larger than that for dehydration, which has been suggested to be equivalent to solvent exchange,[4] be established? Recent work by Nielsen[4,5] indicates that the latter corresponds to the growth activation energy. Solvent exchange in the vicinity of most alkaline-earth ions is known to be very fast.[3] Does it mean that the incorporation of growth units into kinks by dehydration becomes rate-determining and the activation energy of this step determines the growth activation energy? Furthermore, would it be correct to infer that the activation energy for dissolution for HAP is larger than that for crystal growth? If not, why not?

The asymmetry between interfacial controlled crystal growth and dissolution lies in the faster dissolution rate coefficient (for a given type of crystal).[6,7] How can my inference from the suggestion of Christoffersen and Christoffersen, if correct, be consistent with this well established experimental observation?

The overall dissolution is an electroneutral process with no net charge transfer. How can variations of electric potential difference between the crystal surface and the solution have a major effect on the rate of dissolution?

Previous workers[8,9] have often used $m^{2/3}$ (*i.e.* surface area) as the variable representing the activity of the interface. A number of recent studies,[7,10] however, have suggested that such a relationship may hold for only a given type of crystal, regarding their history of preparation and storage. This raises two questions.

(1) What is the reason for selecting a fixed value of m/m_0 at which to compare the rate of dissolution whilst the 'surface activity' of the different crystal samples used in the large number of runs was not explicitly considered?

(2) What is the variation of rate with concentration over the course of a single dissolution run with HAP?

[1] W. K. Burton, N. Cabrera and F. C. Frank, *Philos. Trans. R. Soc. London, Ser. A*, 1951, **243**, 299.
[2] N. Cabrera and V. Coleman, in *Art and Science of Crystal Growth*, ed. J. J. Gilman (Wiley, New York, 1963), p. 3.
[3] S. Petrucci, in *Ionic Interactions, II. Kinetics and Structure*, ed. S. Petrucci (Academic Press, New York, 1971).
[4] A. E. Nielsen, *Pure Appl. Chem.*, 1981, **53**, 2025.
[5] A. E. Nielsen, in *Industrial Crystallisation 81*, ed. S. J. Jancic and E. J. de Jong (North Holland, Amsterdam, 1982).
[6] J. W. Mullin, *Crystallisation* (Butterworths, London, 1972), p. 199.
[7] S. T. Liu, G. H. Nancollas and E. A. Gasiecki, *J. Cryst. Growth*, 1976, **33**, 11.
[8] C. W. Davies and A. L. Jones, *Trans. Faraday Soc.*, 1955, **51**, 812.
[9] D. M. S. Little and G. H. Nancollas, *Trans. Faraday Soc.*, 1970, **66**, 3103.
[10] M. R. Christoffersen, J. Christoffersen, M. P. C. Weijnen and G. M. van Rosmalen, *J. Cryst. Growth*, 1982, **58**, 585.

Prof. J. Christoffersen and Dr. M. R. Christoffersen (*University of Copenhagen, Denmark*) replied: We agree with you that spiral growth and dissolution can be expected to be faster than nucleation-controlled rates close to saturation. In our experiments the rates become too low to be measured using our present technique when $C/C_s > 0.7$.

The factor $C(1)$ in eqn (8) is not simple to determine accurately. We have estimated $C(1)$ in the following way. With $C(1)$ being the density (mole fraction) of surface sites from which an ion is missing, but where none of the lateral neighbours is missing, and $C(0)$ being the corresponding density of sites where the ion is not missing, we may estimate

$$C(1)/C(0) \approx \exp(-4a^2\sigma/kT) \approx 0.02$$

using $a \approx 0.3$ nm, $\sigma \approx 0.045$ J m^{-2} and $kT = 4 \times 10^{-21}$ J. Using $C(0) \approx 1$ we have $C(1) \approx 0.02$. For a polynuclear mechanism the linear rate of growth is proportional to $C(1)^{1/3}$, ≈ 0.25. This factor is not very important. As σ has been found to be nearly independent of pH, we do not expect $C(1)/C(0)$ to vary much with pH or with C/C_s.

The rate of dissolution of HAP can be explained by quite a normal polynuclear mechanism. Edges and apices do not seem to play a special role. As the rate is controlled by a nucleation process, the density of kink sites on the crystal surface is also controlled by the nucleation process. Steps associated with spirals are thus unimportant.

Mr. Cheng's statement about ref. (3) in his question is quite correct, but one may calculate the activation energies from dehydration frequencies using Eyring's formula. The only answers we can suggest to the question why the calculated activation energy for dissolution of HAP is about three times the activation energy for removing one water molecule from a dehydrated calcium ion is either that more calcium–oxygen bonds in the crystal surface have to be broken in order to remove a calcium ion, or that the calcium–oxygen bonds in the crystal surface are stronger than the calcium–oxygen bond in the aqueous solution.

We agree that the activation energy for growth of crystals of simple salts containing alkaline-earth cations according to Nielson's work is less than the activation energy for dissolution of HAP. This may be due to the frequency of detachment being lower than the frequency of complete incorporation in the crystal surface, once a cation like Ca^{2+} is partly dehydrated and has formed at least one bond to an anion in the crystal surface. The rate of growth of HAP is not well known to us, neither are accurate rate constants for growth and dissolution of other crystals with a complex structure like apatites. For crystals of simple electrolytes the rate of growth is normally less than the rate of dissolution, both rates being determined for the same value of the affinity.

The question concerning the effect on the rate of the electrical potential difference between the crystal and the solution touches a very important problem. If we do not distinguish between cations and anions, imagining crystals to be made of identical charge-less building units, this effect cannot be understood. On the other hand, if the density of hydrogen phosphate groups in the crystal surface increases, bonds to Ca^{2+} are weakened and these ions may leave the crystal surface faster than if no hydrogen phosphates were present in the crystal surface. In the model used for HAP dissolution one may think of hydrogen ions as catalysing the rate of dissolution.

The last few questions about the difference between $(m/m_0)^{2/3}$ and the expression $F(m/m_0)$ in eqn (5) of our paper can be answered quite simply. If the crystals used in an experiment are not identical, the surface area can in general not be represented accurately by $(m/m_0)^{2/3}$, but the area can be expressed as an unknown function of (m/m_0) if the linear rate can be expressed by an equation of the type given in eqn (4) of our paper. As far as we know, the rate of growth or dissolution of crystals can only be separated into a concentration term and an area term, if the above eqn (4) can be applied. In order to obtain an accurate expression for the influence of solution composition, we have determined rates, J/m_0, for the same values of pH and m/m_0, but for different values of the dissolution affinity. The rate varies by a factor of ca. 10 from the start to the point where an experiment is terminated, see for example fig. 1 in ref. (6) of our paper.

Dr. R. M. Sellers (*C.E.G.B., Berkeley*) said: The dissolution of ionic crystals generally takes place with a rate law given by

$$R_d = kA(C_{eq} - C_t)^n$$

where A is the surface area, C_{eq} is the salt solubility and C_t is the concentration of salt in solution at time t. n is a constant which apparently takes integral values. Table 1 summarises some of the measured values. Linge[1] comments that $n = 2$ for 1:1 electrolytes and $n = 3$ for 2:1 electrolytes, but the results of Simon[2] seem to run counter to this. There seems to be some underlying order here, but it is not clear what these results mean at a fundamental level. Perhaps Prof. Christoffersen, Dr. Coller or Mr. Cheng would care to comment.

Table 1. Experimentally determined values of n in eqn (1)[a]

n	crystal dissolving
1	NaCl, KCl
2	$PbSO_4$, $SrSO_4$, $CaSO_4$, $BaSO_4$, TlBr, $CaCO_3$
3	Ag_2CrO_4, MgF_2, $Ba(IO_3)_2$.

[a] Data from Linge[1] and Simon.[2]

I should also like to hear their comments on whether in the systems they have studied there is any evidence for migration on the surface (say from kink sites to adatoms) prior to dissolution. There is some evidence for this in the work of Jones et al.[3] on the dissolution of MgO crystals in HCl, but little other information seems to be available.

[1] H. G. Linge, *Adv. Colloid Interface Sci.*, 1981, **14**, 239.
[2] B. Simon, *J. Cryst. Growth*, 1981, **52**, 789.
[3] C. F. Jones, R. L. Segall, R. St. C. Smart and P. S. Tucker, *Proc. R. Soc. London, Ser. A*, 1981, **374A**, 141.

Prof. J. Christoffersen (*University of Copenhagen, Denmark*) replied: As far as I know, there exists no sound theory for crystal growth or dissolution relating the power n to the stoichiometry of the crystals. In general, a simple diffusion-controlled rate will lead to $n = 1$, whereas spiral growth (and dissolution) will show $n \leq 2$. For nucleation-controlled rates the power of n increases as saturation is approached.[1,2]

In regard to Dr. Sellers' question about surface migration, the rate of dissolution of HAP can be accounted for without taking such migration into account. This indicates that we have equilibrium between the solution and adatoms in an adsorbed layer.

[1] A. E. Nielsen, *J. Cryst. Growth*, 1984, in press.
[2] A. E. Nielsen and J. M. Toft, *J. Cryst. Growth*, 1984, in press.

Mr. V. K. Cheng (*Monash University, Australia*) said: Dr. Sellers' first question is of particular relevance to our understanding of the growth and dissolution of ionic crystals at the fundamental level.[1,2] The rate law given is an empirical one. Its concentration factor is expressed as the distance from saturation raised to the nth power (the order) and is distinct from those commonly encountered in homogeneous kinetics or the 'reaction controlled' interfacial kinetics reported by Dr. Sellers and others in this meeting. The parameters in the rate law, among which n is of particular interest, are determined from the fitting of experimental data, and our fundamental objective is to establish the meaning of n in terms of mechanisms.

We highlighted the disagreement between the BCF theory involving a one-component solid/vapour interface and the second order ($n = 2$) dependence of the rate on the distance from equilibrium observed for the growth and dissolution of bivalent metal sulphates. A number of modified BCF theories have been proposed in the past[3,4] which appeared to be capable of providing agreement between theory and experiment.

Linge's view[2] is an alternative mechanism which considers the transfer of ions across the 'double layer' region of the interface as equivalent to that of electron transfer. The electroneutrality and/or stoichiometric composition of ions in this region and the double-layer potential are necessarily involved in the description of growth and dissolution kinetics. This idea was initiated by Davies and Jones[5] and we have also raised its lack of rational physical basis in our paper. In particular, it cannot account for the asymmetry between growth and dissolution. Basic chemistry textbooks[6,7] remind us of the general absence of relationship between the stoichiometry and the empirical rate order of a reaction. I have raised the issue of preservation of electroneutrality at the interface in discussion with Prof. Christoffersen. Under such conditions I am doubtful of any involvement of the double-layer potential in determining the growth and dissolution of ionic crystals.

We must note the apparent regularity of the data given in the table and 'the usefulness of the Davies–Jones *model*'[8] (*i.e.* the rate equation with $n = 2$). After all, the equation is empirical and its usefulness is self-assured. The agreement between stoichiometry and order for binary salts is perhaps a coincidence and can be accounted for in terms of more adequately formulated theories based on step movement. In my opinion, the third-order law reported for the given group of 2:1 crystals and its relation to stoichiometry is disputable. The dissolution of Ag_2CrO_4 is largely controlled by volume diffusion.[9] The third-order kinetics was established from the data measured at C/C_s over 96%. Inspection of the published 'interfacial-controlled' data suggests to me that they can be fitted over a range of n. The published data for $Ba(IO_3)_2$[10] are not conclusive in support of the third-order law and can also be fitted with a second-order law.[11] Perhaps observed orders higher than 2 can be accounted for in terms of the process reported by Christoffersen in this Discussion which gives rise to a very complicated theoretical rate expression. However, nucleation is not a plausible process near equilibrium except with the aid of lattice imperfections[12,13] during dissolution. The second order for 2:1 ionic crystals was established a long time ago for K_2SO_4[14] and very recently for SrF_2.[15] Despite the choice of a Nernst-volume diffusion model unfamiliar to me and the mix up of enthalpy and activation enthalpy in Simon's paper,[16] the first-order law together with the small (activation) enthalpy found for the dissolution of KCl and NaCl was accounted for in terms of volume diffusion.

[1] G. H. Nancollas, *Adv. Colloid Interface Sci.*, 1979, **10**, 215.
[2] H. G. Linge, *Adv. Colloid Interface Sci.*, 1981, **14**, 239.
[3] R. Reich and M. Kahlweit, *Ber. Bunsenges. Phys. Chem.*, 1968, **72**, 66; 75.
[4] A. E. Neilsen, *Pure Appl. Chem.*, 1981, **53**, 2025.
[5] C. W. Davies and A. L. Jones, *Trans. Faraday Soc.*, 1955, **51**, 812.
[6] G. I. Brown, *Introduction to Physical Chemistry* (Longmans, London, 1972), p. 336.
[7] B. H. Mahan, *University Chemistry* (Addison-Wesley, 2nd edn, 1969), p. 356.
[8] W. A. House, *J. Chem. Soc., Faraday Trans. 1*, 1981, **77**, 341.
[9] A. L. Jones, H. G. Linge and I. R. Wilson, *J. Cryst. Growth*, 1974, **26**, 37; 1975, **28**, 254.
[10] A. L. Jones, G. A. Madigan and I. R. Wilson, *J. Cryst. Growth*, 1973, **20**, 93, 99.
[11] G. A. Madigan, Ph.D. Thesis (Monash University, Melbourne, 1969).
[12] B. van der Hoek, J. P. van der Eerden and P. Bennema, *J. Cryst. Growth*, 1982, **56**, 621.
[13] G. Z. Liu, J. P. van der Eerden and P. Bennema, *J. Cryst. Growth*, 1982, **58**, 152.
[14] R. Marc, *Z. Phys. Chem.*, 1908, **61**, 385; 1909, **67**, 470.
[15] R. A. Bochner, A. Abdeul-Raman and G. H. Nancollas, *J. Chem. Soc., Faraday Trans. 1*, 1984, **80**, 217.
[16] B. Simon, *J. Cryst. Growth*, 1981, **52**, 789.

Dr. R. M. Sellers (*C.E.G.B., Berkeley*) said: In much of our work on the dissolution of mixed oxides we have found that one or more components (often iron) dissolves slightly more rapidly than the others. There must therefore be some enrichment of the surface in the other components, and we have demonstrated this with crystals of magnetite or franklinite using surface-sensitive techniques such as X-ray photoelectron spectroscopy[1] or chemically.[2] In general these effects do not cause a dramatic change in kinetics, and the shrinking-core model describes the data to upwards of 75% dissolution.[3] There are exceptions, for instance in the oxidative dissolution of nickel chromium ferrites, where an outer barrier of increasing thickness is formed. Here the Crank, Ginstling and Brounshtein equation:

$$1 - \tfrac{2}{3}\alpha - (1-\alpha)^{2/3} = 1 - kt$$

holds.[4] Does either Prof. Christoffersen or Mr. Cheng have any evidence that the surfaces of his crystals become enriched in one or other of the components during dissolution, and if so, how does this evolve with time?

[1] G. C. Allen, R. M. Sellers and P. Tucker, *Philos. Mag.*, 1983, **48**, L5.
[2] G. V. Buxton, T. Rhodes and R. M. Sellers, *J. Chem. Soc., Faraday Trans. 1*, 1983, **79**, 2961.
[3] M. G. Segal and R. M. Sellers, *J. Chem. Soc., Faraday Trans. 1*, 1982, **78**, 1149.
[4] F. Habashi, *Extractive Metallurgy* (Gordon and Breach, New York, 1969), vol. 1.

Prof. J. and Dr. M. R. Christoffersen (*University of Copenhagen, Denmark*) said: In the case of the dissolution of calcium hydroxyapatite (HAP) we have avoided such effects by studying the rate of dissolution at constant pH and constant calcium to phosphate ratio. We have found that the rate constant increases with increasing concentration of hydrogen ions in solution. In our experiments we have equilibrium between hydrogen ions in solution and on the crystal surface. If this was not the case it would be most difficult to analyse the data. Dissolution of HAP with some of the hydroxy groups substituted by fluoride ions in solutions unsaturated with respect to HAP and FAP (calcium fluorapatite) will cause a surface enrichment with respect to fluoride ions.[1]

[1] M. R. Christoffersen, J. Christoffersen and J. Arends, *J. Cryst. Growth*, 1984, **67**, 107.

Mr. V. K. Cheng (*Monash University, Australia*) said: Continuing my response to Dr. Sellers, I wish to remark that our paper did not consider the problem of surface diffusion. However, the involvement of surface diffusion in crystal growth has been inferred from the fitting of growth data (first-order growth-rate law) for $NaClO_3$ and potash alum with the B.C.F.[1,2] Alternatively it is possible to decide the involvement of surface diffusion in crystal growth and dissolution from the ratio of direct detachment to surface diffusion probabilities of surface units.

For example, in the SOS model for the solid/fluid interface,[3] the evaporation probability of a site with n lateral neighbours is given by

$$k_n^- = \nu \exp\left(\frac{\gamma}{2}(2-n)\right) \qquad (1)$$

where $\gamma = 4\phi/kT$.

From eqn (12), (14) and (15) in ref. (3) of this comment, the probability of surface migration from a site with n lateral neighbours to a neighbouring site with m neighbours is given by

$$k_{nm} = \frac{k_n^- k_m^-}{k_0^-} \exp\left(\frac{m\gamma}{2}\right)\left(\frac{x_s}{a}\right)^2$$

$$= \frac{k_n^- k_m^-}{k_0^-} \exp\left(\frac{m\gamma}{2}\right)\frac{k_{00}}{k_0^-}. \qquad (2)$$

k_{00} can be calculated from eqn (12) of ref. (3) and the B.C.F. theory.[4]

$$k_{00} = \nu \exp(-U_s/kT).$$

The migration along sites which does not result in any change in coordination number gives $U_s = 0$. Thus $k_{00} = \nu$.

Substitution of eqn (1) into (2) with the appropriate choice of n leads to

$$k_{n0} = \nu \exp(-n\gamma/2).$$

Therefore

$$\frac{k_n^-}{k_{n0}} = \exp(\gamma).$$

A decrease in temperature or an increase in γ would make the ratio larger and surface diffusion for less soluble crystals less favourable. However, simulations involving surface diffusion carried out so far[3,5] consider the surface migration frequency as an independent input variable. Such choice would have been effective for the examination of the role of surface diffusion on crystal growth at a given ϕ/kT.

The estimate given above is very useful for describing the growth and dissolution of (non-polar) ionic crystals far from edges and apices. It has been shown[7] that because of the partial cancellation of contributions from lattice ions of opposite charges, adsorbed ions experience short-ranged attraction from the rest of the lattice. The movement of solvent molecules during an elementary event of incorporation or detachment is more complicated to describe, but the assumption given in ref. (3) that the solvent can be treated as a continuum with defined average interactions is probably valid because of the short lifetime of the solvated water. The short-ranged electrostatic repulsion between ad-ions and lattice ions of like charge, however, would probably limit the number of available surface diffusion paths. The ad-ions move like a bishop on a chess board!

[1] P. Bennema, *J. Cryst. Growth*, 1967, **1**, 287.
[2] P. Bennema, *J. Cryst. Growth*, 1969, **5**, 29.
[3] G. H. Gilmer and P. Bennema, *J. Appl. Phys.*, 1972, **13**, 1347.
[4] W. K. Burton, N. Cabrera and F. C. Frank, *Philos. Trans. R. Soc. London, Part A*, 1951, **243**, 299.
[5] J. D. Weeks and G. H. Gilmer, *Adv. Chem. Phys.*, 1979, **40**, 157.
[6] J. E. Lennard-Jones and B. M. Dent, *Trans. Faraday Soc.*, 1928, **24**, 92.

Prof. J. Christoffersen (*University of Copenhagen, Denmark*) said: The rates of dissolution of different preparations of $BaSO_4$ are described as proportional to $(C_s - C)^2$, but in order to obtain this power law a different solubility, C_s, is assigned to each preparation. A plot of log rate against $\log(C_s - C)$ of Mr. Cheng's data given by the points marked ● in his fig. 1, taking C_s to be the accepted equilibrium value of the solubility of $BaSO_4$, 10.4 μmol dm^{-3}, leads to the rate being proportional to $(C_s - C)^{1.6}$. Could this indicate that some diffusion process has to be taken into account?

That the adsorption of Sr^{2+} and Ca^{2+} onto $BaSO_4$ crystals causes an increase in the rate of dissolution is interesting. Has this also been found for the rate of dissolution of microcrystals of $BaSO_4$ in aqueous suspension? Has Mr. Cheng an explanation for the order of effectiveness of $CaCl_2$ and $SrCl_2$ given in table 3: 10 μmol dm^{-3} $SrCl_2$ < 10 μmol dm^{-3} $CaCl_2$ < 54 μmol dm^{-3} $CaCl_2$ < 20 μmol dm^{-3} $SrCl_2$?

Dr. B. A. W. Coller and Mr. V. K. Cheng (*Monash University, Australia*) said: The exponent n in crystal growth and dissolution kinetics can be obtained by means of a number of apparently equivalent presentations of experimental data.[1,2] The solubility of a crystal is involved in all presentations. We have shown[3] that choice of a smaller solubility will lead to a decrease in the value of the exponent in the log–log plot. The 'accepted' equilibrium solubility used by Prof. Christoffersen is smaller than the kinetic solubility obtained in our work. Therefore the exponent of 1.6 is not surprising.

By using an exponent of 2 we obtain a kinetic solubility; a parameter that we interpret in terms of energies of dissolution affected by regions of stress and discontinuity at the surfaces of the crystals. Prof. Christoffersen's use of the accepted equilibrium solubility would lead to exponents depending on the nature of the crystal which could be interpreted in terms of energies of initiation of dissolution, possibly by formation of dissolution nuclei, in stressed or unstressed regions. The ambiguity highlights the lack of a single agreed theoretical model and empirical presentation of rate data for the dissolution of ionic crystals found in the literature at present.

We are confident that the dissolution of barium sulphate is controlled by transfer at the interface because the rates found with the single crystal did not vary with rotation speed and with microcrystalline samples did not depend on stirring speed.

We have not yet studied the effects of additives on the dissolution of microcrystalline samples, nor do we yet have an explanation for the large effect of $SrCl_2$ at 20 μmol dm^{-3} with the single crystal.

[1] G. H. Nancollas, *J. Cryst. Growth*, 1968, **34**, 335.
[2] J. L. Powell, B. A. W. Coller and A. L. Jones, *J. Cryst. Growth*, 1978, **43**, 185.
[3] V. K. Cheng, *Ph.D. Thesis* (Monash University, Australia), to be submitted.

Dr. W. A. House (*Freshwater Biological Association, Dorset*) asked Dr. Coller if there was any systematic trend in the activation energies E_a and enthalpies of dissolution shown in table 1 with ageing and use of the crystals? Could one foresee a situation when after prolonged ageing and allowing the surface to reform that this variability was reduced?

Dr. B. A. W. Coller (*Monash University, Australia*) replied: Activation energies and enthalpies of dissolution, shown in fig. 2 of our paper refer to seven different preparations of barium sulphate crystals mounted on polythene plates. For a given sample, successive values of E_a and $\Delta \bar{H}$ obtained in dissolution trials at intervals of one to three days did not show a systematic trend along the correlation line. This could be the effect of variation in the periods of dissolution ('washing') and storage in contact with saturated solution ('ageing'). However, it is not clear that the slow processes of growth and dissolution during extended ageing will remove the effects of rapid dissolution during washing, particularly in regions of localized stress associated with dislocations, grain boundaries and crystal edges. There is need for further study of the effects of wash-and-age cycles on crystal morphology.

Dr. J. J. M. Binsma (*IRI, Delft, The Netherlands*) said: I would like to ask Dr. Coller whether the barium sulphate crystals of type II and III are still facetted? For dendritic or aggregated crystals it is to be expected that they are bounded by faces which do not have a specific crystallographic orientation and which are rough on an atomic scale. Such faces will have a very low activation energy for dissolution, because almost all atoms (ions) are found in kink or less strongly bound positions and no 'dissolution nuclei' need to be formed.

Dr. B. A. W. Coller (*Monash University, Australia*) replied: Type II crystals of barium sulphate appear to retain many small facets of low Miller index but bounded by edges and multiple steps (see, for example, electron micrographs published by Liu et al.[1]). In qualitative terms, the surfaces of these crystals dissolve more rapidly than those of type I and appear to have lower energies of dissolution, but our experiments indicate that their activation energies for dissolution tend to be higher.

We suggest that the rhombohedral type I crystals are produced by growth around screw dislocations which are centres of stress and serve as ready-made dissolution nuclei. Such crystals appear to dissolve from the central areas of their faces. If dendritic type II crystals grown from more concentrated solutions are relatively free of dislocations, then dissolution of layers on facets will begin at apices and along edges.

Kink units on faces, kink units on edges and kink units at apices may have similar energies of total detachment. However, displacement of a crystal unit from an apex kink or edge kink onto a terrace will involve passage through a transition state in which the unit is more exposed than when an intrafacial kink unit is moved out onto an adjacent terrace.

For molecular crystals, where non-neighbour interactions may be neglected, the two processes can be expected to have similar activation energies. However, with ionic crystal units, next-to-nearest neighbour electrostatic interactions that can affect the potential-energy surface and make a difference between the activation energies. Simple models of movement of ion pairs lead us to expect a higher activation energy for displacement from edge kink to terrace than for displacement from intrafacial kink to terrace. There is need for more detailed calculations.

Type III crystals of barium sulphate have not yet been subjected to scanning electron microscopy to investigate the surface topography, although transmission and diffraction studies by Takiyama[2] indicate a well developed crystal structure even in very small crystals.

[1] S. T. Liu, H. Nancollas and E. A. Gasiecki, *J. Cryst. Growth*, 1976, **33**, 11.
[2] K. Takiyama, *Bull. Chem. Soc. Jpn*, 1959, **32**, 68.

Mr. V. K. Cheng (*Monash University, Australia*) added: The roughening of surfaces depends on the temperature and interaction (ϕ/kT).[1] The roughening transition occurs at $\phi/kT \approx 1.75$. I would expect sparingly soluble salts to have high ϕ/kT, i.e. they have low-temperature surfaces. The absence of macroscopic facets is a characteristic of a roughened crystal. Arrangement of ions at this type of crystal surface will create local polarity, as reflected by the non-zero surface unit-cell dipole moment. The electrostatic energy of such a crystal is known to be enormous and makes formation of roughened ionic crystal very unfavourable.[2] The formation of polar habits, such as the (111) face of KCl, requires the stabilisation provided by added impurities. The surfaces of ionic crystals are therefore likely to be flat. Furthermore the growth and dissolution kinetics of ionic crystals in general do not conform to those of a roughened surface.[3]

[1] H. J. Leamy, G. H. Gilmer and K. A. Jackson, in *Surface Physics of Materials*, ed. C. Blakely (Academic Press, New York, 1975), p. 121.
[2] E. R. Smith, *Proc. R. Soc. London, Ser. A*, 1982, **381**, 241.
[3] J. D. Weeks, G. H. Gilmer and K. A. Jackson, *J. Chem. Phys.*, 1976, **65**, 712.

Dr. B. A. W. Coller (*Monash University, Australia*) said:

(1) Binsma and Kolar's compartment model allows for exchange between solution, surface (terrace), steps and kinks. The numbers of sites available for adsorption in each of these compartments could be very different from one another. Have the capacities of the compartments been estimated from the exchange data?

(2) According to the hard-sphere model of an ionic crystal, the transition states for diffusion of ions across the surface occur at points of zero electrostatic potential, *i.e.* the energy of the transition state in a diffusion step would be equal to the energy of the completely detached ion. However, the presence of solvent at the interface could make a considerable difference. Is Dr. Binsma able to assess the ease of diffusion on the crystal surface, by comparison with the ease of detachment?

(3) Dr. Binsma drew attention to the seemingly irreversible incorporation of some labelled ions into the crystal and supposed them to be contained in layers that had been overlain by several others. It seems to us that dissolution can be as different from the simple reversal of growth as the unravelling of a knitted garment can be different from reversing the sequence of the original knitting. With exchange under equilibrium conditions, the changes of surface configuration should be at their most nearly reversible, but should still be subject to statistical fluctuations in which ions are taken up in surface layer sites and surrounded by other surface units laid down in rows to form patches.

Has Dr. Binsma considered the statistics of incorporation and release of labelled ions, in terms of rows rather than layers?

Dr. J. J. M. Binsma (*IRI, Delft, The Netherlands*) said: The answers to Dr. Coller's questions can be summarized as follows.

(1) The activity against time curves have been fitted to a three-compartment model made up of a central compartment (the solution) and two peripheral ones which are not connected to each other. Within this model it appears that the compartment sizes do not differ very much from each other and are each equal to about half the total amount of exchangeable Ca^{2+} ions.

(2) As an upper limit for the activation energy of surface diffusion one can take the activation energy for detachment of an adatom from the underlying crystal lattice.[1] The jumping frequency for surface diffusion, which can be calculated from the activation energy *via* the Eyring equation, will then be equal to or lower than the real value. For the calculation of the surface-diffusion jumping frequency along this line the required data concerning the dissolution of CaF_2 are lacking, however.

(3) I agree that the kinetics of dissolution may be entirely different from the kinetics of growth. With regard to the seemingly irreversible incorporation of labelled ions it should be kept in mind that the number of relatively weakly bonded ions (*e.g.* in kink or step-like positions) will be very small under equilibrium conditions.[1] Most ions will be more strongly bonded, making the crystal surface very flat. These more strongly bonded ions nevertheless take part in the exchange since the experiments showed that *ca.* one lattice layer of CaF_2 participates in the exchange. The exchange of these ions may proceed *via* one of two alternative pathways, namely either by direct transitions out of the surface into the solution or *via* the fluctuation of kink-like positions. If, starting at a kink position, rows of ions go into the solution, the kink position travels along the surface giving all ions the opportunity to detach from a relatively weakly bonded state. With regard to the statistics it remains in the latter case that the exchange of ions should be reversible.

[1] G. H. Gilmer and P. Bennema, *J. Appl. Phys.*, 1972, **43**, 1347.

Prof. J. Christoffersen (*University of Copenhagen, Denmark*) said: Could Dr. Binsma explain why the different rate constants obtained cannot be explained by two of the three compartments involved being the adsorbed layer and the more solid surface of the crystals?

Dr. J. J. M. Binsma (*IRI, Delft, The Netherlands*) said: Prof. Christoffersen suggested that two of the three compartments involved in the exchange processes may correspond to the adsorption layer and the more solid surface of the crystals. If this were so, the measured rate constants should be of the same order of magnitude as the frequency of adsorption of Ca^{2+} ions from the solution, f_{ad}, and the frequency of integration, f_{in}. The frequency for adsorption is related to the flux of ions F by

$$F = Q_a f_{ad} = V_a c_s f_{ad} = A a c_s f_{ad} \tag{1}$$

where Q_a is the amount of Ca^{2+} ions in the solution which are available for adsorption, V_a the volume of the solution from which adsorption in one 'jump' is possible, c_s the solubility of CaF_2, A the surface area of the crystal in contact with the solution and a the so-called molecular diameter (0.3 nm according to Nielsen[1]). It is thus supposed that adsorption in one jump is possible from a layer of thickness a adjacent to the crystal surface. For a flux of ions of 1×10^{-10} mol m^{-2} s^{-1} (see our paper), the adsorption frequency should be 1.7 s^{-1}, which is much larger than the rate constants found in our experiments. The integration frequency, f_{in}, for incorporation into the more solid surface (kink sites) appears to be related to the cation dehydration frequency f_{dh} by $f_{in} = 10^{-3} f_{dh}$, as checked for CaF_2 by Nielsen.[1] For CaF_2 ($f_{dh} = 1.6 \times 10^8$ s^{-1}) f_{in} should be 1.6×10^5 s^{-1}, which is again much higher than the experimentally observed rate constants.

[1] A. E. Nielsen, *J. Cryst. Growth*, to be published.

Dr. A. R. Flambard (*Free University of Berlin, West Germany*) said: I address my remarks to Dr Binsma. My questions concern your re-exchange studies. In accordance with similar observations by Möller and Sastri you ascribe a loss in exchangability of some 30% of the total amount of the $^{45}Ca^{2+}$ ions as being due to diffusion into the bulk of the crystal. Do you have any evidence for this from other types of measurements?

Have you considered any statistical analyses of the partition of $^{45}Ca^{2+}$ between the solid (surface) and solution phases?

Rather than diffusion into the bulk of the crystal, is it not also possible that your observation of at least two different types of bonding for Ca^{2+} within the solid phase could be explainable if one considers an exchange of Ca^{2+} between the first and second, or possibly third, layers of the crystal's surface? (Then in simple terms one would expect the properties of these layers at, or close to the surface to differ from those in the bulk.)

Finally, during your re-exchange experiments, after the system had reached equilibrium again, did you try replacing the saturated CaF_2 solution with fresh solution and if so, were there any changes in observed activities?

Dr. J. J. M. Binsma (*IRI, Delft, The Netherlands*) said: The following comments can be made regarding Dr. Flambard's questions.

(1) The diffusion of Ca^{2+} from the surface into the bulk of the crystal has not yet been measured independently.

(2) The amounts of $^{45}Ca^{2+}$ in the solid (surface) and in the solution are correlated with the size of the respective compartments. When the solid has gathered its final (equilibrium) amount of activity, the relation between the amounts of $^{45}Ca^{2+}$ and the total amounts of Ca^{2+} in the solid and in the solution is given by eqn (2) of our paper.

(3) It is certainly possible that exchange occurs between the solution and a number of layers of the crystal. More refined experiments are needed in order to be able to draw conclusions in this respect.

(4) The saturated aqueous solution of CaF_2 has not been renewed in the re-exchange experiment. This will not have had any influence on the results because the specific activity of the solution after the system had reached equilibrium was ca. 1000 times lower than that of the solution used in the exchange experiments, whereas the specific activity of the solid had decreased by only 70%.

Dr. W. A. House (*Freshwater Biological Association, Dorset*) asked: Does Dr. Binsma think the ^{45}Ca isotopic exchange method is at all suitable for measuring the specific surface areas of calcium carbonate minerals? The range of f values shown in table 2 of the paper indicates that for {100} face of CaF_2 only part of the surface takes part in the exchange and models based on the entire lattice taking part in the exchange would lead to erroneous estimates of the extent of surface. I raise this point because the ^{45}Ca exchange method[1] is used to estimate specific surface areas.

[1] C. G. Inks and R. B. Hahn, *Anal. Chem.*, 1967, **39**, 625.

Dr. J. J. M. Binsma (*IRI, Delft, The Netherlands*) said: In reply to Dr. House it should be stressed that the so-called isotopic exchange method for determining specific surface areas as used for instance by Inks and Hahn[1] can only be applied to steady-state systems and if a correlation exists between the final value of the activity of the solid phase and its geometrical surface area. The method is carried out generally for populations of small particles of micrometre size or smaller having rather rough surfaces. In that case a relatively fast and complete exchange with the first surface layer can be expected. It should not be excluded, however, that deeper layers are also taking part in the exchange, especially if the surface is very rough. Whether the isotopic-exchange method can be used should therefore be checked for each compound and even for each type of sample because of the influence of the surface properties.

[1] C. G. Inks and R. B. Hahn, *Anal. Chem.*, 1967, **39**, 625.

Mr. V. K. Cheng (*Monash University, Australia*) said: I address my question to Drs. Sellers and Williams.

What is the variable k in fig. 1 of your paper? Its unit apparently involves an area factor. How did you characterise the surface area of the crystals?

I am not able to find an explicit description on the rate measurement. Presumably, in each dissolution run, a known amount of solid was dissolved over a fixed duration and is determined thereafter. Thus each data point in fig. 1-6 is obtained with different initial solid samples. Over a period of many hours of dissolution, as indicated in fig. 2, the large (ca. 20%) amount of solid dissolved would change the surface activity significantly through the decrease in both surface area and mass of crystals.

What is the variation of the initial amount of solid in each run and what would be the relative contributions of the variation of initial crystal mass and crystal imperfections to the rate measured? How was aging characterised?

How many Fe^{3+} ions are there in the solids? Can the absence of autocatalytic paths be related to the Fe^{3+} content in the solid?

At 140 °C pH values of the acids different from that at 25 °C would be expected. Has the variation of pH due to temperature been considered and if so what would be the effect of the variation of pH of the individual acids in the solution at different temperatures on the dissolution rate?

The rate law in general expresses the rate in terms of the concentration of reactants or products raised to the nth power, where n is the rate order. In crystal growth and dissolution, the concentration factor is expressed as the supersaturation or undersaturation and the order for the dissolution of barium sulphate and HAP, as presented in this Discussion, was found to be, respectively, 2 and 4–6. (The order varies over a wide range of values as you have commented earlier.) However, your rate equation [eqn (3)] does not express rate and the concentration variables in the form used by us and Prof. Christoffersen, nor is the index in your rate equation equal to 3. How can a third-order or cubic law be justifiably declared? Furthermore does the rate law, as derived from the 'shrinking-core model' which is not described in your paper, contain any mechanistic information such as the time evolution of the surface topography?

What is the binuclear complex mentioned in the dissolution involving oxalic acid? Do you suggest that a small molecule like the oxalate ion can become a bridging ligand to two neighbouring cation sites and, as a result, the dissolution depend on the electron transfer from the complex to Fe^{3+} ions at the surface?

What is the origin of the values for the variables k and K_4 used in eqn (8)?

The cause of reduction of the dissolution rate of crystals was related to the increasing Cr^{3+} content and the substitutional inertness of the Cr^{3+} ions. Is it true that the Cr^{3+} ions are more difficult to remove by complex formation, compared with Fe^{3+} and other divalent ions in the crystal? Have you proved this point by investigating the stability constants of the complexes between Cr^{3+} and the ligands?

How do you know that NTA is more adsorbable than oxalic acid and that NTA completely covers the oxide particles (or do you mean microcrystals)?

Dr. R. M. Sellers (*C.E.G.B., Berkeley*) said: The variable k in fig. 1 of our paper is a rate constant, and is related to the k_{obs} of eqn (3) by $k_{obs} = k/r_0\rho$, where r_0 is the initial particle radius and ρ the density. As described in the Experimental section the mean initial radii of the various oxides were determined using a Coulter counter. The oxides have low surface roughness, so the mean radius is a reasonable measure of the surface area. In fact the oxides had nearly the same sizes, and did not appear to differ significantly in surface roughness, so the corrections for differences in size were small.

The rate measurements were carried out as described in the Experimental section. *Ca.* 3 mg of oxide were used in each run. Small differences in this amount are taken into account in calculating k_{obs} from the shrinking-core model. It is important only that the other reagents are present in sufficient excess that their concentrations do not vary appreciably during a particular run. We assume that faults are distributed evenly throughout the particles. The ageing of the $Ni_{0.6}Cr_{0.6}Fe_{1.8}O_4$ oxide manifested itself through a reduction in dissolution rate between the initial and final phases of the work. We do not know for certain what the cause of this was, but changes in the nature of the fault structure of the oxide are probably implicated.

The surface area undoubtedly increases in the initial phases of the dissolution, but there appears to be no concomitant increase in rate. We believe the reason for this is that the number of reactive sites does not vary in parallel with the surface area, but rather decreases with time in a manner mathematically the same as the shrinking-core model.

The spinel oxides have a general composition AB_2O_4 where A is a divalent cation and B trivalent. The mixed oxides used in our work have a composition $Ni_{0.6}Cr_nFe_{2.4-n}O_4$. The divalent ions are therefore $Ni_{0.6}+Fe_{0.4}$, and the trivalent Cr_n and $Fe_{2.0-n}$. The absence of autocatalytic pathways is related to (a) failure to dissolve ferric ions reductively and (b) a rate constant for reaction (1) comparable with that for reaction (2):

$$Fe^{II}L_n + {>}Fe^{3+} \rightarrow Fe^{III}L_n + Fe^{2+} \qquad (1)$$

$$L_n + {>}Fe^{3+} \rightarrow Fe^{III}L_n. \qquad (2)$$

The pK_a values of both oxalic acid and NTA at 140 °C are undoubtedly different from the values at room temperature. No data appear to be available on these, however. In view of this we have been unable to calculate the pH values in the solutions at 140 °C, and so throughout this paper pH values are quoted at 25 °C. This introduces some bias to the data (e.g. in fig. 6), but will not alter the general conclusions, for instance, that with oxalic acid the dissolution rate increases as the pH decreases.

Eqn (3) in the paper is an integrated form of the rate equation. Its derivation is given, for instance, by Segal and Sellers[1] or Habashi[2] but starts with the assumption that the rate is proportional to the instantaneous surface area. The shrinking-core model is relatively crude, and does not take into account, for instance, the evolution of surface topography as the reaction proceeds. Nevertheless, we have found it useful in this and previous work on spinel dissolution.[1,3] It seems that the number of active surface sites (kinks?) is the most important factor, and as noted above, the variation of this with time parallels the change in surface area of a dissolving sphere. The description of eqn (3) as a 'cubic' rate law is misleading; the 'shrinking-core' equation is to be preferred.

Small molecules such as oxalic acid can indeed bind simultaneously at two neighbouring sites, as indicated by i.r. measurements.[4] Other species such as picolinic acid,[3] selenious acid, orthophosphoric acid and sulphuric acid[5-10] behave similarly on iron-rich oxides such as haematite or goethite. In the case of oxalic acid electron transfer can be envisaged as occurring from such a 'complex' as follows:

The values quoted for k and K_4 in eqn (8) were deduced from the slope and intercept of plots of k_{obs}^{-1} against $[C_2O_4^{2-}]_T^{-1}$.

Cr^{3+} ions in these mixed spinel oxides are difficult to dissolve by any means.[11] They can be dissolved by oxidative processes but these are inhibited by iron,[12] or with Cr_2O_3 by reductive means if the reductant is sufficiently powerful.[13] The stability constants for many Cr^{3+} complexes in aqueous solution have been measured[14] but the problem is not one of thermodynamics; rather it is the kinetics that are limiting. The well known substitution inertness of Cr^{3+} is unquestionably a key factor.

That NTA is more strongly adsorbed than oxalic acid is inferred from the results (compare fig. 3 and 5). It is also the expected result since NTA is tetradentate, whilst oxalic acid is bidentate. For NTA to be removed from the surface requires that four bonds are broken, a statistically improbable, and therefore slow, process. Other aminocarboxylate ligands, such as EDTA, are known for their ability to bind strongly to oxide surfaces.[15]

[1] M. G. Segal and R. M. Sellers, *J. Chem. Soc., Faraday Trans. 1*, 1982, **78**, 1149.
[2] F. Habashi, *Extractive Metallurgy* (Gordon and Breach, New York, 1969), vol. 1.
[3] D. Bradbury, M. G. Segal, R. M. Sellers, T. Swan and C. J. Wood, Electric Power Research Institute Report No. EPRI NP-3177 (1983).
[4] R. L. Parfitt, V. C. Farmer and J. D. Russell, *J. Soil Sci.*, 1977, **28**, 29.
[5] J. D. Russell, R. L. Parfitt, A. R. Fraser and V. C. Farmer, *Nature (London)*, 1974, **248**, 220.
[6] R. J. Atkinson, R. L. Parfitt and R. St. C. Smart, *J. Chem. Soc., Faraday Trans. 1*, 1974, **70**, 1472.
[7] R. L. Parfitt, R. J. Atkinson and R. St. C. Smart, *Soil. Sci. Soc. Am. Proc.*, 1975, **39**, 837.
[8] J. D. Russell, E. Paterson, A. R. Fraser and V. C. Farmer, *J. Chem. Soc., Faraday Trans. 1*, 1975, **71**, 1623.
[9] R. L. Parfitt, J. D. Russell and V. C. Farmer, *J. Chem. Soc., Faraday Trans. 1*, 1976, **72**, 1082.
[10] R. L. Parfitt and R. St. C. Smart, *J. Chem. Soc., Faraday Trans. 1*, 1977, **73**, 796.
[11] M. G. Segal and T. Swan, in *Water Chemistry of Nuclear Reactor Systems 3* (British Nuclear Energy Society, London, 1983), p. 187.
[12] W. J. Williams, unpublished data.
[13] S. Bennett, D. Bradbury, B. Daniel, R. M. Sellers, M. G. Segal and T. Swan, in *Water Chemistry of Nuclear Reactor Systems 3* (British Nuclear Energy Society, London, 1983), p. 361.
[14] (a) Chemical Society Special Publication No. 17 (The Chemical Society, London, 1964); (b) Chemical Society Special Publication No. 25 (The Chemical Society, London, 1971).
[15] E.g. H-C. Chang, T. W. Healy and E. Matijević, *J. Colloid Interface Sci.*, 1982, **92**, 469.

Prof. J. Christoffersen (*University of Copenhagen, Denmark*) said: Integrated rate laws are in general not very accurate in use. Could the authors, despite the complexity of their system, apply the differential rate law corresponding to their eqn (3), *i.e.*

$$-\frac{d(1-\alpha)}{dt} = 3k_{obs}(1-\alpha)^{2/3}$$

or is this not possible for their system?

The term $(1-\alpha)^{2/3}$ in the above equation indicates that the rate of dissolution is proportional to the surface area of the remaining solid; the rate constant k_{obs} includes a term representing the solution composition. The above equation can hardly be used to take into account that the surface of the solid may become richer in Ni during the dissolution process. Could one apply a rate expression of the type in our paper, eqn (5)? If so, do Drs. Sellers and Williams agree that the possible effect of surface enrichment with Ni can to a first approximation be included in the term $F(m/m_0)$?

Dr. R. M. Sellers (*C.E.G.B., Berkeley*) said: No doubt the differential form of the rate equation could be used, but it does not offer any particular advantages over the more simply applied integrated form. Similarly eqn (5) in Prof. Christoffersen's paper could also be used. His suggestion that surface enrichment in Ni could, to a first approximation, be included in the term $F(m/m_0)$ is a good one and deserves further investigation.

Dr. W. J. Williams (*C.E.G.B., Berkeley*) said: In addition to Dr. Sellers' comments on eqn (3), it is also worthwhile re-emphasising the possible stages involved in heterogeneous reactions, *viz*.: (i) diffusion of reactants to the surface, (ii) adsorption of reactants on the surface, (iii) reaction at the surface to yield adsorbed product, (iv) desorption of product from the surface and (v) diffusion of product away from the surface. Any one of these steps may be rate controlling. We have found in most of our systems the reaction rate is consistent with a control step under category (iii). Thus, provided the reactant concentrations are kept in excess and thus essentially constant, the rate will be directly proportional to the instantaneous surface area of the particle provided also that the concentration of reactive sites per unit area of surface remains constant as the particle dissolves; hence

$$-\frac{dR}{dt} = kA.$$

Eqn (3) follows directly from this expression if the loss of mass by dissolution is translated into the increase of product concentration in the dissolving media. The 1/3 order is independent of the shape of the particles; it is a natural corollary of solid volume being dissolved through a surface-controlled process.

Presumably in the systems studied by Dr. Coller, Mr. Cheng and Prof. Christoffersen, steps (iv) and (v) play an increasingly important role as implied by the incorporation of the $(C_s - C)^n$ term.

Mr. A. M. Creeth (*Imperial College, London*) said: This contribution links the ideas of Dr. Spiro's paper with the kinetics of dissolution of solids, discussed in previous papers.

Linear free-energy relationships (LFER) can be derived for the diffusion-controlled dissolution of solids. The situation is depicted in fig. 1, where c_S is the concentration of A at the surface, c_B that in the bulk and x_D is the diffusion-layer thickness.

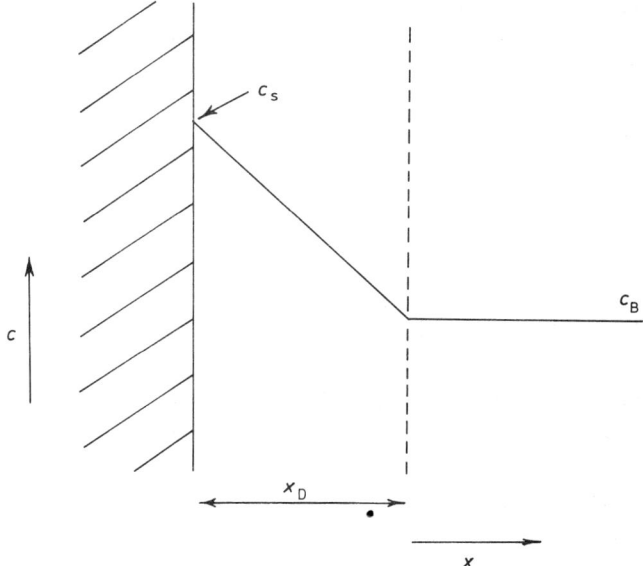

Fig. 1. Concentration profile of solute for the diffusion-controlled dissolution of solids.

If the solid A is dissolving so quickly that, at the surface, solid A is in equilibrium with dissolved A, then c_S is the solubility of A and we may write

$$A(s) \rightleftharpoons A(aq) \xrightarrow{k_D} A(aq).$$

Applying Fick's first law for the flux J (mol m^{-2} s^{-1}) of A(aq) away from the surface

$$J = D(c_S - c_B)/x_D.$$

Initially $c_B = 0$, whereupon

$$\frac{dc_B}{dt} = \frac{ADc_S}{x_D V}$$

where A is here the area of the solid surface exposed to the liquid, V is the volume of the liquid phase and D is the diffusion coefficient of A in solution.

For a molecular solid $c_S = K_s$; thus

$$\frac{dc_B}{dt} = \frac{ADK_s}{x_D V} = k_{obs}$$

where k_{obs} is the initial observed zero-order rate constant. Despite being diffusion-controlled, the rate is proportional to the concentration of A at the surface and hence to the equilibrium constant. Taking logarithms

$$\log k_{obs} = \log\left(\frac{AD}{x_D V}\right) + \log K_s.$$

This is an LFER with slope $\alpha = 1$. A plot of $\log k_{obs}$ against $\log K_s$ will yield a straight line with intercept $\log(AD/x_D V)$. For fixed hydrodynamic conditions comparisons between compounds may be made. These are meaningful because diffusion coefficients vary by at most a factor of 10, whereas the solubility constants vary by several powers of 10.

An electrolyte produces slightly more complicated equations. Consider the following salt, which is completely dissociated in solution:

$$A_a B_b \rightleftharpoons aA^{z_A} + bB^{z_B}$$

$$c_s = \left(\frac{K_s}{a^a b^b}\right)^{1/(a+b)}.$$

This leads to

$$\log k_{obs} = \log\left(\frac{DA}{x_D V}\right) - \left(\frac{1}{a+b}\right)\log(a^a b^b) + \left(\frac{1}{a+b}\right)\log K_s.$$

In this case the slope α equals $1/(a+b)$.

Examples of solids for which dissolution has been found to be diffusion-controlled are
(1) AgCl,[1,2] $K_s = 1.77 \times 10^{-10}$ mol^2 dm^{-6}
(2) CaSO$_4$ in the gypsum form,[3] $K_s = 4 \times 10^{-5}$ mol^2 dm^{-6} and
(3) Ca(H$_2$PO$_4$)$_2$·H$_2$O,[4] $K_s = 9 \times 10^{-5}$ mol^3 dm^{-9}.

The experimental results are not sufficiently detailed to allow a graph to be plotted.

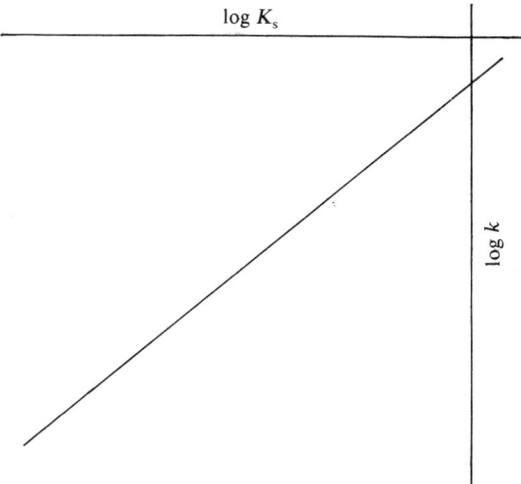

Fig. 2. Plot of log k against log K_s.

One interesting aspect of this situation is the absence of a plateau region in the plot of log k against log K_s (fig. 2). This is unlike the cases mentioned by Dr. Spiro because here there is only one type of diffusion control.

[1] A. L. Jones, *Trans. Faraday Soc.*, 1963, **59**, 2355.
[2] *Stability Constants of Metal Ion Complexes*, ed. L. G. Sillén and A. E. Martell (Special Publication, The Chemical Society, London, 1964).
[3] A. F. M. Barton and N. M. Wilde, *Trans. Faraday Soc.*, 1971, **67**, 3590.
[4] A. F. M. Barton and S. R. McConnel, *J. Chem. Soc., Faraday Trans. 1*, 1974, **70**, 2355.

Dr. D. Leahy (*ICI Pharmaceuticals, Macclesfield*) said: I would like to make a few comments as an addendum to Dr. Spiro's interesting discussion of LFER in interfacial kinetics.

Exactly the same kind of Brönsted-Eigen relationship described by Dr. Spiro (fig. 1 in his paper) exists between the partitioning rates and equilibria of organic molecules between water and organic solvents. Similarly, the diffusion-controlled rate plateau can be raised by stirring (fig. 4). More interestingly, work by Brodin in which diffusion across unstirred layers is eliminated has shown a linear log–log relationship between intrinsic water/oil partitioning rate and equilibrium with slopes around 0.9. This suggests that the solute has lost most of its aqueous solvation shell at the transition state, a quite reasonable and predictable result.

[1] J T. M. van de Waterbeemd, in *Quantitative Approaches to Drug Design*, ed. J. C. Dearden (Elsevier, Amsterdam, 1983), p. 183.
[2] (*a*) A. Brodin and M. I. Nilsson, *Acta Pharm. Suecica*, 1973, **10**, 187; (*b*) A. Brodin, *Acta Pharm. Suecica*, 1974, **11**, 141.

Dr. M. Spiro (*Imperial College, London*) (*partly communicated*): I am grateful to Dr. Leahy and to Mr. Creeth for pointing out further applications of linear free-energy relationships to interfacial reactions in solution. I had in fact referred briefly in my paper to van de Waterbeemd's work on interfacial drug transfer for which, as Dr. Leahy has pointed out, the plots of log(rate) against log(partition coefficient, P) correspond in shape to those for the heterogeneous catalysis of redox reactions between irreversible couples in my fig. 4. The work of Brodin, however, is new to me: it illustrates how LFER can provide structural insights into interfacial processes. It might be worth adding that the log P values themselves possess an additive–constitutive nature and can be correlated with Hammett and Taft parameters.[1]

Mr. Creeth's interesting contribution concerning the dissolution rates of solids that dissolve rapidly has brought to light another set of diffusion-controlled interfacial processes whose LFER is of finite slope. In fact, perusal of several of the papers in this Discussion shows that analogous LFER can be written for certain other types of interfacial process. Thus it follows from eqn (20) of Savage *et al.* and eqn (12) of House *et al.* that for transport-controlled gas dissolution in liquids

$$\ln J = \ln (Dp/\delta) + \ln H$$

where J is the initial mass-transfer flux (mol m^{-2} s^{-1}), p the partial pressure of the gas over the liquid, δ the effective thickness of the boundary layer and H the Henry's law constant or solubility coefficient. For different gases under similar flow conditions, plots of $\ln J$ should therefore vary linearly with $\ln H$, the slope α being unity. The term $\ln (Dp/\delta)$ should remain relatively constant because, following Mr. Creeth's argument, the values of D will vary much less than will the values of H. The kinetics of various transport-controlled chemical reactions at liquid/liquid interfaces can also lead to LFER with $\alpha = 1$. For example, by an extension of eqn (10) in the paper by Crooks and Chisholm, one can obtain

$$\ln J = \ln (Dc/\delta) + \ln P.$$

In this situation the rate-determining step is the diffusion through the Nernst layer of the reactant (toluene) from the organic phase, where its concentration is c, into the aqueous phase where it reacts (is nitrated); P is the partition coefficient c_{aq}/c_{org}. The rates of different organic substrates reacting under similar physical conditions should therefore again fit a LFER. Although this linear log(rate) against log P relation formally resembles that for surface-controlled drug transfer between phases, the different underlying mechanisms are revealed experimentally by the effect of stirring and numerically by the value of the slope α.

It is now clear that the rates of many types of interfacial process will fit LFER with finite slopes. These processes may be catalytic, chemical or transfer in nature. Provided they are fast enough to have become diffusion-controlled, the values of α are determined by the stoichiometry of the process and are frequently unity. This behaviour may be contrasted with that of diffusion-controlled homogeneous reactions, which exhibit log(rate) against log K plots of zero slope.

[1] T. Fujita, in *Progress in Physical Organic Chemistry*, ed. R. W. Taft (Wiley–Interscience, New York, 1983), vol. 14, p. 75.

Closing Remarks

By M. Spiro

Department of Chemistry, Imperial College of Science and Technology, London SW7 2AY

Scientists working on liquid/liquid extraction processes normally gather together in little groups of their own, as do those concerned with drug transfer, those studying the dissolution of gases, those whose interests lie in the crystallisation and dissolution of solids or in ion exchange and those involved with various catalytic phenomena. It was the aim of the organisers of this Discussion to bring them together. We hoped that this would highlight the common ground and also allow new experimental techniques and new theoretical ideas to be more freely disseminated. One criterion by which we can judge the success of this aim is by the institutions from which the 52 authors have come. Fewer than half worked in departments of chemistry; the others came from industry, various kinds of research institute, departments of pharmacy and departments of chemical engineering or chemical technology. The participants at the meeting have equally varied backgrounds, with over 36% from overseas. I believe that the resulting interactions have been both stimulating and fruitful, not only at the formal discussion sessions but also in the informal gatherings in between.

Turning to the scientific content of the papers, we have seen that reactions occurring at or through interfaces may be kinetically controlled by one of three steps: transport through the diffusion layer or layers, reaction in one of the bulk phases or in a reaction layer, or reaction at the interface itself. Examples of all three types have been given in this Discussion, with frequent observations of intermediate control. Since diffusion is always to be reckoned with, it must be either identified as the rate-controlling step or else extrapolated out to yield surface or bulk rate constants. Several different designs of cell have been employed for this purpose and, as someone who does not work in this field, I have been struck by the close resemblance between them and the standard methods of determining diffusion coefficients. Thus cells in which the bulk phases on each side of the interface are stirred, such as the Lewis cell and its variants, bear a likeness to the Stokes diaphragm cell.[1] The Albery rotating diffusion cell is clearly based on two rotating discs to which the Levich equation applies. The rising- or falling-drop method[2] shows a superficial similarity to the dropping mercury electrode, although the hydrodynamics of the former are far less well understood.[3] Laminar jets, too, have been employed both to study interfacial reactions[4] and to measure diffusion coefficients.[5] In this meeting Dr. Guy has described a radiotracer capillary technique which is modelled on the idea of Anderson and Saddington[6] for measuring self-diffusion coefficients. This line of thought suggests the possibility of applying other diffusion methods (recently described in considerable detail by Tyrrell and Harris[7]) to interfacial kinetics. Optical methods involving interference patterns, for example, appear not to have been tried except in the study of Marangoni effects. Moreover, I believe the time has come for us to heed Prof. Nitsch's plea for a proper comparison between different types of apparatus, particularly between those employing free and those employing supported interfaces. He has suggested the zinc/dithizone system as a suitable one for normalising procedures because it exhibits both transport and chemical steps and shows no interfacial instabilities. Certainly the few comparisons available so far do not all agree with the ±10% claimed by the proponents of the various techniques, and I therefore hope that his proposal will be taken up.

In many practical situations, transport processes at interfaces are speeded up by mechanical instabilities—the Marangoni effects. We have all been impressed by the beautiful pictures of these in several of the papers and especially by the remarkable phenomena in the film shown by Dr. Nakache. Prof. Meares has rightly urged experimenters to be aware of these effects and drawn attention to the theories of Drs. Sanfeld and Steinchen which should allow us to predict under what conditions such instabilities can be avoided or encouraged. However, it is still not clear to what extent Marangoni effects will manifest themselves in various kinds of supported liquid/liquid interface, and experiments designed to test this point would be welcome. It is a pity that Marangoni phenomena are rarely treated in standard courses of chemistry although chemical engineers do learn about them. One chemical engineer who has made significant contributions in this area is Prof. Sawistowski, who was also a member of our organising committee. You will all share our regret that due to illness he was unable to be present with us.

Reactions become completely transport controlled when the chemical or surface step is sufficiently fast. This can occur when the interfacial reaction has come to equilibrium (as in some heterogeneously catalysed reactions) or when the surface concentrations in one phase are in equilibrium with the bulk concentrations in the other phase (which we have learnt happens with CO_2 transfer into certain solutions, or the dissolution of soluble solids). The resulting rates of reaction then exhibit the interesting property of being a function of thermodynamic and hydrodynamic parameters only. These rates can therefore be completely calculated from ancillary data provided the flow conditions are sufficiently well defined. All the relevant examples at this meeting have referred to transport control in the liquid phase, but this is not a necessary restriction. Liquid/solid systems showing transport control in the solid phase are met with in extraction processes. Here it has been shown that the rate-determining step is frequently the diffusion of the soluble material from the interior of the plant product to its periphery, as in the extraction of caffeine from swollen coffee beans to make decaffeinated coffee.[8] The rate then depends only on the diffusion coefficient within the bean and on geometrical and thermodynamic properties.[9]

Several authors have reported rates which are completely surface-controlled or else they were able to determine such rates by suitably extrapolating out the transport contributions. These rates exhibit the expected sensitivity to steric factors. This aspect was beautifully illustrated in the pioneering work of Alexander and coworkers[10] for reactions taking place at the water/air interface. Compression of films of substrate (e.g. ethyl palmitate) at the surface forced the molecules to pack more tightly, and their changed orientations produced a decrease in the rate of attack by ions (e.g. OH^-) in the water phase. The deliberate introduction of chemically inert surfactant species into the films also affected the rates markedly, especially if these species were charged.[11] In several of the papers presented at this Discussion we have seen similar features. The important effect of surfactants at liquid/liquid interfaces has been mentioned a number of times and this aspect deserves further study. Adsorption of foreign ions or molecules was also found to affect the kinetics of exchange and dissolution processes at solid/liquid interfaces, and Cheng, Coller and Powell have made progress in understanding this phenomenon by carrying our Monte Carlo simulations. There can be no doubt, either, that spatial requirements both on and within the structure of the solid play an important role, as several contributors have pointed out. We must remember that as yet we cannot predict the rate constants of surface-controlled reactions although we do understand transport processes sufficiently well to predict diffusion-controlled rates in an appropriately designed cell.

This brings me to the question of specialist physicochemical techniques that have been applied in these researches. Take electrochemical ones. The electroanalytical methods employed include conductance, ion-selective electrodes and the pH-stat, as well as an ingenious arc–ring electrode affixed onto the disc of a rotating diffusion cell. The structure of the liquid/liquid interface itself has been probed by cyclic voltammetry and electrical pulse methods, an activity in which the Prague school has been prominent. Interfacial potentials will particularly affect the rates of ionic interfacial reactions. Current–voltage curves at metal electrodes, moreover, have allowed us to predict the catalytic effect of the metal on redox reactions in solution. However, if electrochemical techniques, and also radiochemical ones, have been widely used, the same cannot be said for spectroscopic ones. This is rather surprising. Nowadays more and more spectroscopic probes are being turned on to catalyst and electrode surfaces to study their own structure and that of species adsorbed on them, yet in this Discussion such investigative means have rarely been mentioned except in connection with Marangoni studies. Perhaps there is a lesson here for us.

Moving from techniques to systems one cannot help noticing the emphasis in this Discussion on geochemical and mineral systems. Metal extraction, salt and gas dissolution, and to a lesser extent drug transfer, have clearly been the major processes of interest. Only one paper, that of Crooks and Chisholm, deals with an organic chemical reaction. Nevertheless, the experimental methods and theories so lucidly described by various contributors should apply also to other types of interfacial process, such as interfacial polymerisation and phase-transfer catalysis. The last mentioned, in particular, has proved of increasing utility to practical synthetic chemists but has received less than its fair share of attention from physical chemists.

Finally, I would like to draw your attention to some other broader aspects. Interfacial reactions, whether transfer or chemical in nature, display many features in common with catalytic processes that are either truly heterogeneous or microheterogeneous. By the latter I mean reactions at dissolved entities like enzymes or micelles, which are so large in comparison with substrate molecules that they act as if they were part of another phase. All these processes, transfer or chemical or catalytic, occur at or through an interface of limited area and therefore display saturation phenomena. These have several times been mentioned here. Moreover, the reacting species must compete for sites at the interface with other reactants, with products, with solvent molecules and with impurity species. The presence of the latter may render reproducibility more difficult and can lead to poisoning or blocking of the interface. The effect of surfactants, already referred to, is particularly important in this respect. Another consequence of the physical resemblance between all interfacial processes is the strikingly similar form of many of the resulting rate equations. Once any transport contributions have been extrapolated out or removed by use of sufficiently strong forced convection, we find repeatedly equations of the Langmuir type:

$$\frac{1}{v} = \frac{1}{a} + \frac{b}{c}$$

where v is the rate or flux and c the concentration of a reactant. Such equations are found not only in the heterogeneous catalysis of gas reactions but also with micellar and enzyme catalysis, and indeed a derivative of the Michaelis–Menten equation was specifically named by Hadgraft *et al.* in their analysis of the flux of dye transferred across a liquid membrane.

At every session the discussion has been lively, frequently illuminating, and at times controversial. The question as to what is meant by an interfacial reaction provoked particularly vigorous comment. Since at one stage the argument was described as one of semantics, it reminded me of an apt little verse by my favourite poet, Ogden Nash:

> I give you now Professor Twist,
> A conscientious scientist.
> Trustees exclaimed, 'He never bungles!'
> And sent him off to distant jungles.
> Camped on a tropic riverside,
> One day he missed his loving bride.
> She had, the guide informed him later,
> Been eaten by an alligator.
> Professor Twist could not but smile.
> 'You mean,' he said 'a crocodile'.

This seems a fitting point on which to end, for we shall all now be swallowed up by the outside world on leaving this comfortable conference centre. Dr. Aveyard and his helpers at Hull University deserve our grateful thanks for their hospitality; their efforts have played a significant role in making this Discussion such a successful one.

[1] R. H. Stokes, *J. Am. Chem. Soc.*, 1950, **72**, 763.
[2] W. Nitsch, *Ber. Bunsenges. Phys. Chem.*, 1965, **69**, 884.
[3] R. J. Whewell, M. A. Hughes and C. Hanson, *J. Inorg. Nucl. Chem.*, 1975, **37**, 2303.
[4] C. Hanson and H. A. M. Ismail, *Chem. Eng. Sci.*, 1977, **32**, 775.
[5] C. Hanson and H. A. M. Ismail, *J. Appl. Chem. Biochem.*, 1976, **26**, 111.
[6] J. S. Anderson and K. Saddington, *J. Chem. Soc.*, 1949, S 381.
[7] H. J. V. Tyrrell and K. R. Harris, *Diffusion in Liquids* (Butterworths, London, 1984), chap. 5.
[8] B. Bichsel, *Food Chem.*, 1979, **4**, 53.
[9] M. Spiro and R. M. Selwood, *J. Sci. Food Agric.*, 1984, **35**, 915.
[10] A. E. Alexander and J. H. Schulman, *Proc. R. Soc. London, Ser. A*, 1937, **161**, 115; A. E. Alexander and E. K. Rideal, *Proc. R. Soc. London, Ser. A*, 1937, **163**, 70.
[11] J. T. Davies, *Adv. Catal.*, 1954, **6**, 1.

INDEX OF NAMES*

Albery, W. J., **53**, 139, 144, 148, 151
Amantea, M., **127**
Astarita, G., **17**, 48
Barker, N., **97**
Binsma, J. J. M., **257**, 297, 299, 300, 301
Cheng, V. K., 51, **243**, 290, 293, 295, 296, 298, 301
Chisholm, J. M., **105**
Christoffersen, J., **235**, 288, 289, 291, 293, 295, 296, 300, 304
Christoffersen, M. R., **235**, 288, 289, 291, 295
Choudhery, R. A., **53**
Coller, B. A. W., **243**, 288, 296, 297, 298, 299
Creeth, A. M., 305
Crooks, J. E., **105**, 151
Dickel, G., **157**, 217
Du, G., **209**
Dupeyrat, M., **189**, 218
Fisk, P. R., **53**, 151
Flambard, A. R., 300
Gu, Z. M., **67**
Guy, R. H., **127**
Hadgraft, J., **97**, 149, 150
Hinz, R. S., **127**
Homolka, D., **197**
House, W. A., **33**, 47, 49, 50, 51, 289, 297, 301
Howard, J. R., **33**
Hughes, M. A., **75**, 139, 146, 147, 148
Ikeda, T., **223**
Kolar, Z., **257**
Koryta, J., **209**, 221

Kreevoy, M. M., 140, 147, 148
Li, N. N., **67**
Leahy, D., 149, 307
Linde, H., 47, **181**, 218
Mareček, V., **197**
Meares, P., **7**, 139, 217
Nitsch, W., **85**, 147
Nakache, E., **189**, 218, 219, 221
Noble, R. D., 143, 146
Powell, J. L., **243**
Robinson, B. H., 140, 149, 155, 287
Rod, V., **75**
Ruth, W., **209**
Samec, Z., **197**, 219, 220
Sanfeld, A., 140, **169**
Sartori, G., **17**, 288
Sasaki, M., **223**
Savage, D. W., **17**
Sellers, R. M., **265**, 292, 294, 302, 304
Skirrow, G., **33**
Spiro, M., 50, 150, 219, 220, **275**, 307, **309**
Steinchen, A., 140, 148, **169**
Tondre, C., 47, **115**, 143, 151, 155, 219
Vanýsek, P., **209**
Vignes-Adler, M., **189**, 218
Wasan, D. T., **67**
Williams, W. J., **265**, 305
Wotton, P. K., **97**
Xenakis, A., **115**
Yasunaga, T., **223**, 287, 288

* The page numbers in heavy type indicate papers submitted for discussion.

GENERAL DISCUSSIONS OF THE FARADAY SOCIETY/FARADAY DISCUSSIONS OF THE CHEMICAL SOCIETY

Date	Subject	Volume
1907	Osmotic Pressure	Trans. 3*
1907	Hydrates in Solution	3*
1910	The Constitution of Water	6*
1911	High Temperature Work	7*
1912	Magnetic Properties of Alloys	8*
1913	Colloids and their Viscosity	9*
1913	The Corrosion of Iron and Steel	9*
1913	The Passivity of Metals	9*
1914	Optical Rotatory Power	10*
1914	The Hardening of Metals	10*
1915	The Transformation of Pure Iron	11*
1916	Methods and Appliances for the Attainment of High Temperatures in a Laboratory	12*
1916	Refractory Materials	12*
1917	Training and Work of the Chemical Engineer	13*
1917	Osmotic Pressure	13*
1917	Pyrometers and Pyrometry	13*
1918	The Setting of Cements and Plasters	14*
1918	Electrical Furnaces	14*
1918	Co-ordination of Scientific Publication	14*
1918	The Occlusion of Gases by Metals	14*
1919	The Present Position of the Theory of Ionization	15*
1919	The Examination of Materials by X-Rays	15*
1920	The Microscope: Its Design, Construction and Applications	16*
1920	Basic Slags: Their Production and Utilization in Agriculture	16*
1920	Physics and Chemistry of Colloids	16*
1920	Electrodeposition and Electroplating	16*
1912	Capillarity	17*
1921	The Failure of Metals under Internal and Prolonged Stress	17*
1921	Physico-Chemical Problems Relating to the Soil	17*
1921	Catalysis with special reference to Newer Theories of Chemical Action	17*
1922	Some Properties of Powders with special reference to Grading by Elutriation	18*
1922	The Generation and Utilization of Cold	18
1923	Alloys Resistant to Corrosion	19*
1923	The Physical Chemistry of the Photographic Process	19
1923	The Electronic Theory of Valency	19*
1923	Electrode Reactions and Equilibria	19
1923	Atmospheric Corrosion. First Report	19*
1924	Investigation on Oppau Ammonium Sulphate-Nitrate	20*
1924	Fluxes and Slags in Metal Melting and Working	20*
1924	Physical and Physico-Chemical Problems relating to Textile Fibres	20*
1924	The Physical Chemistry of Igneous Rock Formation	20*
1924	Base Exchange in Soils	20*
1925	The Physical Chemistry of Steel-Making Processes	21*
1925	Photochemical Reactions in Liquids and Gases	21
1926	Explosive Reactions in Gaseous Media	22
1926	Physical Phenomena at Interfaces, with special reference to Molecular Orientation	22
1927	Atmospheric Corrosion. Second Report	23*
1927	The Theory of Strong Electrolytes	23*
1927	Cohesion and Related Problems	24
1928	Homogeneous Catalysis	24
1929	Crystal Structure and Chemical Constitution	25*
1929	Atmospheric Corrosion of Metals. Third Report	25*
1929	Molecular Spectra and Molecular Structure	26*
1930	Colloid Science Applied to Biology	26

Date	Subject	Volume
1931	Photochemical Processes	27
1932	The Adsorption of Gases by Solids	28
1932	The Colloid Aspect of Textile Materials	29
1933	Liquid Crystals and Anisotropic Melts	29*
1933	Free Radicals	30
1934	Dipole Moments	30
1934	Colloidal Electrolytes	31*
1935	The Structure of Metallic Coatings, Films and Surfaces	31*
1935	The Phenomena of Polymerization and Condensation	32*
1936	Disperse Systems in Gases: Dust, Smoke and Fog	32*
1936	Structure and Molecular Forces in (a) Pure Liquids, and (b) Solutions	33*
1937	The Properties and Functions of Membranes, Natural and Artificial	33*
1937	Reaction Kinetics	34*
1938	Chemical Reactions Involving Solids	34*
1938	Luminescence	35*
1939	Hydrocarbon Chemistry	35*
1939	The Electrical Double Layer (owing to the outbreak of war the meeting was abandoned, but the papers were printed in the *Transactions*)	35*
1940	The Hydrogen Bond	36*
1941	The Oil-Water Interface	37*
1941	The Mechanism and Chemical Kinetics of Organic Reactions in Liquid Systems	37*
1942	The Structure and Reactions of Rubber	38
1943	Modes of Drug Action	39
1944	Molecular Weight and Molecular Weight Distribution in High Polymers (Joint Meeting with the Plastics Group, Society of Chemical Industry)	40*
1945	The Application of Infra-red Spectra to Chemical Problems	41*
1945	Oxidation	42*
1946	Dielectrics	42 A
1946	Swelling and Shrinking	42 B
1947	Electrode Processes	Disc. 1*
1947	The Labile Molecule	2
1947	Surface Chemistry (Jointly with the Société de Chimie Physique at Bordeaux) Published by Butterworths Scientific Publications, Ltd	
1947	Colloidal Electrolytes and Solutions	Trans. 43*
1948	The Interaction of Water and Porous Materials	Disc. 3
1948	The Physical Chemistry of Process Metallurgy	4*
1949	Crystal Growth	5
1949	Lipo-proteins	6
1949	Chromatographic Analysis	7
1950	Heterogeneous Catalysis	8*
1950	Physico-chemical Properties and Behaviour of Nuclear Acids	Trans. 46*
1950	Spectroscopy and Molecular Structure and Optical Methods of Investigating Cell Structure	Disc. 9
1950	Electrical Double Layer	Trans. 47
1951	Hydrocarbons	Disc. 10
1951	The Size and Shape Factor in Colloidal Systems	11
1952	Radiation Chemistry	12*
1952	The Physical Chemistry of Proteins	13
1952	The Reactivity of Free Radicals	14
1953	The Equilibrium Properties of Solutions on Non-electrolytes	15
1953	The Physical Chemistry of Dyeing and Tanning	16*
1954	The Study of Fast Reactions	17
1954	Coagulation and Flocculation	18
1955	Microwave and Radio-frequency Spectroscopy	19
1955	Physical Chemistry of Enzymes	20
1956	Membrane Phenomena	21
1956	Physical Chemistry of Processes at High Pressures	22
1957	Molecular Mechanism of Rate Processes in Solids	23
1958	Interactions in Ionic Solutions	24
1957	Configurations and Interactions of Macromolecules and Liquid Crystals	25
1958	Ions of the Transition Elements	26
1959	Energy Transfer with special reference to Biological Systems	27
1959	Crystal Imperfections and the Chemical Reactivity of Solids	28
1960	Oxidation-Reduction Reactions in Ionizing Solvents	29
1960	The Physical Chemistry of Aerosols	30
1961	Radiation Effects in Inorganic Solids	31
1961	The Structure and Properties of Ionic Melts	32

Date	Subject	Volume
1962	Inelastic Collisions of Atoms and Simple Molecules	33*
1962	High Resolution Nulcear Magnetic Resonance	34
1963	The Structure of Electronically Excited Species in the Gas Phase	35
1963	Fundamental Processes in Radiation Chemistry	36
1964	Chemical Reactions in the Atmosphere	37
1964	Dislocations in Solids	38
1965	The Kinetics of Proton Transfer Processes	39
1965	Intermolecular Forces	40
1966	The Role of the Adsorbed State in Heterogeneous Catalysis	41
1966	Colloid Stability in Aqueous and Non-aqueous Media	42
1967	The Structure and Properties of Liquids	43
1967	Molecular Dynamics of the Chemical Reactions of Gases	44
1968	Electrode Reactions of Organic Compounds	45
1968	Homogeneous Catalysis with Special Reference to Hydrogenation and Oxidation	46
1969	Bonding in Metallo-organic Compounds	47
1969	Motions in Molecular Crystals	48
1970	Polymer Solutions	49
1970	The Vitreous State	50
1971	Electrical Conduction in Organic Solids	51
1971	Surface Chemistry of Oxides	52
1972	Reactions of Small Molecules in Excited States	53
1972	The Photoelectron Spectroscopy of Molecules	54
1973	Molecular Beam Scattering	55
1973	Intermediates in Electrochemical Reactions	56
1974	Gels and Gelling Processes	57
1974	Photo-effects in Adsorbed Species	58
1975	Physical Adsorption in Condensed Phases	59
1975	Electron Spectroscopy of Solids and Surfaces	60
1976	Precipitation	61
1977	Potential Energy Surfaces	62
1977	Radiation Effects in Liquids and Solids	63
1977	Ion–Ion and Ion–Solvent Interactions	64
1978	Colloid Stability	65*
1978	Structure and Motion in Molecular Liquids	66
1979	Kinetics of State Selected Species	67
1979	Organization of Macromolecules in the Condensed Phase	68
1980	Phase Transitions in Molecular Solids	69
1980	Photoelectrochemistry	70
1981	High Resolution Spectroscopy	71
1981	Selectivity in Heterogeneous Catalysis	72
1982	Van der Waals Molecules	73
1982	Electron and Proton Transfer	74
1983	Intramolecular Kinetics	75
1983	Concentrated Colloidal Dispersions	76

* *Not available; for current information on prices, etc., of available volumes, please contact the Marketing Officer, Royal Society of Chemistry, Burlington House, London W1V 0BN stating whether or not you are a member of the Society.*